T0367474

Quickest Detection

The problem of detecting abrupt changes in the behavior of an observed signal or time series arises in a variety of fields, including climate modeling, finance, image analysis, and security. Quickest detection refers to real-time detection of such changes as quickly as possible after they occur. Using the framework of optimal stopping theory, this book describes the fundamentals underpinning the field, providing the background necessary to design, analyze, and understand quickest detection algorithms.

For the first time the authors bring together results that were previously scattered across disparate disciplines, and provide a unified treatment of several different approaches to the quickest detection problem. This book is essential reading for anyone who wants to understand the basic statistical procedures for change detection from a fundamental viewpoint, and for those interested in theoretical questions of change detection. It is ideal for graduate students and researchers in engineering, statistics, economics, and finance.

H. Vincent Poor is the Michael Henry Strater University Professor of Electrical Engineering, and Dean of the School of Engineering and Applied Science, at Princeton University, from where he received his Ph.D. in 1977. Prior to joining the Princeton faculty in 1990, he was on the faculty of the University of Illinois at Urbana-Champaign, and has held visiting positions at a number of other institutions, including Imperial College, Harvard University, and Stanford University. He is a Fellow of the IEEE, the Institute of Mathematical Statistics, and the American Academy of Arts and Sciences, as well as a member of the US National Academy of Engineering.

Olympia Hadjiliadis is an Assistant Professor in the Department of Mathematics at Brooklyn College of the City University of New York, where she is also a member of the graduate faculty of the Department of Computer Science. She was awarded her M.Math in Statistics and Finance in 1999 from the University of Waterloo, Canada. After receiving her Ph.D. in Statistics with distinction from Columbia University in 2005, Dr. Hadjiliadis joined the Electrical Engineering Department at Princeton as a Postdoctoral Fellow, where she was subsequently appointed as a Visiting Research Collaborator until 2008.

Quickest Detection

Quickest Detection

H. VINCENT POOR
Princeton University

OLYMPIA HADJILIADIS
City University of New York

CAMBRIDGE
UNIVERSITY PRESS

University Printing House, Cambridge CB2 8BS, United Kingdom

Cambridge University Press is part of the University of Cambridge.

It furthers the University's mission by disseminating knowledge in the pursuit of education, learning and research at the highest international levels of excellence.

www.cambridge.org
Information on this title: www.cambridge.org/9780521621045

First published 2009

A catalogue record for this publication is available from the British Library

ISBN 978-0-521-62104-5 Hardback

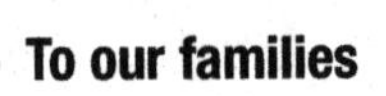
To our families

Contents

Figures

Preface

Change detection is a fundamental problem arising in many fields of engineering, in finance, in the natural and social sciences, and even in the humanities. This book is concerned with the problem of change detection within a specific context. In particular, the framework considered here is one in which changes are manifested in the statistical behavior of quantitative observations, so that the problem treated is that of *statistical change detection*. Moreover, we are interested in the on-line problem of *quickest detection*, in which the objective is to detect changes in real time as quickly as possible after they occur. And, finally, our focus is on formulating such problems in such a way that optimal procedures can be sought and found using the tools of stochastic analysis.

Thus, the purpose of this book is to provide an exposition of the extant theory underlying the problem of quickest detection, with an emphasis on providing the reader with the background necessary to begin new research in the field. It is intended both for those familiar with basic statistical procedures for change detection who are interested in understanding these methods from a fundamental viewpoint (and possibly extending them to new applications), and for those who are interested in theoretical questions of change detection themselves.

The approach taken in this book is to cast the problem of quickest detection in the framework of optimal stopping theory. Within this framework, it is possible to provide a unified treatment of several different approaches to the quickest detection problem. This approach allows for exact formalism of quickest detection problems, and for a clear understanding of the optimality properties they enjoy. Moreover, it provides an obvious path to follow for the researcher interested in going beyond existing results.

This treatment should be accessible to graduate students and other researchers at a similar level in fields such as engineering, statistics, economics, finance, and other fields with comparable mathematical content, who have a working knowledge of probability and stochastic processes at the level of a first-tier graduate course. Although the book begins with an overview of necessary background material in probability and stochastic processes, this material is included primarily as a review and to establish notation, and is not intended to be a first exposure to these subjects. Otherwise, the notes are relatively self-contained and can be read with a knowledge only of basic analysis. Some previous exposure to statistical inference is useful, but not necessary, in interpreting some of the results described here.

The material presented in this book is comprised primarily of results published previously in journals. However, the literature in this area is quite scattered across both time and disciplines, and a unique feature of this treatment is the collection of these results into a single, unified work that also includes the necessary background in optimal stopping theory and probability. Although this book is intended primarily as a research monograph, it would also be useful as a primary text in a short course or specialized graduate course, or as a supplementary text in a more basic graduate course in signal detection or statistical inference.

This book grew out of a series of lectures given by the first author at the IDA Center for Communications Research (CCR) in Princeton, New Jersey, and much of the material here, including primarily the discrete-time formalism treated in Chapters 2–6, originally served as lecture notes for that series. On the other hand, most of the material on continuous-time models in Chapters 2–6, and all of the material in Chapter 7, was completed while the second author was a post-doctoral associate at Princeton University under the support of the US Army Pantheon Project.

The writing of this book has benefited from a number of people other than the authors. Among these are David Goldschmidt, former Director of CCR, who suggested and encouraged the series of lectures that led to the book, and George Soules, former Deputy Director of CCR, who provided many useful suggestions on the subject matter and presentation of this material. Early drafts of this material also benefited considerably from the comments and suggestions of the many members of the technical staff at CCR who attended the lectures there. We are very grateful to these colleagues, as well as to many other colleagues at Princeton University and elsewhere who have made useful suggestions on this treatment. Of these, we mention in particular Richard Davis of Colorado State University, Savas Dayanik and Stuart Schwartz of Princeton, Robert Grossman of the University of Illinois, Chris Heyde of the Australian National University, Kostas Kardaras of Boston University, George Moustakides of the University of Patras, and Alexander Tartakovsky of the University of Southern California. Finally, the authors are also grateful to Phil Meyler of Cambridge University Press for his encouragement of this project and for his patience in awaiting its completion.

Frequently used notation

$\mathbb{R}$ is the set of all real numbers
$\mathcal{Z}$ is the set of all integers
$\mathbb{R}^\infty$ is the set of (one-sided) sequences of real numbers
$C[0, \infty)$ is the set of continuous functions mapping $[0, \infty)$ to $\mathbb{R}$
$D[0, \infty)$ is the set of functions mapping $[0, \infty)$ to $\mathbb{R}$ that are right-continuous and have left limits

1 Introduction

The problem of detecting abrupt changes in the statistical behavior of an observed signal or time series is a classical one, whose provenance dates at least to work in the 1930s on the problem of monitoring the quality of manufacturing processes [224]. In more recent years, this problem has attracted attention in a wide variety of fields, including climate modeling [15], econometrics [4,5,7,8,34,51], environment and public health [110,115,170,201], finance [7,30,193], image analysis [17,214], medical diagnosis [59,84,85,171,229], navigation [148,159], network security [53,55,125,164,208,213], neuroscience [60,66,217,232], other security applications such as fraud detection and counter-terrorism [86,87,125,202], remote sensing (seismic, sonar, radar, biomedical) [104,143,172], video editing [126,134], and even the analysis of historical texts [50,95,182]. This list, although long, is hardly exhaustive, and other applications can be found, for example, in [6,8,18,19,45,52,87,114,128,131,149,161,162,163,165,230]. These cited references only touch the surface of a very diverse and vibrant field, in which this general problem is known variously as *statistical change detection*, *change-point detection*, or *disorder detection*.

Many of these applications, such as those in image analysis, econometrics, or the analysis of historical texts, involve primarily off-line analyses to detect a change in statistical behavior during a pre-specified frame of time or space. In such problems, it is of interest to estimate the occurrence time of a change, and to identify appropriate statistical models before and after the change. However, it is not usually an objective of these applications to perform these functions in real time.

On the other hand, there are many applications of change detection in which it is of interest to perform on-line (i.e. real-time) detection of such changes in a way that minimizes the delay between the time a change occurs and the time it is detected. This latter type of problem is know as the *quickest detection problem*, and this problem arises in many of the above-noted applications. For example, in seismology, quickest detection can be used to detect the onset of seismic events that may presage earthquakes. It is important that such events be detected as quickly as possible so that emergency action can be taken. Similar issues arise in the monitoring of cardiac patients, the monitoring of the radio spectrum for opportunistic wireless transmission, the analysis of financial indicators to detect fundamental shifts in sector performance or foreign exchange trends, the monitoring of computer networks for faults or security breaches, etc. Again, the list of such problems is quite long and diverse.

This book describes a theoretical basis for the design, analysis and understanding of quickest detection algorithms. There are six chapters (plus bibliography) beyond the current one. These are described briefly as follows.

(2) Probabilistic framework
This chapter provides an overview of the elements of probability theory needed to place the quickest detection problem in a mathematical setting. The topics reviewed include probability spaces, random variables, expectations, Radon–Nikodym derivatives, conditional expectations and independence, properties of random sequences, martingales, stopping times, Brownian motion, and Poisson processes. It is assumed that the reader has prior exposure to most of these ideas, and this chapter is intended primarily as a review and as a mechanism for establishing notation and vocabulary.

(3) Markov optimal stopping theory
This chapter develops the concepts and tools of Markov optimal stopping theory that are necessary to derive and understand optimal procedures for quickest detection. These concepts include the general characterization of optimal stopping procedures in terms of the Snell envelope, and the explicit techniques (e.g. dynamic programming) for computing solutions to Markov optimal stopping problems. Three cases are treated: finite-horizon discrete time, infinite-horizon discrete time, and infinite-horizon continuous time. The emphasis here is on the first two of these cases, with the third case being treated only briefly. Several examples are used to illustrate this theory, including the classical selection problem, option trading, etc.

(4) Sequential detection
This chapter formulates and solves the classical sequential detection problem as an optimal stopping problem. This problem deals with the optimization of decision rules for deciding between two possible statistical models for an infinite, statistically homogeneous sequence of random observations. The optimization is carried out by penalizing, in various ways, the probabilities of error and the average amount of time required to reach a decision. By optimizing separately over the error probabilities with the decision time fixed, this problem becomes an optimal stopping problem that can be treated using the methods of the preceding chapter. As this problem is treated in many sources, the primary motivation for including it here is that it serves as a prototype for developing the tools needed in the related problem of quickest detection.

With this in mind, both Bayesian and non-Bayesian, as well as discrete-time and continuous-time formulations of this problem are treated as well as models that combine both a discrete and a continuous nature. In the course of this treatment, a set of analytical techniques is developed that will be useful in the solution and performance analysis of problems of quickest detection to be treated in subsequent chapters. Specific topics included are Bayesian optimization, the Wald–Wolfowitz theorem, the fundamental identity of sequential analysis, Wald's approximations, and diffusion approximations.

A basic conclusion of this chapter is the general optimality of the sequential probability ratio test (SPRT), which, in its various forms, is a central algorithm for the problem of sequential detection.

(5) Bayesian quickest detection

This chapter treats the 'disorder' problem, first posed by Kolmogorov and Shiryaev, in which the distribution of an observed random sequence changes abruptly at an unknown time (the change point). This change point is assumed to have a known geometric prior distribution, and hence this problem provides a Bayesian framework for quickest detection. The choice of a geometric prior is mathematically convenient, but it also provides a reasonable model for a number of practical applications.

The objective of a detection procedure in this situation is to react as quickly as possible to the change in distribution, within a constraint on the probability of reacting before the change occurs. Thus, the design of such procedures involves the satisfaction of optimization criteria comprised of two performance indices: the mean delay until detection, and the probability of false alarm (i.e. premature detection). As with the classical sequential detection problem, this problem can also be formulated as an optimal stopping problem after a suitable transformation.

We also describe various other formulations of this problem starting with its continuous-time analog of detecting a change in the drift of a Brownian motion, where the prior for the change point is assumed to be exponential. We subsequently consider a Poisson model which combines both continuous-time and discrete-time features in its nature and treatment. We further include a different treatment of this problem that focuses on devising a stopping rule that, with high probability, is as close as possible to the change point. This problem is also formulated as an optimal stopping problem and the tools developed in Chapter 3 are used for its treatment.

After a discussion of several alternative optimization criteria comprised of trade-offs similar to the ones mentioned above, we finally conclude this chapter with a game theoretic approach to the problem of Bayesian quickest detection, in which the change point is viewed as having been selected by an opponent ("nature") playing a competitive game with the designer of the detection procedure.

A central theme of this chapter is the general Bayesian optimality of procedures that announce the presence of a change point at the first upcrossing of a threshold by the posterior probability of a change, given the past and present observations. An exception to this general optimality arises in the game theoretic formulation, for which the optimal solution is the so-called cumulative sum (CUSUM) procedure, also known as Page's test, which plays a central role in non-Bayesian formulations of the quickest detection problem. This latter formalism thus provides a bridge between Bayesian and non-Bayesian problems, and a segue to the next chapter of this book.

(6) Non-Bayesian quickest detection

This chapter treats a non-Bayesian formulation of the quickest detection problem, first proposed by Lorden, in which no prior knowledge of the change point is known. For many applications, this formulation is more useful than the Shiryaev formulation, since the assumption of a prior distribution for the change point is sometimes unrealistic. Without a prior, however, the performance indices used in the Shiryaev formulation of this problem – namely, mean detection delay and false-alarm probability – are not meaningful since there is an infinite set of possible distributions for the observations, one for each possible value of the change point.

Lorden's formulation deals with this difficulty by replacing the mean detection delay with a worst-case conditional delay, where the conditioning is with respect to the change point, and the worst case is taken over all possible values of the change point and all realizations of the measurements leading up to the change point. False alarms are controlled by placing a lower bound on the allowable mean time between false alarms. Here, considering first the discrete-time case, we present a solution to the problem of minimizing delay within this constraint, again by appealing to a related optimal stopping problem. The above-noted CUSUM algorithm is the optimal solution here, and it is in fact the central (although not the only) algorithm arising in non-Bayesian quickest detection problems.

Results from renewal theory are also used here to relate the performance of optimal detection procedures for this problem to that of the classical sequential detection procedures described in Chapter 4. Through this connection, a number of approximations and bounds for the relevant performance indices are developed.

The non-Bayesian quickest detection problem is also treated under the assumption of several continuous-time models for the observations. The problem is seen once again as an optimal stopping problem, but now we introduce a different approach, based on establishing global lower bounds for the performance of all relevant stopping times, in solving it. In the case of a specific continuous-time model we also discuss the practically important problem of an unknown change after the change point.

We finally give several asymptotic results that are useful in the analysis of the CUSUM algorithm, and in its generalization. We also apply this asymptotic analysis to treat the problem of two-sided alternatives (i.e. changes in the mean of unknown sign) in the context of a specific continuous-time observation model.

(7) Additional topics

This final chapter considers the problem of sequential and quickest detection in several settings that arise from practical considerations not treated in the previous chapters. These include decentralized, robust, and adaptive methods for quickest detection. Decentralized problems arise, for example, in applications involving sensor networks or distributed databases. Robust and adaptive methods are generically of interest when there is uncertainty in the statistical models used to describe observations. Other generalizations and alternative formulations of the quickest-detection problem are also described, notably in connection with problems in which the observations do not form an independent sequence. Such problems arise in applications involving the analysis of time series, for example. All the results of this chapter are cast in a discrete-time framework.

The basic idea in the treatment of the decentralized detection problems is to again formulate them as optimal stopping problems and use the results of Chapter 3 to solve them. Adopting a Bayesian model for the change point, the similarity of the problems of decentralized sequential and quickest detection to the problems treated in Chapters 4 and 5, in both formulation and solution, is easily seen. However, an additional feature here that does not arise in the earlier formulations is the need of optimizing local decisions in addition to global ones.

As noted above, the problems of robust and adaptive quickest detection treat situations of modeling uncertainty. The two approaches are quite different, as robust procedures seek to provide guaranteed performance in the face of small, but potentially damaging, non-parametric uncertainties in statistical models, whereas adaptive procedures are based on the on-line estimation of parametrized models. In the former case, we describe and solve a minimax formulation of quickest detection, whereas the latter problem is solved using combined detection–estimation procedures. In both case, the approach is essentially non-Bayesian.

Finally, we present the problem of quickest detection in the case of more general dependence models than the independent-sampling models used in earlier chapters, again using a non-Bayesian formulation. After first giving a precise generalization of the optimality of the CUSUM to a class of dependent observation processes, we then turn to a more general approach to change detection in time series models based on a general asymptotic local formulation of change detection, which makes heavy use of diffusion approximations to develop asymptotically optimal procedures.

2 Probabilistic framework

2.1 Introduction

Probability theory provides a useful mathematical setting for problems of optimal stopping and statistical change detection. This chapter provides a brief overview of the concepts from probability that will be used in the sequel. This overview is organized into five sections, a review of basic probability (Section 2.2), a collection of results about martingales and stopping times (Section 2.3), some introductory material on Brownian motion and Poisson processes (Section 2.4), an overview of semimartingales (Section 2.5), and the definition and properties of the stochastic integral (Section 2.6). Those who are already familiar with the basic definitions of probability, expected value, random variables, and stochastic convergence, may want to skip Section 2.2.

2.2 Basic setting

In this section, we define some essential notions from probability theory that will be useful in the sequel. Most of this material can be found in many basic books, including [37,46], or in the first chapter of [225].

2.2.1 Probability spaces

The basic notion in a probabilistic model is that of a *random experiment*, in which outcomes are produced according to some chance mechanism. From a mathematical point of view, this notion is contained in an abstraction – a *probability space*, which is a triple $(\Omega, \mathcal{F}, P)$ consisting of the following elements:

- a *sample space* Ω of elemental outcomes of the random experiment;
- an *event class* $\mathcal{F}$, which is a nonempty collection of subsets of Ω to which we wish to assign probabilities; and
- a *probability measure* (or *probability distribution*) P, which is a real-valued set function that assigns probabilities to the events in $\mathcal{F}$.

In order to be able to manipulate probabilities, we do not allow the event class to be arbitrary, but rather we assume that it is a σ-*field* (or σ-*algebra*); that is, we assume that $\mathcal{F}$ is closed under complementation and under countable unions. The usual algebra of set operations then assures that $\mathcal{F}$ is closed under arbitrary countable sequences of the operations: union, intersection, and complementation. Such a class necessarily contains the sample space Ω and the null set $\emptyset$. The elements of $\mathcal{F}$ are called *events*. A pair $(\Omega, \mathcal{F})$ consisting of a sample space and event class is called a *measurable space* or a *pre-probability space*.

The probability measure P is constrained to have the following properties, which axiomatize the intuitive notion of what probability means:

- $P(\Omega) = 1$;
- $P(F) \geq 0, \ \forall F \in \mathcal{F}$; and
- $$P\left(\bigcup_{n=1}^{\infty} F_n\right) = \sum_{n=1}^{\infty} P(F_n),$$
  for all sequences $\{F_k; k = 1, 2, \ldots\}$ of elements of $\mathcal{F}$ satisfying $F_m \cap F_n = \emptyset, \ \forall\, m \neq n$.

That is, P is constrained to be non-negative, normalized, and countably additive.

2.2.2 Random variables

The probability space is a useful abstraction that allows us to think of a chance mechanism underlying more concrete, observable phenomena that we wish to model as being random. Such concrete things can be modeled as being *random variables*.

Mathematically, a random variable is defined to be a measurable mapping from the sample space Ω (endowed with the event class $\mathcal{F}$) to the real line $\mathbb{R}$ (endowed with the usual Borel σ-field $\mathcal{B}$).[1] That is, $X : \Omega \to \mathbb{R}$ is a random variable if

$$X^{-1}(B) \in \mathcal{F}, \ \forall\, B \in \mathcal{B}, \tag{2.1}$$

where $X^{-1}(B)$ denotes the pre-image under X of $B : \{\omega \in \Omega | X(\omega) \in B\}$.

The measurability of X assures that probabilities can be assigned to all Borel subsets of $\mathbb{R}$ via the obvious assignment:

$$P_X(B) = P\left(X^{-1}(B)\right), \ \forall\, B \in \mathcal{B}. \tag{2.2}$$

In this way, X induces a probability measure P_X on $(\mathbb{R}, \mathcal{B})$ so that $(\mathbb{R}, \mathcal{B}, P_X)$ is also a probability space. Once P_X is known, the structure of the underlying probability space $(\Omega, \mathcal{F}, P)$ is irrelevant in describing the probabilistic behavior of the random variable X.

[1] Recall that the Borel σ-field in $\mathbb{R}$ is the smallest σ-field that contains all intervals. This is a natural event class to consider on the real line, since the intervals, their complements, unions, and intersections, are the sets of most interest in the context of describing the behavior of observed real phenomena.

The information contained in the probability measure P_X is more succinctly described in terms of the *cumulative probability distribution function* (cdf) of X, defined as

$$F_X(x) = P(X \leq x) = P_X((-\infty, x]), \; x \in \mathbb{R}. \tag{2.3}$$

Either of the two functions F_X or P_X determines the other.

The family of all cdf's is the set of all non-decreasing, right-continuous functions with left limit zero and right limit one. That is, all cdf's satisfy these properties; and, given any function with these properties, one can construct a random variable having that function as its cdf. Random variables are classified according to the nature of their cdf's. Two distinct types of interest are: *continuous* random variables, whose cdf's are absolutely continuous functions; and *discrete* random variables, whose cdf's are piecewise constant.

It is sometimes of interest to generalize the notion of a random variable slightly to permit the values $\pm\infty$ in the range of the random variable. To preserve measurability, the sets of outcomes in Ω for which the variable takes on the values $+\infty$ and $-\infty$ must be in $\mathcal{F}$. This generalization of a random variable is known as an *extended* random variable.

2.2.3 Expectation

The cdf of a random variable completely describes its probabilistic behavior. A coarser description of this behavior can be given in terms of the *expected value* of the random variable.

A *simple* random variable is one taking on only finitely many values. The expected value of a simple random variable X taking on the values $x_1, x_2, \ldots, x_n$, is defined as

$$E\{X\} = \sum_{k=1}^{n} x_k P(F_k), \tag{2.4}$$

where $F_k = \{X = x_k\}$. The expected value of a general non-negative random variable X is defined as the (possibly infinite) value

$$E\{X\} = \sup_{\{\text{simple } Y | P(Y \leq X) = 1\}} E\{Y\}. \tag{2.5}$$

The expected value of an arbitrary random variable is defined if at least one of the non-negative random variables, $X^+ = \max\{0, X\}$ and $X^- = (-X)^+$, has a finite expectation, in which case

$$E\{X\} = E\{X^+\} - E\{X^-\}. \tag{2.6}$$

Otherwise $E\{X\}$ is undefined. The interpretation of $E\{X\}$ is as the average value of X, where the average is taken over all values in the range of X weighted by the probabilities with which these values occur.

When $E\{X\}$ exists, we write it as the integral

$$\int_{\Omega} X(\omega)P(\mathrm{d}\omega) = \int_{\Omega} X\mathrm{d}P. \tag{2.7}$$

If $E\{|X|\} < \infty$, we say that X is *integrable*.

The simplest possible non-trivial random variable is the indicator function of an event, say F, which is defined as

$$1_F(\omega) = \begin{cases} 1 & \omega \in F \\ 0 & \omega \notin F \end{cases}. \tag{2.8}$$

Since we have $E\{1_F\} = P(F)$, it follows that knowledge of expectations of all random variables is equivalent to knowledge of the probability distribution P. For an event F and a random variable X whose expectation exists, we write

$$E\{X1_F\} = \int_F X(\omega)P(\mathrm{d}\omega) = \int_F X\mathrm{d}P. \tag{2.9}$$

The integral (2.7) is a Lebesgue–Stieltjes integral, and it equals the Lebesgue–Stieltjes integral

$$\int_{\mathbf{R}} xP_X(\mathrm{d}x), \tag{2.10}$$

which in turn equals the Riemann–Stieltjes integral

$$\int_{-\infty}^{\infty} x\mathrm{d}F_X(x), \tag{2.11}$$

whenever this integral converges. For a continuous random variable we thus have

$$E\{X\} = \int_{-\infty}^{\infty} xf_X(x)\mathrm{d}x, \tag{2.12}$$

where $f_X(x) = \mathrm{d}F_X(x)/\mathrm{d}x$ is the *probability density function* (pdf) of X. Similarly, for a discrete random variable X taking the values $x_1, x_2, \ldots$, we have

$$E\{X\} = \sum_{k=1}^{\infty} x_k p_X(x_k), \tag{2.13}$$

where $p_X(x) = P(X = x)$ is the *probability mass function* (pmf) of X.

If X is a random variable and g is a measurable function from $(\mathbb{R}, \mathcal{B})$ to $(\mathbb{R}, \mathcal{B})$, then the composite function $Y = g(X)$ is also a random variable, and its expectation (assuming it exists) is given by

$$E\{Y\} = \int_{\Omega} g(X(\omega))P(\mathrm{d}\omega) = \int_{\mathbf{R}} g(x)P_X(\mathrm{d}x). \tag{2.14}$$

(The right-hand integral also must equal $\int_{\mathbf{R}} yP_Y(\mathrm{d}y)$, of course.) The following quantities are of interest.

- The *moments* of a random variable:

$$E\{X^n\},\ n = 1, 2, \ldots \tag{2.15}$$

The first moment (which is the expected value) is called the *mean* of X.

- The *central moments* of a random variable:

$$E\left\{(X - E\{X\})^n\right\},\ n = 1, 2, \ldots \tag{2.16}$$

The second central moment is the *variance* of X.

- The function $M_X(t) = E\{e^{tX}\}$ for t complex, which is known as the *moment generating function* if $t \in \mathbb{R}$, and the *characteristic function* if $t = \mathrm{i}u$ with $\mathrm{i} = \sqrt{-1}$ and $u \in \mathbb{R}$. The characteristic function is sometimes written as $\phi_X(u) = M_X(\mathrm{i}u)$. Note that P_X and ϕ_X form a unique pair.

A useful result involving expectations of functions of random variables is *Jensen's inequality:*

$$E\{g(X)\} \geq g\left(E\{X\}\right), \tag{2.17}$$

which holds for convex functions g such that the left-hand side exists. If g is strictly convex, then the inequality in Jensen's inequality is strict unless X is almost surely constant.

2.2.4 Radon–Nikodym derivatives

Suppose P and Q are two probability measures on a measurable space $(\Omega, \mathcal{F})$. Then, we have the following theorems.

- *Lebesgue decomposition theorem.* There exists a random variable f (unique up to sets of P-probability zero), and an event H satisfying $P(H) = 0$, such that

$$Q(F) = \int_F f \mathrm{d}P + Q(H \cap F),\ \forall\ F \in \mathcal{F}. \tag{2.18}$$

We say that Q is *absolutely continuous with respect to P* (or that *P dominates Q*) if $P(F) = 0$ implies $Q(F) = 0$. We write $Q \ll P$. If $Q \ll P$ and $P \ll Q$, we say that P and Q are *equivalent* and we write $P \equiv Q$.
A trivial corollary to the Lebesgue decomposition theorem is the following.

- *Radon–Nikodym theorem.* Suppose $Q \ll P$. Then there exists a random variable f such that

$$Q(F) = \int_F f \mathrm{d}P,\ \forall\ F \in \mathcal{F}. \tag{2.19}$$

The function f appearing in (2.19) is called the *Radon–Nikodym derivative of Q with respect to P*, and we write

$$f(\omega) = \frac{\mathrm{d}Q}{\mathrm{d}P}(\omega). \tag{2.20}$$

If μ is a third probability measure on $(\Omega, \mathcal{F})$, and we have $Q \ll P \ll \mu$, then we can write[2]

$$\frac{dQ}{dP} = \frac{dQ/d\mu}{dP/d\mu}. \tag{2.21}$$

If P does not dominate Q, then, for any μ that dominates both P and Q, the ratio on the right-hand side of (2.21) is called the *generalized* Radon–Nikodym derivative of Q with respect to P.[3] The generalized Radon–Nikodym derivative is an extended random variable. In (2.18), the random variable f is the generalized Radon–Nikodym derivative of Q with respect to P, and $H = \{\omega \in \Omega | f(\omega) = \infty\}$. For any two probability measures Q and P, there is a common dominating probability measure μ (e.g. $\mu = (P+Q)/2$).

The Lebesgue decomposition theorem (and, hence, the Radon–Nikodym theorem) remains valid if the normalization axiom ($P(\Omega) = Q(\Omega) = 1$) is replaced with the condition that P and Q be *σ-finite*; i.e. that Ω can be written as the countable union of events, each of which has finite measure under P, and similarly for Q.

As a simple example of a Radon–Nikodym derivative, suppose $(\Omega, \mathcal{F}) = (\mathbb{R}, \mathcal{B})$ and P and Q have pdf's p and q, respectively. Then, the condition $Q \ll P$ is equivalent to the condition that the support of q is contained in the support of p, and in this case we have

$$\frac{dQ}{dP} = \frac{q}{p}. \tag{2.22}$$

This is a restatement of (2.21) for the case in which μ is the Lebesgue measure on $(\mathbb{R}, \mathcal{B})$.

2.2.5 Conditional expectation and independence

Conditional expectation is one of the most important fundamental concepts needed in optimal stopping theory. To introduce this notion, we consider again a probability space $(\Omega, \mathcal{F}, P)$, an integrable random variable X, and an event F with $P(F) > 0$.

The *conditional expectation* of X given F is the constant

$$E\{X|F\} = \frac{\int_F X dP}{P(F)}. \tag{2.23}$$

Note that this constant satifies the condition

$$\int_F E\{X|\mathcal{F}\} dP = \int_F X dP; \tag{2.24}$$

[2] Note that Radon–Nikodym derivatives are only unique up to sets of P-probability zero. From here on, whenever we equate two random variables, it is understood that equality holds *almost surely*; i.e. that the set of experimental outcomes for which equality does not hold has zero probability under the basic probability P. A similar interpretation should be used when interpreting inequalities or other statements involving random variables.

[3] The ratio can be defined arbitrarily in the ambiguous case in which $dQ/d\mu(\omega) = dP/d\mu(\omega) = 0$, since the set of ω for which this holds has zero probability under both measures P and Q.

that is, it has the same P-weighted integral that X does over the event F.

For our purposes, we would like to generalize this notion of conditional expectation to allow for conditioning on groups (in particular σ-fields) of events. To do so we begin with a few preliminary definitions.

- A σ-field $\mathcal{G}$ is a *sub-σ-field* of $\mathcal{F}$ if each element of $\mathcal{G}$ is also in $\mathcal{F}$. We write $\mathcal{G} \subseteq \mathcal{F}$.
- A random variable X is *measurable with respect to the sub-σ-field* $\mathcal{G}$ if $X^{-1}(B) \in \mathcal{G}$, $\forall\, B \in \mathcal{B}$.
- The *σ-field generated by a random variable* X (denoted by $\sigma(X)$) is the smallest σ-field with respect to which X is measurable.

Heuristically, $\sigma(X)$ can be thought of as the set of events on which the random variable X is constant. Further refinement of $\mathcal{F}$ beyond $\sigma(X)$ is not observable through X. This interpretation is exact for simple random variables.

Now suppose we have an integrable random variable X and a sub-σ-field $\mathcal{G}$. The conditional expectation of X given $\mathcal{G}$ is defined as any random variable Z, measurable with respect to $\mathcal{G}$, such that

$$\int_G Z\mathrm{d}P = \int_G X\mathrm{d}P,\ \forall\, G \in \mathcal{G}. \tag{2.25}$$

Such a random variable Z always exists, and we write $E\{X|\mathcal{G}\}$ to denote any such random variable. Any two versions of $E\{X|\mathcal{G}\}$ differ only on sets of probability zero. Of course, if X itself is $\mathcal{G}$-measurable, then $E\{X|\mathcal{G}\} = X$.

The interpretation of $E\{X|\mathcal{G}\}$ is that it is a random variable that behaves like X does to the extent that is consistent with the constraint that it be $\mathcal{G}$-measurable. In this sense, $E\{X|\mathcal{G}\}$ can be thought of as a projection of X onto $\mathcal{G}$.[4] Again appealing to a heuristic interpretation, $E\{X|\mathcal{G}\}$ is a random variable formed by replacing X by its centroid (with respect to P) on each set in $\mathcal{G}$. This can be illustrated with a very simple example. In particular, choose an event F with $0 < P(F) < 1$, and consider the sub-σ-field

$$\mathcal{G} = \{F, F^c, \Omega, \emptyset\}. \tag{2.26}$$

Then, for any random variable X we have

$$E\{X|\mathcal{G}\} = \begin{cases} E\{X|F\} & \omega \in F \\ E\{X|F^c\} & \omega \in F^c \end{cases}. \tag{2.27}$$

The existence of conditional expectation follows as a corollary to the Radon–Nikodym theorem. To see this, suppose first that X is non-negative. Consider the measurable space $(\Omega, \mathcal{G})$ and define two measures Q and P' on $(\Omega, \mathcal{G})$ by

$$Q(G) = \int_G X\mathrm{d}P \quad \text{and} \quad P'(G) = P(G),\ \forall\, G \in \mathcal{G}. \tag{2.28}$$

[4] This notion of conditional expectation as a projection is precise if we assume that $E\{X^2\} < \infty$. In this case, it is easily shown that $E\{X|\mathcal{G}\}$ minimizes $E\{(Z - X)^2\}$ over all $\mathcal{G}$-measurable random variables Z.

Note that Q is a finite measure (i.e. $Q(\Omega) < \infty$), but it is not a probability measure unless $E\{X\} = 1$. The probability measure P' is the *restriction of P to $\mathcal{G}$*. We have trivially that $Q \ll P'$ and thus we can write

$$Q(G) = \int_G f \mathrm{d}P', \ \forall \, G \in \mathcal{G}, \tag{2.29}$$

with $f = \mathrm{d}Q/\mathrm{d}P'$. Since f is $\mathcal{G}$-measurable and P and P' agree on $\mathcal{G}$, (2.29) can be rewritten as

$$Q(G) = \int_G f \mathrm{d}P, \ \forall \, G \in \mathcal{G}, \tag{2.30}$$

Comparing this relationship with the definition of $E\{X|\mathcal{G}\}$, we see that $E\{X|\mathcal{G}\} = \mathrm{d}Q/\mathrm{d}P'$. If X is not non-negative, then the existence of $E\{X|\mathcal{G}\}$ follows straightforwardly by decomposing X as $X^+ - X^-$, and proceeding as above.

If $\mathcal{H}$ is a sub-σ-field of $\mathcal{G}$, then it is easy to see that

$$E\left\{E\{X|\mathcal{G}\}\Big|\mathcal{H}\right\} = E\{X|\mathcal{H}\}. \tag{2.31}$$

The interpretation of this property is that the projection of X onto the σ-field $\mathcal{H}$ can be found by first projecting onto the finer σ-field $\mathcal{G}$ and then projecting the result onto $\mathcal{H}$. So, for example, since $E\{X\}$ is the conditional expectation of X given the trivial σ-field $(\Omega, \emptyset)$, we have

$$E\{E\{X|\mathcal{G}\}\} = E\{X\}, \tag{2.32}$$

for any $\mathcal{G} \subseteq \mathcal{F}$.

If Y is another random variable, then the conditional expectation of X given Y, is defined by

$$E\{X|Y\} = E\{X|\sigma(Y)\}. \tag{2.33}$$

The quantity $E\{X|Y\}$ has the interpretation of being the (probabilistically-weighted) average value of X, given that Y is fixed. It can be shown that X is $\sigma(Y)$-measurable if and only if there is a measurable function $g : (\mathbb{R}, \mathcal{B}) \to (\mathbb{R}, \mathcal{B})$ such that $X = g(Y)$. So, $E\{X|Y\}$ can be considered to be a measurable function of Y. Thus, we can think of a measurable function $g : \mathbb{R} \to \mathbb{R}$ such that $E\{X|Y\} = g(Y)$; we can write this function as $g(y) = E\{X|Y = y\}$.

Conditional expectations are closely related to the notion of (statistical) *independence*, defined in various circumstances as follows.

- Two events F and G are said to be independent if $P(F \cap G) = P(F)P(G)$.
- Two σ-fields $\mathcal{G} \subseteq \mathcal{F}$ and $\mathcal{H} \subseteq \mathcal{F}$ are said to be independent if all elements of $\mathcal{G}$ are independent of all elements of $\mathcal{H}$.
- A random variable X is said to be independent of a σ-field $\mathcal{G}$ if $\sigma(X)$ is independent of $\mathcal{G}$.
- Two random variables X and Y are said to be independent if $\sigma(X)$ and $\sigma(Y)$ are independent.

If X is independent of $\mathcal{G}$, then $E\{X|\mathcal{G}\} = E\{X\}$; and so if X and Y are independent, then $E\{X|Y\} = E\{X\}$. This condition is the opposite extreme of the condition that X be $\sigma(Y)$-measurable, which means that X (and $E\{X|Y\}$) is a function of Y.

Example 2.1: $(\mathbb{R}^2, \mathcal{B}^2)$. As an example to illustrate the above concepts, consider the case in which the measurable space of interest is the plane equipped with its Borel σ-field;[5] i.e. $(\Omega, \mathcal{F}) = (\mathbb{R}^2, \mathcal{B}^2)$. Consider the sub-$\sigma$-field $\mathcal{G}$ consisting of those sets of the form

$$\{\omega = (\omega_1, \omega_2) \in \mathbb{R}^2 | \omega_2 \in B\}, \;\; B \in \mathcal{B}, \tag{2.34}$$

where, as usual, $\mathcal{B}$ is the Borel σ-field of $\mathbb{R}$. It is straightforward to see that the set of $\mathcal{G}$-measurable random variables is the set of random variables that are functions of only the second coordinate of ω.

So, for an integrable random variable X, the random variable $E\{X|\mathcal{G}\}$ is any measurable function of ω_2 satisfying the equation

$$\int_{\mathbb{R}\times B} E\{X|\mathcal{G}\}(\omega_2) P(\mathrm{d}\omega) = \int_{\mathbb{R}\times B} X(\omega_1, \omega_2) P(\mathrm{d}\omega), \;\; \forall\, B \in \mathcal{B}. \tag{2.35}$$

Now suppose that the probability measure P assigns probabilities via the integral

$$P(F) = \int_F p(\omega_1, \omega_2)\mathrm{d}\omega_1 \mathrm{d}\omega_2, \;\; \forall\, F \in \mathcal{F}, \tag{2.36}$$

where p is a non-negative integrable function on $\mathbb{R}^2$, with total integral 1. Then, (2.35) becomes

$$\int_B E\{X|\mathcal{G}\}(\omega_2) p_2(\omega_2)\mathrm{d}\omega_2 = \int_B \int_{\mathbb{R}} X(\omega_1, \omega_2) p(\omega_1, \omega_2)\mathrm{d}\omega_1 \mathrm{d}\omega_2, \tag{2.37}$$

where

$$p_2(\omega_2) = \int_{\mathbb{R}} p(\omega_1, \omega_2)\mathrm{d}\omega_2. \tag{2.38}$$

It follows that the random variable

$$\frac{\int_{\mathbb{R}} X(\omega_1, \omega_2) p(\omega_1, \omega_2)\mathrm{d}\omega_1}{p_2(\omega_2)} \tag{2.39}$$

is a version of $E\{X|\mathcal{G}\}$. Note that the two measures Q and P' defined above are given here by

$$Q(B) = \int_B \int_{\mathbb{R}} X(\omega_1, \omega_2) p(\omega_1, \omega_2)\mathrm{d}\omega_1 \mathrm{d}\omega_2, \;\; B \in \mathcal{B}, \tag{2.40}$$

and

$$P'(B) = \int_B p_2(\omega_2)\mathrm{d}\omega_2, \;\; B \in \mathcal{B}. \tag{2.41}$$

[5] The Borel σ-field in the plane is the smallest σ-field containing all rectangles.

To illustrate the conditioning of one random variable on another, consider the random variable Y given by $Y(\omega) = \omega_2$. Then, $\sigma(Y) = \mathcal{G}$, and so

$$E\{X|Y\} = E\{X|\mathcal{G}\}(Y); \tag{2.42}$$

in other words

$$E\{X|Y = y\} = \frac{\int_{\mathbb{R}} X(\omega_1, y)p(\omega_1, y)d\omega_1}{p_2(y)}. \tag{2.43}$$

To illustrate independence, consider the sub-σ-field of sets of the form

$$\{(\omega_1, \omega_2) \in \mathbb{R}^2 | \omega_1 \in B\}, \; B \in \mathcal{B}. \tag{2.44}$$

Then $\mathcal{G}$ and $\mathcal{H}$ being independent means that P is a *product measure*; i.e. that

$$P(\{\omega|\omega_1 \in A\} \cap \{\omega|\omega_2 \in B\}) = P_1(A)P_2(B), \; \forall \, A, B \in \mathcal{B}, \tag{2.45}$$

where P_1 and P_2 are the restrictions of P to $\mathcal{G}$ and $\mathcal{H}$, respectively. Note that $\mathcal{H}$ is the σ-field generated by the random variable $X(\omega) = \omega_1$, so this X is independent of the above Y if and only if P is a product measure. In the case in which P has the representation (2.36), this is equivalent to the condition that $p(\omega_1, \omega_2)$ factor into the product of two functions, one of which depends only on ω_1 and the other of which depends only on ω_2.

Having defined conditional expectations allows for the definition of conditional probabilities by considering the conditional expectations of indicator functions. (Such conditional probabilities obey the same axioms that P does.) For example, the conditional cdf of a random variable X given another random variable Y is defined as

$$F_{X|Y}(x|Y) = E\left\{1_{\{X \leq x\}}|Y\right\}, \; x \in \mathbb{R}. \tag{2.46}$$

Since this random variable is, for each x, a function of Y, we can define a function $g(x, y)$ of two real variables such that $F_{X|Y}(x|Y) = g(x, Y)$. So, we can write $g(x, y)$ as $F_{X|Y}(x|y)$ and think of this function as the conditional cdf of X given the event $Y = y$. As a function of x, $F_{X|Y}(x|y)$ has the properties of an (unconditional) cdf. Similarly, conditional pdf's, pmf's, etc., can easily be defined. Such quantities allow for the straightforward computation of conditional expectations and probabilities.

2.2.6 Random sequences

A *stochastic process* (or *random process*) on a probability space $(\Omega, \mathcal{F}, P)$ is an indexed collection of random variables on $(\Omega, \mathcal{F})$, in which the set of indices is a subset of the reals. That is, a stochastic process is a collection $\{X_k; k \in \mathcal{K}\}$, where $\mathcal{K} \subseteq \mathbb{R}$ and where, for each $k \in \mathcal{K}$, X_k is a random variable on $(\Omega, \mathcal{F})$.

In this book, we are interested primarily in stochastic processes with one of two index sets, the non-negative integers (discrete-time processes, or *random sequences*), and the non-negative reals (continuous-time processes.) In either case, a stochastic process can be viewed as a mapping from the set Ω to the set of real-valued functions with domain $\mathcal{K}$. The required measurability of X_k for each $k \in \mathcal{K}$ constrains this mapping. For the case of random sequences, this mapping can be any measurable mapping from $(\Omega, \mathcal{F})$ to $(\mathbb{R}^\infty, \mathcal{B}^\infty)$, where the set $\mathbb{R}^\infty$ is the set of all sequences of real numbers. The σ-algebra $\mathcal{B}^\infty$ contains all cylinder sets, that is sets of the form

$$E = \{x \in \mathbb{R}^\infty; \left(x_{k_1}, \ldots, x_{k_n}\right) \in A\}, \tag{2.47}$$

where A ranges through[6] $\mathcal{B}^n$, $\{k_1, \ldots, k_n\}$ ranges through n-long subsets of the non-negative integers, and $n = 1, 2, \ldots$

Stochastic processes are essentially random functions, and so analytical properties are of interest. This general area is a very broad topic, and here we will discuss only one aspect of it: asymptotic behavior (i.e. notions of convergence). We limit attention to the case of discrete time, although the analogous continuous-time concepts are largely identical.

There are a number of useful modes of convergence that are of interest in analyzing problems involving random sequences. The four most common of these are defined as follows. In each case, we assume that $X, X_0, X_1, \ldots$ are random variables on the probability space $(\Omega, \mathcal{F}, P)$.

- $\{X_k; k = 0, 1, \ldots\}$ converges to X *almost surely* (or *with probability 1*) if

$$P\left(\lim_{k\to\infty} X_k = X\right) = 1.$$

  In this case we write $X_k \stackrel{\text{a.s.}}{\to} X$.
- Suppose $p \geq 1$. $\{X_k; k = 0, 1, \ldots\}$ converges to X *in the* pth *mean* (or *in* L^p) if

$$\lim_{k\to\infty} E\left\{|X_k - X|^p\right\} = 0.$$

  In this case we write $X_k \stackrel{L^p}{\to} X$.
- $\{X_k; k = 0, 1, \ldots\}$ converges to X *in probability* if

$$\lim_{k\to\infty} P\left(|X_k - X| > \epsilon\right) = 0, \forall \epsilon > 0.$$

  In this case we write $X_k \stackrel{\text{i.p.}}{\to} X$.
- $\{X_k; k = 0, 1, \ldots\}$ converges to X *in distribution* (or *in law*) if

$$\lim_{k\to\infty} F_{X_k}(x) = F_X(x),$$

  for all x at which F_X is continuous. In this case we write $X_k \stackrel{D}{\to} X$.

Either of almost-sure or L^p convergence implies convergence in probability, which in turn implies convergence in distribution. None of these implications can be reversed in general, although some limited converses hold. For example, convergence in

[6] Analogously with $\mathcal{B}$ and $\mathcal{B}^2$, $\mathcal{B}^n$ denotes the Borel σ-field in $\mathbb{R}^n$; that is, the smallest σ-field containing all n-dimensional rectangles.

distribution to a constant implies convergence in probability to this constant; convergence in probability of a dominated sequence implies convergence in L^p (see below); and every convergent-in-probability sequence has an almost surely convergent subsequence.

Some basic results relating the interchange of expectation and passage to the limit are listed here.

- *Fatou's lemma.* Suppose $X_k \geq 0, \ \forall k$. Then
$$\limsup_{k\to\infty} E\{X_k\} \leq E\left\{\limsup_{k\to\infty} X_k\right\}.$$
- *Dominated convergence theorem.* Suppose $\{X_k\}$ converges in probability to X, and there exists a random variable Y such that $|X_k| \leq Y, \ \forall k$ and $E\{Y^p\} < \infty$. Then
$$\lim_{k\to\infty} E\{|X_k - X|^p\} = 0.$$
- *Monotone convergence theorem.* Suppose $\{X_k\}$ converges monotonically up to X with probability one. Then, we have $E\{X_k\} \uparrow E\{X\}$.

The dominated convergence theorem[7] is a corollary to the more general result that the order of expectation and passage to the limit can be interchanged for a non-negative random sequence converging with probability 1, if and only if the sequence is *uniformly integrable* (u.i.):

$$\lim_{x\to\infty} \sup_k \int_{\{|X_k|\geq x\}} |X_k| \mathrm{d}P = 0. \tag{2.48}$$

In fact the non-negativity condition is not necessary in the case of uniform integrability. In particular, suppose that $X_k \overset{\text{a.s.}}{\to} X$. If $\{X_k\}$ is u.i. then $E\{X\} < \infty$ and $E\{X_k\} \to E\{X\}$. (See for comparison p. 185 [37].)

Moreover, there is the following relationship between u.i. L^p convergence and convergence in probability.

(1) If the random sequence $\{|X_k|^p\}$ is u.i. for some $p > 0$ and $X_k \overset{\text{i.p.}}{\to} X$, then $E\{|X|^p\} < \infty$ and $X_n \to X$ in L^p.
(2) If $E\{|X_k|^p\} < \infty$ and $X_k \overset{L^p}{\to} X$, then $\{|X_k|^p\}$ is u.i.

For a proof of the above result please refer to [58].

Similarly, the conclusion of Fatou's lemma holds under the more general hypothesis that $\{X_k^+\}$ is u.i. and $E\{\limsup_{k\to\infty} X_k\}$ exists.

Note that all of the above results remain valid if the expectations are conditional expectations.

As with pairs of events, σ-fields, and random variables, we can define independence among larger groups of these same objects.

[7] The case of the dominated convergence theorem in which the dominating random variable Y is a constant is also known as the *bounded convergence theorem*.

- A set of events $\{F_k, k \in \mathcal{K}\}$ is independent if

$$P\left(\bigcap_{k\in\mathcal{I}} F_k\right) = \bigcap_{k\in\mathcal{I}} P(F_k), \tag{2.49}$$

for all subsets $\mathcal{I}$ of $\mathcal{K}$.
- A set of sub-σ-fields $\{\mathcal{F}_k, k \in \mathcal{K}\}$ is independent if all sets of events of the form $\{F_k \in \mathcal{F}_k, k \in \mathcal{K}\}$ are independent.
- A set of random variables $\{X_k; k \in \mathcal{K}\}$ is independent if the set of σ-fields $\{\sigma(X_k); k \in \mathcal{K}\}$ is independent.

Some well-known basic results involving limits of sequences of random variables are given in the following list.

- *Kolmogorov's strong law of large numbers.* Suppose $\{X_k; k = 1, 2, \ldots\}$ is a sequence of independent and identically distributed (i.i.d.) random variables, each having finite mean μ. Then

$$\frac{1}{n}\sum_{k=1}^{n} X_k \stackrel{\text{a.s.}}{\to} \mu. \tag{2.50}$$

- *Chebychev's central limit theorem.* Suppose $\{X_k; k = 1, 2, \ldots\}$ is a sequence of i.i.d. random variables, each having finite mean μ and variance σ^2 satisfying $0 < \sigma^2 < \infty$. Then

$$\frac{1}{\sqrt{n}\sigma}\sum_{k=1}^{n}(X_k - \mu) \stackrel{D}{\to} X, \tag{2.51}$$

where X is a standard (zero mean and unit variance) Gaussian random variable.
- *Borel–Cantelli lemmas.* Suppose $F_1, F_2, \ldots$ is a sequence of events. Consider the event consisting of those experimental outcomes that are in infinitely many of the F_k's; i.e. consider

$$\{F_k \text{ i.o.}\} = \bigcap_{n=1}^{\infty}\bigcup_{k=n}^{\infty} F_k. \tag{2.52}$$

(Here 'i.o.' stands for 'infinitely often'.)
  - If $\sum_{k=1}^{\infty} P(F_k) < \infty$, then $P(F_k \text{ i.o.}) = 0$.
  - If $F_1, F_2, \ldots,$ are independent and $\sum_{k=1}^{\infty} P(F_k) = \infty$, then $P(F_k \text{ i.o.}) = 1$.

The first two of the above limit theorems can be generalized considerably.

2.3 Martingales and stopping times

In this section, we define the notions of martingales and stopping times, and we give some general properties relating to these objects. We do so primarily in the discrete-time case, leaving the continuous-time case to be developed at the end of the section.

Thus, except at the end, the time set for all sequences in this section is the set of non-negative integers. A good source for the material in this section is [157]. Reference [191] also contains most of these results.

2.3.1 Martingales

We consider throughout a probability space $(\Omega, \mathcal{F}, P)$, and we begin with some definitions.

- A *filtration* $\{\mathcal{F}_k; k = 0, 1, \ldots\}$ is an increasing sequence of sub-σ-fields of $\mathcal{F}$.
- A random sequence $\{X_k\}$ on $(\Omega, \mathcal{F}, P)$ is *adapted* to $\{\mathcal{F}_k\}$ if, for each k, X_k is $\mathcal{F}_k$-measurable.
- Suppose $\{X_k\}$ is a sequence of integrable random variables adapted to a filtration $\{\mathcal{F}_k\}$. Then, $\{X_k, \mathcal{F}_k\}$ is a *submartingale* if

$$E\{X_k|\mathcal{F}_\ell\} \geq X_\ell \; \forall \, \ell \leq k. \tag{2.53}$$

  (Again, as noted above, (in)equality of random variables is taken to mean almost-sure (in)equality with respect to the measure P.) When the filtration is understood, we will simply say that such a sequence $\{X_k\}$ is a submartingale. Note that it is sufficient that (2.53) hold for all k with $\ell = k - 1$.
- $\{X_k, \mathcal{F}_k\}$ is a *supermartingale* if $\{-X_k, \mathcal{F}_k\}$ is a submartingale. Submartingales and supermartingales are sometimes referred to as *semimartingales* as they are special cases of them.
- $\{X_k, \mathcal{F}_k\}$ is a *martingale* if it is both a submartingale and a supermartingale; i.e. if

$$E\{X_k|\mathcal{F}_\ell\} = X_\ell \; \forall \, \ell \leq k.$$

One way of thinking of a martingale is as the fortune accumulated by a gambler in playing a sequence of fair games. In particular, if X_ℓ is the gambler's fortune after the ℓth play, then the gambler's expected fortune at any time k in the future given all that has transpired up until time ℓ is just $E\{X_k|\mathcal{F}_\ell\} = X_\ell$. This interpretation is reinforced if we note that any martingale $\{X_k\}$ can be decomposed into the cumulative sum of a *fair sequence:*

$$X_k = \sum_{\ell=0}^{k} \Delta_\ell, \tag{2.54}$$

where the sequence $\Delta_0 = X_0$, $\Delta_k = X_k - X_{k-1}$ if *fair;* i.e. it satisfies the property

$$E\{\Delta_k|\mathcal{F}_\ell\} = 0, \; \forall \, 0 \leq \ell < k. \tag{2.55}$$

In the context of the gambling analogy, the fair sequence represents the sequence of winnings on each play of the game. Note that a submartingale can be thought of as the gambler's fortune when the game is biased in favor of the gambler, and a supermartingale is the gambler's fortune when the game is biased against the gambler.

A filtration can be viewed as describing the evolution in time of observable information in a given model. For example, $\{\mathcal{F}_k\}$ could be generated by a sequence $\{X_k\}$ of random variables, viz.,

$$\mathcal{F}_k = \sigma(X_0, X_1, \ldots, X_k), \; k = 0, 1, 2, \ldots, \tag{2.56}$$

where $\sigma(X_0, X_1, \ldots, X_k)$ denotes the smallest σ-algebra with respect to which all of the random variables $X_0, X_1, \ldots, X_k$, are measurable. Note that this filtration is the minimal filtration with respect to which the sequence $\{X_k\}$ is measurable. In this case, the martingale property is

$$E\{X_k|X_0, X_1, \ldots, X_\ell\} = X_\ell, \; \forall \, \ell \leq k. \tag{2.57}$$

It is easily seen that, if a sequence $\{X_k\}$ is a martingale with respect to any filtration, then it must be a martingale with respect to its natural filtration. However, there are often reasons for using filtrations other than the natural one (for example, there may be more than one martingale arising in a given model, each of which has its own natural filtration).

As a concrete example of a filtration, consider the measurable space $(\Omega, \mathcal{F}) = (\mathbb{R}^\infty, \mathcal{B}^\infty)$. Define a sequence of random variables $\{X_k\}$ via $X_k(\omega) = \omega_k, \; k = 0, 1, \ldots,$ where $\omega = (\omega_0, \omega_1, \omega_2, \ldots)$. Then, $\mathcal{F}_k$ in the filtration $\{\mathcal{F}_k = \sigma(X_0, X_1, \ldots, X_k)\}$ is the class of all $k+1$-dimensional cylinder sets in $\mathbb{R}^\infty$, as they appear in (2.47). Note that these sub-σ-fields are analogous to the sub-σ-field $\mathcal{G}$ discussed in Example 2.1 of Section 2.2.5.

Three major results concerning martingales are the following. In each case, we assume that $\{X_k, \mathcal{F}_k\}$ is a submartingale.

- *Kolmogorov's inequality.*

$$P\left(\max_{1\leq k\leq n} X_k \geq \alpha\right) \leq \frac{E\{X_n^+\}}{\alpha}, \; \forall \, \alpha > 0. \tag{2.58}$$

- *Martingale convergence theorem.* Suppose $\sup_k E\{X_k^+\} < \infty$. Then there is a random variable X finite a.s. on the set $\{X_1 > -\infty\}$ such that $X_k \stackrel{\text{a.s.}}{\to} X$. Moreover, if $\sup_k E\{|X_k|\} < \infty$, then $E\{|X|\} < \infty$.
- *Doob decomposition theorem.* $\{X_k\}$ can be written in a unique way as

$$X_k = M_k + A_k, \; k = 0, 1, 2, \ldots, \tag{2.59}$$

where $\{M_k, \mathcal{F}_k\}$ is a martingale and $\{A_k\}$ is an *increasing process;* that is, $\{A_k\}$ satisfies the following conditions:

(i) A_0 is $\mathcal{F}_0$-measurable, and A_{k+1} is $\mathcal{F}_k$-measurable for all $k \geq 0$; and
(ii) $0 = A_0 \leq A_1 \leq A_2 \leq \cdots$.

A process satisfying condition (i) is said to be *predictable.*

Kolmogorov's inequality can be compared with Markov's inequality, given by

$$P(X \geq \alpha) \leq \frac{E\{X^+\}}{\alpha}, \; \forall \, \alpha > 0 \tag{2.60}$$

for arbitrary random variables X.

Note that, if $\{X_k, \mathcal{F}_k\}$ is a non-negative supermartingale, then $\{-X_k, \mathcal{F}_k\}$ is a non-positive submartingale, which trivially satifies the hypothesis of the martingale convergence theorem. We can conclude that *all non-negative supermartingales are almost-surely convergent*. Since a martingale is also a supermartingale, this conclusion also applies, of course, to all non-negative martingales.

It is easily seen that the processes $\{M_k\}$ and $\{A_k\}$ arising in the Doob decomposition are given by

$$M_{k+1} - M_k = X_{k+1} - E\{X_{k+1}|\mathcal{F}_k\},\ k = 0, 1, \ldots,\ \ M_0 = X_0; \tag{2.61}$$

and

$$A_{k+1} - A_k = E\{X_{k+1}|\mathcal{F}_k\} - X_k,\ k = 0, 1, \ldots,\ \ A_0 = 0. \tag{2.62}$$

We now consider some examples of martingales and semimartingales.

Example 2.2: *(the likelihood ratio under the null hypothesis).* Consider a sequence $\{Y_k; k = 0, 1, \ldots\}$ of i.i.d. random variables, each of which has pdf p, and let $\{\mathcal{F}_k\}$ be the minimal filtration associated with $\{Y_k\}$. Suppose q is another pdf such that the *likelihood ratio*, $L = q/p$, is finite a.s. Define a random sequence $\{X_k; k = 0, 1, \ldots\}$ by

$$X_k = \prod_{\ell=0}^{k} L(Y_\ell),\ k = 0, 1, \ldots \tag{2.63}$$

Then, since $X_k = X_{k-1}L(Y_k)$, we have

$$\begin{aligned} E\{X_k|\mathcal{F}_{k-1}\} &= E\{X_{k-1}L(Y_k)|\mathcal{F}_{k-1}\} \\ &= X_{k-1}E\{L(Y_k)|\mathcal{F}_{k-1}\} \\ &= X_{k-1}E\{L(Y_k)\}, \end{aligned}$$

where the second equality follows because X_{k-1} is $\mathcal{F}_{k-1}$-measurable, and the third equality follows because $L(Y_k)$ is independent of $\mathcal{F}_{k-1}$. Now, since Y_k has pdf p, we have

$$E\{L(Y_k)\} = \int_{-\infty}^{\infty} \frac{q(x)}{p(x)} p(x)\mathrm{d}x = \int_{-\infty}^{\infty} q(x)\mathrm{d}x = 1. \tag{2.64}$$

So we conclude that $\{X_k, \mathcal{F}_k\}$ is a martingale. Since it is non-negative, it is therefore almost-surely convergent. (In fact $X_k \overset{\text{a.s.}}{\to} 0$.)

Example 2.3: *(the likelihood ratio under the alternative hypothesis).* Continuing with the set-up of Example 2.2, define $\{Y_k\}$, $\{\mathcal{F}_k\}$, and $\{X_k\}$ as above, with the exception that we now assume that each Y_k has pdf q instead of p. Now, we have

$$\begin{aligned} E\{L(Y_k)\} &= \int_{-\infty}^{\infty} \frac{q(x)}{p(x)} q(x)\mathrm{d}x \\ &= \int_{-\infty}^{\infty} \left(\frac{q(x)}{p(x)}\right)^2 p(x)\mathrm{d}x \end{aligned}$$

$$\geq \left(\int_{-\infty}^{\infty} \frac{q(x)}{p(x)} p(x) \mathrm{d}x \right)^2 = 1,$$

where the inequality is Jensen's inequality. Assuming p and q are distinct, this inequality will be strict, and so $\{X_k, \mathcal{F}_k\}$ is a submartingale. Here, we have

$$E\{|X_k|\} = (E\{L(Y_1)\})^k, \quad (2.65)$$

which grows exponentially if p and q are distinct, and the sufficient condition for submartingale convergence fails. (In this case, it happens that $\{X_k\}$ diverges almost surely.)

Example 2.4: *(the moment-generating function of cumulative sums).* Suppose $\{Y_k\}_{k=0}^{\infty}$ is a sequence of i.i.d. random variables, each of which has moment-generating function:

$$M(t) = E\left\{ \mathrm{e}^{tY_k} \right\}, t \in \mathbb{R}. \quad (2.66)$$

Let $\{\mathcal{F}_k\}$ be the minimal filtration generated by $\{Y_k\}$. Consider $t \in \mathbb{R}$ such that $M(t) < \infty$, and define a random sequence $\{X_k\}$ by

$$X_k = \frac{\mathrm{e}^{tS_k}}{[M(t)]^k}, k = 0, 1, \ldots, \quad (2.67)$$

where $S_k = \sum_{\ell=0}^{k} Y_\ell,\ k = 0, 1, \ldots$. Then, since

$$X_k = X_{k-1} \frac{\mathrm{e}^{tY_k}}{M(t)}, \quad (2.68)$$

an argument similar to that used in Example 2.2 shows that $\{X_k, \mathcal{F}_k\}$ is a martingale.

Since $X_k \geq 0$ for each k it follows that $\{X_k\}$ is almost surely convergent. In fact, it is readily seen that this example is a particular case of Example 2.2, in which the likelihood ratio is given by $L(y) = \mathrm{e}^{ty}/M(t)$. Thus, $X_k \stackrel{\text{a.s.}}{\to} 0$.

Example 2.5: Suppose $\{X_k, \mathcal{F}_k\}$ is a martingale and $g : \mathbb{R} \to \mathbb{R}$ is a convex function such that $g(X_k)$ is integrable for all k. Then, the sequence $\{Y_k\}$ with $Y_k = g(X_k)$ satifies, for each $k > 0$,

$$\begin{aligned} E\{Y_k|\mathcal{F}_{k-1}\} &= E\{g(X_k)|\mathcal{F}_{k-1}\} \\ &\geq g\left(E\{X_k|\mathcal{F}_{k-1}\}\right) \\ &= g(X_{k-1}) = Y_{k-1}, \end{aligned}$$

where the inequality is Jensen's inequality. So, $\{Y_k, \mathcal{F}_k\}$ is a submartingale.

As a specific example, consider the set-up of Example 2.2. Since $g(x) = -\log(x)$ is convex, the accumulated "log-odds" process $Z_k = \sum_{k=1}^{n} \log L(Y_k)$ is a supermartingale. Alternatively, under the set-up of Example 2.3, this process is a submartingale.

Example 2.6: Consider a filtration $\{\mathcal{F}_k; k = 0, 1, \ldots\}$, an integrable random variable X, and the sequence of random variables defined by

$$X_k = E\{X|\mathcal{F}_k\}, k = 0, 1, \ldots. \tag{2.69}$$

Since we have that

$$E\{|X_k|\} = E\left\{\left|E\{X|\mathcal{F}_k\}\right|\right\} \leq E\left\{E\{|X|\,|\mathcal{F}_k\}\right\} = E\{|X|\} < \infty, \tag{2.70}$$

where the inequality is Jensen's inequality, we can take X to be non-negative. Moreover, by the basic properties of conditional expectation:

$$E\{X_k|\mathcal{F}_\ell\} = E\{E\{X|\mathcal{F}_k\}|\mathcal{F}_\ell\} = E\{X|\mathcal{F}_\ell\} = X_\ell,\ \ell \leq k, \tag{2.71}$$

so that $\{X_k, \mathcal{F}_k\}$ is a martingale. Since (2.70) implies $\sup_k E\{|X_k|\} < \infty$, the martingale convergence theorem implies that $\{X_k\}$ is almost-surely convergent. It can be shown that, if $\{X_k\}$ can be written in the form (2.69), then it can be written in this form with X taken to be the almost-sure limit of $\{X_k\}$, since in that case $X_\infty = \lim_{n\to\infty} X_n$ a.s. is such that $E\{|X_\infty|\} < \infty$.

Martingales that can be generated as in Example 2.6 by conditioning on their (integrable) almost-sure limits are known as *regular* martingales. A necessary and sufficient condition for regularity of a martingale is that it be uniformly integrable. (See for comparison Proposition IV-2-3 of [157].) An equivalent condition is that the martingale converges in L^1.

Example 2.7: *(square-integrable martingales).* Consider the set-up of Example 2.5 with $g(x) = x^2$. (A martingale with finite second moments is called a *square-integrable* martingale.) Then the increasing process $\{A_k\}$ appearing in the Doob decomposition of the submartingale $\{X_k^2, \mathcal{F}_k\}$ is called a *quadratic variation* process and is given by

$$A_{k+1} - A_k = E\left\{(X_{k+1} - X_k)^2|\mathcal{F}_k\right\},\ k = 0, 1, 2, \ldots \tag{2.72}$$

Square-integrable martingales have some useful properties. Among these are the following: (Here $\{A_k\}$ is the increasing process associated with $\{X_k^2, \mathcal{F}_k\}$.)

- $\{X_k\}$ converges almost surely on $\{A_\infty < \infty\}$; and
- if $E\{A_\infty\} < \infty$ then $\{X_k\}$ converges in L^2 and is therefore regular.

2.3.2 Stopping times

Consider a probability space $(\Omega, \mathcal{F}, P)$ and a filtration $\{\mathcal{F}_k; k = 0, 1, \ldots\}$. A *stopping time* (or *Markov time*)[8] is an extended random variable T taking values in the set $\{0, 1, 2, \ldots\} \cup \{\infty\}$, with the property that

$$\{\omega \in \Omega | T(\omega) \leq k\} \in \mathcal{F}_k, \ \forall\, k \geq 0. \tag{2.73}$$

Note that this condition is equivalent to

$$\{\omega \in \Omega | T(\omega) = k\} \in \mathcal{F}_k, \ \forall\, k \geq 0, \tag{2.74}$$

or that $\{1_{\{T=k\}}\}$ be an adapted sequence.

Thus, a stopping time associated with a filtration is an extended random variable taking values in the time set of the filtration, with the property that it can assume the value k only on events that are measurable with respect to the filtration at k. If the filtration is the minimal filtration generated by a random sequence $\{X_k\}$, then whether $T = k$ or not can be determined by observing $X_0, X_1, \ldots, X_k$.

Stopping times play an important role in the theory of martingales. We now give some basic results relating these two concepts.

For a filtration $\{\mathcal{F}_k; k = 0, 1, \ldots\}$ define the σ-field $\mathcal{F}_\infty$ as the smallest σ-field containing $\cup_{k=0}^{\infty} \mathcal{F}_k$. For a stopping time T associated with $\{\mathcal{F}_k\}$ define the σ-field $\mathcal{F}_T$ as

$$\mathcal{F}_T = \{F \in \mathcal{F}_\infty | F \cap \{T \leq k\} \in \mathcal{F}_k \ \forall\, k = 0, 1, \ldots\}. \tag{2.75}$$

(Events in $\mathcal{F}_T$ are said to be *prior* to T.) It can be shown that, if $\{X_k\}$ is a random sequence adapted to $\{\mathcal{F}_k\}$, then X_T (i.e., $X_{T(\omega)}(\omega)$) is an $\mathcal{F}_T$-measurable random variable.

We have the following results, known as the *Optional Sampling theorem*.

- Suppose $\{X_k, \mathcal{F}_k\}$ is a submartingale, and that there is an integrable random variable X such that

  $$X_k \leq E\{X | \mathcal{F}_k\}, \ \forall\, k = 0, 1, \ldots \tag{2.76}$$

  Then for any stopping time T, the random variable X_T is integrable; and for any two stopping times S and T satisfying $P(S \leq T) = 1$, we have

  $$X_S \leq E\{X_T | \mathcal{F}_S\}. \tag{2.77}$$

  (Note that this results generalizes the submartingale property to randomly stopped submartingales.)
- Suppose $\{X_k, \mathcal{F}_k\}$ is a regular martingale. Then the above conclusions hold with (2.77) becoming

  $$X_S = E\{X_T | \mathcal{F}_S\}. \tag{2.78}$$

  In particular, for any stopping time T we have $E\{X_T\} = E\{X_0\}$.

[8] Some authors (e.g. Shiryaev [191]) reserve the term "stopping time" for stopping times that are almost surely finite, and use the term "Markov time" for extended stopping times. We will use the term "stopping time" as defined above.

Note that the optional sampling theorem implies that the above conclusions hold for any non-negative supermartingale with (2.77) being replaced by

$$X_S \geq E\{X_T|\mathcal{F}_S\}. \tag{2.79}$$

The optional sampling theorem holds for arbitrary stopping times T. However, its conclusions hold under weaker conditions for specific stopping times. To discuss this idea, suppose $\{X_k, \mathcal{F}_k\}$ is a martingale. Then, for every stopping time T associated with $\{\mathcal{F}_k\}$, the "stopped" sequence $\{X_{k\wedge T}, \mathcal{F}_k\}$ is also a martingale.[9] A stopping time is said to be *regular* for $\{X_k, \mathcal{F}_k\}$, if $\{X_{k\wedge T}, \mathcal{F}_k\}$ is a regular martingale.

A variety of useful conditions assuring the regularity of stopping times can be found in [157]. For example, if $\{X_k, \mathcal{F}_k\}$ is square integrable and $\{A_k\}$ is given by (2.72), then every stopping time T satisfying $E\{\sqrt{A_T}\} < \infty$ is regular for $\{X_k, \mathcal{F}_k\}$. Also, any bounded stopping time is trivially regular. The notion of a regular stopping time allows us to generalize the optional sampling theorem as follows.

- Suppose S, T, and U are stopping times such that U is regular for the martingale $\{X_k, \mathcal{F}_k\}$ and $P(S \leq T \leq U) = 1$. Then S and T are also regular for $\{X_k, \mathcal{F}_k\}$, $k = 0, 1, \ldots$ and

$$X_S = E\{X_T|\mathcal{F}_S\}. \tag{2.80}$$

- Suppose $\{X_k, \mathcal{F}_k\}$ is a submartingale with Doob decomposition $X_k = M_k + A_k$, and suppose the stopping time U is regular for $\{M_k, \mathcal{F}_k\}$. Then X_U is integrable if and only if A_U is, and in this case

$$X_S \leq E\{X_T|\mathcal{F}_S\} \tag{2.81}$$

holds for all stopping times S and T satisfying $P(S \leq T \leq U) = 1$.

Note that the second of these two results implies that, if T is regular for the martingale $\{M_k, \mathcal{F}_k\}$ in the Doob decomposition of the submartingale $\{X_k, \mathcal{F}_k\}$, then

$$E\{X_T\} = E\{X_0\} + E\{A_T\}. \tag{2.82}$$

An application of regular stopping times is found in the following set of results, known collectively as *Wald's identities*.

Suppose $\{Y_k; k = 1, 2, \ldots\}$ is an i.i.d. sequence adapted to a filtration $\{\mathcal{F}_k\}$, and let $\{S_k\}$ denote the sequence of cumulative sums, $S_k = \sum_{\ell=1}^{k} Y_\ell$. Then we have the following results.

- Suppose $\mu = E\{Y_1\}$ is finite. Then the sequence $\{S_k - k\mu, \mathcal{F}_k\}$ is a martingale, but it is not generally regular. However, every stopping time T satisfying $E\{T\} < \infty$, is regular for $\{S_k - k\mu, \mathcal{F}_k\}$. So, for any such stopping time

$$E\{S_T\} = \mu E\{T\}. \tag{2.83}$$

[9] $x \wedge y = \min\{x, y\}$.

(Note that, if $E\{Y_1^2\} < \infty$, then $\{S_k, \mathcal{F}_k\}$ is square integrable, and thus $E\{\sqrt{T}\} < \infty$ is sufficient for T to be regular for $\{S_k, \mathcal{F}_k\}$.)

- Suppose $E\{Y_1^2\}$ is finite, and let σ^2 denote the variance of Y_1. Then the sequence $\{[S_k - k\mu]^2 - k\sigma^2, \mathcal{F}_k\}$ is a martingale, but again, this martingale is not generally regular. However, similarly to the previous case, every stopping time T satisfying $E\{T\} < \infty$ is regular for $\{[S_k - k\mu]^2 - k\sigma^2, \mathcal{F}_k\}$, and so

$$E\{[S_T - T\mu]^2\} = \sigma^2 E\{T\}. \tag{2.84}$$

- Suppose M is the moment generating function of Y_1. If $t \in \mathbb{R}$ is such that $M(t) < \infty$, then the sequence

$$\left\{e^{tS_k}[M(t)]^{-k}, \mathcal{F}_k\right\} \tag{2.85}$$

is a martingale. (See Example 2.4 of Section 2.3.1.) With regard to this martingale, we have the following results.

  - For scalars $a, b > 0$, define the stopping time

$$T_{-b}^{a} = \inf\{k | S_k \notin (-b, a)\}. \tag{2.86}$$

Suppose $t \neq 0$ is such that $M(t) < \infty$. Then for all $0 < a, b < \infty$, T_{-b}^{a} is regular for the martingale (2.85). Thus, the relationship

$$E\left\{e^{tS_T}[M(t)]^{-T}\right\} = 1, \tag{2.87}$$

holds for any stopping time T such that $P(T \leq T_{-b}^{a}) = 1$.

  - Suppose $a > 0$, and $t \neq 0$ is such that $M(t) < \infty$ and $M'(t) \geq 0$. Then (2.87) holds for any stopping time T such that $P(T \leq T_{-\infty}^{a}) = 1$.

Wald's identities will be useful in analyzing the performance of statistical change detection algorithms.

2.3.3 Continuous-time analogs

We now consider extensions of the basic definitions and results of the preceding sections for processes for which the time set $\mathcal{K} = [0, \infty)$. Because the time set is now an uncountable set, issues of continuity of sample paths of the processes arise. In particular, a sample path is a realization $\{X_t(\omega); t \geq 0\}$ of the process for a fixed $\omega \in \Omega$. It is said that the process $\{X_t(\omega); t \geq 0\}$ has continuous sample paths if the functions $t \to X_t(\omega)$ seen as functions of t (for fixed ω) are continuous. Similarly, it is said that $\{X_t; t \geq 0\}$ has càdlàg paths if the functions $t \to X_t(\omega)$ seen as a functions of t (for fixed ω) are right-continuous with the existence of left limits. These two properties give rise to two very important sample spaces Ω, namely the space of continuous functions $C[0, \infty)$ and the space of functions with càdlàg paths $D[0, \infty)$. These sample spaces are equipped with natural σ-algebras $\mathcal{B}(C[0, \infty))$ and $\mathcal{B}(D[0, \infty))$, respectively, the details of which can be found in [118]. These σ-algebras host filtrations $\{\mathcal{F}_t\}$ which are defined in exactly the same way as in the discrete-time case. Also, the same definition of adaptivity of the process $\{X_t\}$ carries over, as do the definitions of a supermartingale, a submartingale and a martingale. Kolmogorov's inequality also holds with the maximum replaced by the supremum, and the martingale convergence theorem holds as well. Much of this material as well as the material that follows can be found, for example, in [118].

Because of the continuity of the time set, some additional concepts arise in the continuous time case. For example, for a given filtration $\{\mathcal{F}_t\}$ we define

(1) $\mathcal{F}_{t^+} = \cap_{s>t}\mathcal{F}_s$; and
(2) $\mathcal{F}_{t^-} = \sigma\{\mathcal{F}_s,\ 0 \leq s < t\}$.

The filtration $\{\mathcal{F}_t\}$ is said to be right-continuous if $\mathcal{F}_{t^+} = \mathcal{F}_t$ for all $t \geq 0$, and it is said to be left-continuous if $\mathcal{F}_{t^-} = \mathcal{F}_t$ for all $t \geq 0$.

The same definitions as in the discrete-time case also hold for stopping times, but now the notion of an optional time becomes relevant. A random time is said to be an *optional* time with respect to the filtration $\{\mathcal{F}_t\}$ if $\{T < t\} \in \mathcal{F}_t$ for all $t \geq 0$. It is easy to see that an optional time with respect to $\{\mathcal{F}_t\}$ is a stopping time with respect to $\{\mathcal{F}_{t^+}\}$.

We continue by developing a notion of predictability in continuous time. In particular, a process that is $\mathcal{F}_{t^-}$-measurable is predictable. Also, the process $\{X_{t^-}\}$ is $\mathcal{F}_t$-adapted, where $X_{t^-} = \lim_{s\uparrow t} X_s = \sup_{0\leq s<t} X_s$. If a process $\{X_t, \mathcal{F}_t\}$ is predictable and T is an $\mathcal{F}_t$ stopping time, then the random variable $X_T I_{\{T<\infty\}}$ is $\mathcal{F}_{T^-}$-measurable. The notion of predictability will prove useful in the definition of Poissonian martingales in Sections 2.4.2 and 2.6 below.

The Doob decomposition holds in this setting (where it is known as the Doob–Meyer decomposition), but with uniqueness only up to indistinguishability for $\{M_t\}$ and $\{A_t\}$. Also, the process $\{A_t\}$ satisfies additional technical conditions that we will not need here except as noted in the sequel (see for comparison [106,118,180]).

We can also restate the optional sampling theorem for continuous-time processes as follows. Suppose that $\{X_t, \mathcal{F}_t\}$ is a right-continuous submartingale and let S, T, and U be stopping times such that $P(S \leq T \leq U) = 1$. If either one of the following two conditions is satisfied:

(1) U is regular; or
(2) $\{X_t\}$ has a last element X_∞ with $E\{|X_\infty|\} < \infty$,

then

$$E\{X_T|\mathcal{F}_S\} \geq X_S. \tag{2.88}$$

2.4 Brownian motion and Poisson processes

In this section we introduce two of the most basic and important examples of continuous-time random processes: Brownian motion and the basic Poisson process.

2.4.1 Brownian motion

As noted in the preceding section, the theory of martingales can be extended to stochastic processes in which the time set $\mathcal{K}$ is a continuum. Many of the basic results are the same as those cited in the preceding section, although new issues involving the continuity properties of the filtration also arise as we have seen above. Rather than give a further exposition of this somewhat technical area, we will limit attention now to a specific type of continuous-time semimartingale, namely integrable processes with stationary, independent increments.

Throughout this section, we consider a probability space $(\Omega, \mathcal{F}, P)$, the index set $\mathcal{K} = [0, \infty)$, and a filtration $\{\mathcal{F}_t; t \geq 0\}$ of sub-σ-fields of $\mathcal{F}$.

Suppose $\{X_t; t \geq 0\}$ is a stochastic process adapted to $\{\mathcal{F}_t; t \geq 0\}$. We say that $\{X_t, \mathcal{F}_t; t \geq 0\}$ has *independent increments* if $X_t - X_s$ is independent of $\mathcal{F}_s$ for all $s < t$. We further say that $\{X_t, \mathcal{F}_t; t \geq 0\}$ has *stationary increments* if the distribution of $X_{s+t} - X_s$ does not depend on s for all $s, t > 0$.

If $\{X_t, \mathcal{F}_t\}$ has stationary and independent increments (s.i.i.), and X_1 is integrable, then $\{X_t, \mathcal{F}_t\}$ is a semimartingale with

$$E\{X_t|\mathcal{F}_s\} = X_s + \mu(t-s), \ \forall\, t > s \tag{2.89}$$

where $\mu = E\{X_1\}$. The parameter μ is called the *drift* of $\{X_t, \mathcal{F}_t\}$.

If $\{X_t, \mathcal{F}_t\}$ has s.i.i. and also satisfies the continuity condition

$$\lim_{t \downarrow 0} \frac{P\left(|X_t - X_0| > \epsilon \middle| \mathcal{F}_0\right)}{t} = 0, \ \forall\, \epsilon > 0, \tag{2.90}$$

then the conditional distribution of $X_t - X_s$ is Gaussian for all $t > s$. A process of this type is called a *Brownian motion*. Without significant loss of generality we assume that $X_0 = 0$, in which case $X_t \sim \mathcal{N}(\mu t, \sigma^2 t)$, $\forall\, t$,[10] where μ is the drift and $\sigma^2 > 0$ is the *variance parameter.* (σ^2 is the variance of X_1, which is finite.) A Brownian motion satisfying $\mu = 0$ and $\sigma^2 = 1$ is called a *standard* Brownian motion.

Like standard Gaussian random variables, standard Brownian motion is the limit in distribution of certain appropriately centered and scaled sums of random sequences. In particular, consider a sequence of i.i.d. zero-mean random variables $\{Y_k; k = 1, \ldots\}$ each of which has finite, non-zero variance σ^2. For each positive integer n, define the stochastic process $\{X_t^{(n)}\}$ by

$$X_t^{(n)} = \frac{1}{\sigma\sqrt{n}} Z_{nt}, \ t \geq 0, \tag{2.91}$$

[10] A continuous random variable X is Gaussian with mean μ and variance σ^2 if its probability density function f_X is given by

$$f_X(x) = \frac{1}{\sqrt{2\pi\sigma^2}} e^{-\frac{(x-\mu)^2}{2\sigma^2}}, \ x \in \mathbb{R}.$$

In this case we write $X \sim \mathcal{N}(\mu, \sigma^2)$.

where $\{Z_t\}$ is a stochastic process obtained by linearly interpolating the cumulative sums $S_k = \sum_{\ell=1}^{k} Y_\ell$ as[11]

$$Z_t = S_{\lfloor t \rfloor} + (t - \lfloor t \rfloor) Y_{\lfloor t \rfloor + 1}, \; t \geq 0. \tag{2.92}$$

Then, for any times $0 \leq t_1 \leq t_2 \leq \cdots \leq t_d < \infty$, we have[12]

$$\left(X_{t_1}^{(n)}, X_{t_2}^{(n)}, \ldots, X_{t_d}^{(n)} \right) \xrightarrow{\mathcal{D}} (W_{t_1}, W_{t_2}, \ldots, W_{t_d}), \tag{2.94}$$

where $\{W_t, \mathcal{F}_t^W\}$ is a standard Brownian motion and $\{\mathcal{F}_t^W\}$ is the minimal filtration generated by $\{W_t\}$.

Brownian motion can be viewed as a measurable mapping from the underlying probability space to the measurable space $(C[0, \infty), \mathcal{B}(C[0, \infty)))$ where $C[0, \infty)$ denotes the set of all continuous, real-valued functions on $[0, \infty)$ and $\mathcal{B}(C[0, \infty))$ is the smallest σ-field containing all of the open sets in $C[0, \infty)$ generated by the metric

$$\rho(x, y) = \sum_{n=1}^{\infty} \frac{1}{2^n} \max_{0 \leq t \leq n} (|x(t) - y(t)| \wedge 1). \tag{2.95}$$

A standard Brownian motion induces a measure P^* on $(C[0, \infty), \mathcal{B}(C[0, \infty)))$ known as *Wiener measure*. If we denote by $P^{(n)}$ the measure induced on $(C[0, \infty), \mathcal{B}(C[0, \infty)))$ by the process $\{X_t^{(n)}\}$ defined above, then it can be shown that $P^{(n)}$ converges *weakly* to P^*, where weak convergence means that

$$\int f dP^{(n)} \to \int f dP^*, \tag{2.96}$$

for every bounded continuous real-valued function f on $C[0, \infty)$. Weak convergence of probability measures on $\mathbb{R}$ is the same as convergence in distribution, so we again see a role for standard Brownian motion (in this case Wiener measure) that is analogous to that of the standard Gaussian distribution on $\mathbb{R}$.

Brownian motion plays a central role in both the theory and application of continuous-time stochastic processes, and the literature dealing with its properties is vast. A good, although not exhaustive, treatment is found in [118].

Some basic properties of Brownian motion are that its sample paths (i.e. the set of functions $\{X_t(\omega); t \geq 0\}$ with $\omega \in \Omega$) are almost surely continuous, but are almost surely not of bounded variation. The quadratic variation of $\{X_s(\omega); 0 \leq s \leq t\}$ (the quadratic variation will be defined below; compare Example 2.9 of Section 2.5) is almost surely equal to $\sigma^2 t$.

[11] $\lfloor t \rfloor$ denotes the greatest integer not larger than t.

[12] The joint distribution of a set $X_1, X_2, \ldots, X_n$ of random variables is the function

$$F(x) = P(X_1 \leq x_1, X_2 \leq x_2, \ldots, X_n \leq x_n), \; x = (x_1, x_2, \ldots, x_n) \in \mathbb{R}^n. \tag{2.93}$$

Convergence in distribution for sequences of vectors of random variables means that the joint distribution function of the components of the sequence of vectors converge to the joint distribution of the components of the limiting vector at points of continuity of the latter.

If $\{X_t, \mathcal{F}_t\}$ is a Brownian motion, then $\{X_t - \mu t, \mathcal{F}_t\}$ is a square-integrable martingale, and the process $\{(X_t - \mu t)^2, \mathcal{F}_t\}$ has the decomposition

$$(X_t - \mu t)^2 = M_t + \sigma^2 t, \; t \geq 0, \tag{2.97}$$

where $\{M_t, \mathcal{F}_t\}$ is a martingale satisfying $M_0 = 0$.

Analogously with the discrete-time case, Wald identities hold. In particular, we have the following.

- Suppose $\{X_t, \mathcal{F}_t\}$ is a Brownian motion with drift μ and variance parameter σ^2, and T is an integrable stopping time with respect to the filtration $\{\mathcal{F}_t\}$. Then

$$E\{X_T\} = \mu E\{T\}; \tag{2.98}$$

  and

$$E\{(X_T - \mu T)^2\} = \sigma^2 E\{T\}. \tag{2.99}$$

- Consider a family $\{P^{(\mu)};\ \mu \in \mathbb{R}\}$ of measures on $(\Omega, \mathcal{F})$ such that, under $P^{(\mu)}$, $\{X_t, \mathcal{F}_t\}$ is a Brownian motion with drift μ and unit variance. Suppose further that T is a stopping time with respect to the minimal filtration generated by $\{X_t\}$ such that $P^{(0)}(T < \infty) = 1$. Then

$$E^{(0)}\left\{ e^{\mu X_T - \frac{1}{2}\mu^2 T} \right\} = 1, \tag{2.100}$$

  if and only if $P^{(\mu)}(T < \infty) = 1$.

2.4.2 Poisson processes

If we relax the continuity condition (2.90), then a process having s.i.i. is no longer necessarily Gaussian. However, its distribution must still be among the family of *infinitely divisible distributions,* which consists of those distributions that can arise as the limits in distribution of sums of i.i.d. random variables. A distribution is infinitely divisible if its characteristic function is of the form

$$\log \phi(u) = i\alpha u - \frac{\beta^2 u^2}{2} + \gamma \int \left(e^{iux} - 1 - \frac{iux}{1+x^2} \right) \frac{1+x^2}{x^2} \nu(dx), \tag{2.101}$$

where ν is a probability measure on the reals that assigns zero probability to the origin, and where α, β, and $\gamma \geq 0$ are constants. Note that, in the case of Brownian motion, X_t has the above characteristic function with $\alpha = \mu t$, $\beta^2 = \sigma^2 t$, and $\gamma = 0$.

Another interesting case is that in which X_t has the above characteristic function with $\beta = 0, \alpha = \gamma = \lambda t$, where $\lambda > 0$ is a constant, and with ν placing unit mass at $x = 1$. In this case, X_t has the Poisson distribution[13] with parameter λt for all t,

[13] The discrete random variable X has a Poisson distribution with parameter λ if its probability mass function p_X is given by

$$p_X(k) = e^{-\lambda} \frac{\lambda^k}{k!}, \; k = 0, 1, \ldots$$

and $\{X_t; t \geq 0\}$ is a *homogeneous Poisson counting process* (HPCP) with *rate* λ. Since $E\{|X_t|\} = \lambda t$, $\{X_t, \mathcal{F}_t\}$ is a semimartingale (in this case, a submartingale), and as such it shares many of the properties of Brownian motion. However, the sample paths of $\{X_t\}$ are non-decreasing and piecewise constant, taking unit jumps at random times whose number in any given time interval (s, t) is Poisson with parameter $\lambda(t - s)$. Thus, the value of X_t is a count of all of the jumps occurring up to time t.

To get a better sense of this simplest of counting processes, namely the homogeneous Poisson process, and how it arises consider a sequence of events, say arrivals of customers in a queue. Let T_1 be the waiting time to the first event, T_2 the waiting time between the first and the second event, and so on. And let $S_n = \sum_{i=1}^{n} T_i$ be the time of occurrence of the nth event. Assume that $\{T_n\}$ is a sequence of random variables defined on a probability space $(\Omega, \mathcal{F}, P)$. The number of events that occur in any time interval $[0, t]$ is

$$X_t = \sup\{n | S_n \leq t\}.$$

Moreover the number of events in the interval $(s, t]$ is the increment $X_t - X_s$. If, in addition, the sequence of random times $\{T_n\}$ are independent and identically exponentially distributed with parameter λ,[14] then the process $\{X_t\}$ is a homogeneous Poisson counting process with rate λ.

It can easily be verified that if the process $\{X_t\}$ is adapted to a filtration $\{\mathcal{F}_t\}$, then the process $\{X_t - \lambda t\}$ is an $\mathcal{F}_t$- martingale.

An inhomogeneous Poisson process is the càdlàg process that is identical to a homogeneous Poisson process but with the property that,

- for $0 < t_1 < \cdots < t_k$ the increments $X_{t_1}, X_{t_2} - X_{t_1}, \ldots, X_{t_k} - X_{t_{k-1}}$ are independent and
-

$$P(X_t - X_s = n) = \frac{e^{\int_s^t \lambda_u du} (\int_s^t \lambda_u du)^n}{n!}, \qquad n = 0, 1, 2, \ldots,$$

where $\lambda_t > 0$ for all $t \in [0, \infty)$ is a deterministic function, called the *intensity* of $\{X_t\}$.

It can easily be verified that if such a process is adapted to a given filtration $\mathcal{F}_t$, then $X_t - \int_0^t \lambda_s ds$ is an $\mathcal{F}_t$-martingale.

Other processes of interest include the doubly stochastic Poisson process which is a point process with an intensity function $\{\lambda_t; t \geq 0\}$ that is $\mathcal{F}_0$ measurable. A discussion of this type of process can be found in [47]. A further class of interesting processes are the self-exciting Poisson processes, in which the intensity λ_t is $\mathcal{F}_t$ measurable for each $t \geq 0$ [200].

[14] A random variable X is exponentially distributed with parameter λ if its cumulative probability distribution function F_X is given by

$$F_X(x) = 1 - e^{-\lambda x}, \ x \geq 0.$$

A good treatment of the martingale properties of point processes such as Poissonian martingales can be found in [47]. We will also briefly mention them in the sequel as needed.

2.5 Continuous-time semimartingales

In Section 2.4 we introduced a special class of continuous-time semimartingale processes $\{X_t, \mathcal{F}_t\}$ with stationary and independent increments (s.i.i.) and X_1 integrable, with a specific reference to Brownian motion with drift and to the HPCP. We also mentioned that supermartingales, submartingales, and martingales are all special cases of semimartingales. It is natural to expect that the class of continuous-time semimartingale processes is much wider than such a special class of processes. Although we will not go into all the details involved in defining continuous-time semimartingale processes here, we will give some brief insight into their properties. The motivation for this will be made clear in the next section when we define the stochastic integral.

To begin, we will first introduce the local martingale process and the quadratic variation process. A local martingale, as the name suggests, is a process that behaves as a martingale locally. That is,

DEFINITION 2.1. *A real-valued, right-continuous, adapted process* $\{(X_t, \mathcal{F}_t); t \in [0, \infty)\}$ *on a probability space* $(\Omega, \mathcal{F}, P)$ *is a local martingale if there exists a sequence of* $\mathcal{F}_t$*- stopping-times* $\{T_n\}$ *such that*

- $0 \leq T_n(\omega) \leq T_{n+1}(\omega)\ \forall\ \omega \in \Omega,\ n = 1, 2, \ldots$;
- $P(\lim_{n\to\infty} T_n = \infty) = 1$; *and*
- $\{X_{t\wedge T_n}, \mathcal{F}_t); t \in [0, \infty)\}$ *is a martingale for each* n.

The sequence of stopping times $\{T_n\}$ is called a localizing sequence for the local martingale $\{X_t, \mathcal{F}_t\}$. It is obvious that if a process is a martingale then it is also a local martingale, but that the converse is not true. In fact, it is possible to find u.i. local martingales that are not martingales. An example of such a process is given on page 168 of [118]. It is important to mention however that if a given local martingale $\{(X_t, \mathcal{F}_t); t \in [0, \infty)\}$ with localizing sequence $\{T_n\}$ gives rise to the u.i. sequence $\{(X_{t\wedge T_n}, \mathcal{F}_t); t \in [0, \infty)\}$, then $\{(X_t, \mathcal{F}_t); t \in [0, \infty)\}$ is a martingale. Finally, an interesting connection between local martingale processes and supermartingales is that every local martingale that is a.s. non-negative is a supermartingale.

We now define processes of finite variation. We then will use these two ingredients (i.e. the finite variation process and the local martingale) to define a semimartingale.

DEFINITION 2.2. *A random process* $\{A_t; t \geq 0\}$ *is of finite variation on the interval* $[s, t]$ *if*

$$V[A, s, t] = \sup \sum_{k=1}^{n} |A_{t_k} - A_{t_{k-1}}| < \infty, \ a.s., \tag{2.102}$$

where the sup *is taken over all partitions* $s = t_0 < t_1 < \cdots < t_n = t$.

Examples of such processes include any non-increasing or non-decreasing processes. Any process of finite variation can be written as the difference of two non-decreasing processes.

We are now in a position to give a characterization of semimartingales. In particular, an adapted càdlàg process $\{X_t, \mathcal{F}_t\}$ is a semimartingale if it can be written in the form

$$X_t = X_0 + M_t + A_t, \ t \geq 0,$$

where X_0 is an $\mathcal{F}_0$ measurable random variable, $\{M_t\}$ is a square integrable local martingale (i.e. such that $E\{X_t^2\} < \infty$ for all $t > 0$), and $\{A_t\}$ is a càdlàg finite variation process.

It is worth mentioning that both of the above processes $\{M_t\}$ and $\{A_t\}$ are unique to within indistinguishability, which in a loose sense means that they are unique everywhere except on subsets of Ω that have probability 0.

The motivation for the above definitions will be made clear in the next section.

Before concluding this section we motivate the idea of a quadratic variation process. A quadratic variation process is a finite variation process that can be subtracted from the square of a square-integrable continuous (local) martingale in order to make this new process a square-integrable continuous (local) martingale. In Example 2.7 of Subsection 2.3.1 for example, the process $\{A_t\}$ is the quadratic variation process of the square-integrable martingale $\{M_t\}$. We now give a more precise characterization of this process.

Suppose that $\{X_t, \mathcal{F}_t\}$ is a local martingale with continuous paths on the probability space $(\Omega, \mathcal{F}, P)$. A real-valued process $\{A_t; t \in [0, \infty)\}$ on the same space is a quadratic variation process if

(1) $\{A_t, \mathcal{F}_t\}$ is an adapted process,
(2) The mappings $t \to A_t(\omega)$ are continuous and non-decreasing on $[0, \infty)$ for all $\omega \in \Omega$,
(3) $A_0 = 0$, and
(4) $\{X_t^2 - A_t, \mathcal{F}_t\}$ is a continuous local martingale.

Example 2.8: Let $\{X_t, \mathcal{F}_t\}$ be a square-integrable continuous martingale. Then, its associated quadratic variation process is the natural increasing process that arises in the continuous version of the Doob–Meyer decomposition of $\{(X_t)^2\}$.

Example 2.9: The quadratic variation process of the standard Brownian motion process $\{W_t\}$ is t since $E\{W_t^2 - t|\mathcal{F}_s\} = W_s^2 - s$.

In the case in which $\{X_t, \mathcal{F}_t\}$ is a square-integrable martingale (i.e. $E\{X_t^2\} < \infty$ for all $t \in [0, \infty)$), we also have that $E\{A_t\} < \infty$ for all $t \in [0, \infty)$. Moreover, the local martingale $\{X_t, \mathcal{F}_t\}$ with $E\{X_t^2\} < \infty$ is a martingale if and only if $E\{A_t\} < \infty$ for all $t \in [0, \infty)$ (see [106,180]).

REMARK 2.3. The natural increasing process that appears in the Doob–Meyer decomposition of the square of a right-continuous (but not continuous) square-integrable (local) martingale is not the same as the quadratic variation process. In fact the former is the *compensator* of the latter (i.e. the process that needs to be subtracted from the quadratic variation process so that the resulting process is a (local) martingale).

Example 2.10: Let $\{X_t, \mathcal{F}_t\}$ be a homogeneous Poisson counting process with rate λ. The quadratic variation process of the martingale $(X_t - \lambda t)$ is X_t, with compensator λt. Notice that $\{(X_t - \lambda t)^2 - \lambda t\}$ is an $\mathcal{F}_t$-martingale as can be easily verified from

$$E\{(X_t - \lambda t)^2 - \lambda t | \mathcal{F}_s\} = (X_s - \lambda s)^2 - \lambda s,$$

which follows by using $E\{X_t - X_s | \mathcal{F}_s\} = \lambda(t - s)$.

REMARK 2.4. It is interesting to note that any continuous local martingale of finite variation has the same intervals of constancy as its quadratic variation process. This implies that if the quadratic variation process of a martingale is constant everywhere then so will be the local martingale.

Note that the continuity of the sample paths of the local martingale in the above remark is essential. To see this, consider the $\mathcal{F}_t$-martingale $\{X_t - \lambda t\}$ from Example 2.10. While it has no intervals of constancy, its quadratic variation process $\{X_t\}$ is constant everywhere except for the points of jumps.

2.6 Stochastic integration

Brownian motion plays a central role in the theory of stochastic calculus, as we will review here briefly. We begin this discussion by considering how to properly define the stochastic integral

$$\int_0^t \Phi_s dW_s, \tag{2.103}$$

where $\{\Phi_t\}$ is a stochastic process and $\{W_t\}$ is a standard Brownian motion with $W_0 = 0$. It is evident that the above integral does not exist for many types of integrands (e.g., continuous ones) in the Lebesgue–Stieltjes sense, as the paths of Brownian motion are not of finite variation (see [106,180]). That is, the stochastic integral (2.103) cannot exist for such integrands in the "pathwise" sense for individually chosen ω, and thus can only exist as the limit in one of the senses defined in Section 2.2 of sums of appropriately chosen integrands.

To define integrals of the form (2.103), let us start with the elementary processes $\{\Phi_t\}$ of the form

$$\Phi_t = \sum_{i=0}^{2^n - 1} c_i 1_{\{\frac{i}{2^n}, \frac{i+1}{2^n}\}}(t). \tag{2.104}$$

For such functions it would be reasonable to define, for each $\omega \in \Omega$,

$$\int_0^t \Phi_s \mathrm{d}W_s = \sum_{i=0}^{2^n-1} c_i \left[W_{\frac{i+1}{2^n}t} - W_{\frac{i}{2^n}t} \right]. \tag{2.105}$$

However, it is easy to check that with $c_i = W_{\frac{i}{2^n}t}$ the expected value of the right-hand side of (2.105) is equal to 0, while with $c_i = W_{\frac{i+1}{2^n}t}$ the expected value of the right-hand side of (2.105) is equal to t. Thus, we must be careful in situations where the integrand and integrator are dependent. It turns out that the first choice of Φ_t, which was suggested by Itô, preserves important martingale properties of the stochastic integral that will be discussed in the sequel. The characteristic of such integrands is that they are not forward looking in time. That is, for each t, $\{\Phi_t < u\} \in \mathcal{F}_t = \sigma\{W_s, s \leq t\}$ for any $u \in \mathbf{R}$; i.e., $\{\Phi_t\}$ is $\mathcal{F}_t$-adapted.

To consider this issue further, we consider with the following class of integrands $\mathcal{C}$:

$$\mathcal{C} = \Bigg\{ \Phi : [0, \infty) \times \Omega \to \mathbf{R};\ \Phi \text{ is jointly measurable in } t \text{ and } \omega,$$

$$\Phi_t \text{ is } \mathcal{F}_t \text{ adapted, and } E\left\{ \int_0^t [\Phi_s]^2 \, \mathrm{d}s \right\} < \infty \Bigg\}. \tag{2.106}$$

Using the elementary form for Φ_t in (2.104), with $\{c_i\}$ chosen so that $\Phi \in \mathcal{C}$ along with the definition of the stochastic integral in (2.105), we can easily check that

$$E\left\{ \left(\int_0^t \Phi_s \mathrm{d}W_s \right)^2 \right\} = E\left\{ \int_0^t [\Phi_s]^2 \, \mathrm{d}s \right\}, \tag{2.107}$$

a property known as Itô's isometry.

The next step is to generalize the definition in (2.105) to all $\Phi \in \mathcal{C}$ using Itô's isometry. This is achieved in three main steps that involve going first from a sequence $\{\Phi^{(n)}\}$ of elementary integrands to bounded and continuous integrands in $\mathcal{C}$ with mean-square convergence of the sequence $\int_0^t \Phi_s^{(n)} \mathrm{d}W_s$; second from a sequence of bounded and continuous integrands in $\mathcal{C}$ to bounded integrands in $\mathcal{C}$; and finally from a sequence of bounded integrands in $\mathcal{C}$ to any integrand in $\mathcal{C}$, again using mean-square convergence. As a result of these steps we can now define the stochastic integral

$$\int_0^t \Phi_s \mathrm{d}W_s,$$

for any $\Phi \in \mathcal{C}$ as the limit in mean-square of the sequence of stochastic integrals

$$\int_0^t \Phi_s^{(n)} \mathrm{d}W_s,$$

where $\{\Phi^{(n)}\}$ is a sequence of elementary integrands in $\mathcal{C}$ (see [106,160]).

It is possible to show (see, e.g., [106,160]) that for any integrand $\Phi \in \mathcal{C}$ there exists a continuous modification of the stochastic integral of (2.103). It is also possible to show that this stochastic integral is a martingale. In particular, we have the following result (see, e.g., [106,118,160]).

THEOREM 2.5. *The stochastic integral (2.103), with integrand of the class $\mathcal{C}$ as in (2.106), is a continuous square-integrable martingale.*

It turns out that a "converse" to the above result also holds. That is, if $\{M_t\}$ is a continuous square-integrable $\mathcal{F}_t$-martingale, then there exists a unique $g \in \mathcal{C}$, such that

$$M_t = E\{M_0\} + \int_0^t g_s \mathrm{d}W_s \text{ a.s. for all } t \geq 0. \tag{2.108}$$

It is noteworthy that, if the last condition in the integrands of class $\mathcal{C}$ is replaced by the weaker condition $P(\int_0^t [\Phi_s]^2 \, \mathrm{d}s < \infty) = 1$, then the stochastic integral (2.103) still exists as the limit in probability of the sequence of stochastic integrals with elementary integrands (see for comparison, Chapter 3 of [160] or Chapter 5 of [106]), but it is only guaranteed to be a local martingale.

We now consider the class of one-dimensional Itô processes, which are processes of the form

$$X_t = X_0 + \int_0^t \mu_s \mathrm{d}s + \int_0^t \sigma_s \mathrm{d}W_s, \tag{2.109}$$

where the processes $\{\mu_s\}$ and $\{\sigma_s\}$ are $\mathcal{F}_t$-adapted (in which $\{\mathcal{F}_t\}$ can be a larger filtration than $\{\mathcal{F}_t^W\}$ as long as $\{W_t\}$ maintains its martingale property), and are such that

$$P\left(\int_0^t \sigma_s^2 \mathrm{d}s < \infty\right) = 1, \tag{2.110}$$

and

$$P\left(\int_0^t |\mu_s| \mathrm{d}s < \infty\right) = 1. \tag{2.111}$$

Notice that conditions (2.110) and (2.111) are weaker than the condition satisfied by processes of the class $\mathcal{C}$. However, in view of the discussion following equation (2.108), condition (2.110) is sufficient in order to guarantee that the stochastic integral $\int_0^t \sigma_s \mathrm{d}W_s$ is a local martingale. An Itô process thus consists of the sum of a finite variation term (compare Definition 2.2) and a local martingale. In particular, Itô processes are semimartingales but with continuous sample paths.

For (2.109) we use the shorthand notation

$$\mathrm{d}X_t = \mu_t \mathrm{d}t + \sigma_t \mathrm{d}W_t. \tag{2.112}$$

As noted above Itô processes (2.109) are semimartingales. In fact every continuous semimartingale is an Itô process [180]. The space of Itô processes (and in fact of all semimartingales continuous or with discontinuities) is closed under smooth maps. In other words, if $\{X_t\}$ is an Itô process and $g(t, x)$ is a once continuously differentiable function of the first variable and a twice continuously differentiable function of the second variable, then the process $Y_t = g(t, X_t)$ is also an Itô process. We now state the celebrated Itô formula.

THEOREM 2.6. *Suppose $\{X_t\}$ is an Itô process given by $\mathrm{d}X_t = \mu_t \mathrm{d}t + \sigma_t \mathrm{d}W_t$. If $g(t, x)$ is a once continuously differentiable function of the first variable and a* *twice continuously differentiable function of the second variable, then the process* $Y_t = g(t, X_t)$ *is again an Itô process and*

$$dY_t = \frac{\partial g}{\partial t}(t, X_t)dt + \frac{\partial g}{\partial x}(t, X_t)dX_t + \frac{1}{2}\frac{\partial^2 g}{\partial x^2}(t, X_t)(dX_t)^2,$$

where $(dX_t)^2 = dX_t dX_t$ *is computed according to the rules* $dtdt = dtdW_t = dW_t dt = 0$ *and* $dW_t dW_t = dt$.

For a detailed proof of the above theorem we refer the reader to [106], [118] or [160].

In fact the space of all semimartingales is closed under smooth maps $g(t, x)$ described above. This is summarized in the more general version of Theorem 2.6 which includes all semimartingale processes.

THEOREM 2.7. *Suppose* $\{X_t\}$ *is a càdlàg semimartingale and* $g(t, x)$ *is a once continuously differentiable function in the first variable and a twice continuously differentiable function in the second variable. Then the process* $\{Y_t\}$ *defined by* $Y_t = g(t, X_t)$ *is also a semimartingale, and the following formula holds:*

$$\begin{aligned} g(t, X_t) - g(0, 0) = &\int_{0+}^{t} \frac{\partial g}{\partial t}(s, X_{s-})ds \\ &+ \int_{0+}^{t} \frac{\partial g}{\partial x}(s, X_{s-})dX_s \\ &+ \frac{1}{2}\int_{0+}^{t} \frac{\partial^2 g}{\partial x^2}(s, X_s)(dA_s) \\ &+ \sum_{0<s\le t}\left[g(s, X_s) - g(s, X_{s-}) - \frac{\partial g}{\partial x}(s, X_{s-})\Delta X_s\right], \end{aligned}$$

where $\Delta X_s = X_s - X_{s-}$ *and* $\{A_s\}$ *is the quadratic variation process associated with the process* $\{X_s\}$.

For a detailed proof of the above theorem we refer the reader to [180]. We now state another celebrated result, the innovation theorem (see [191]).

THEOREM 2.8. *Suppose that* $\{W_t\}$ *is a standard Brownian motion on the probability space* $(\Omega, \mathcal{F}, P)$, *and that* $\{\mu_t\}$ *is an independent measurable random process such that, for all* $t \ge 0$,

(1) $E\{|\mu_t|\} < \infty$, and
(2) $E\left\{\int_0^t \mu_s^2 ds\right\} < \infty$.

Suppose also that for the process $\{X_t\}$ *we have*

$$dX_t = \mu_t dt + \sigma dW_t,\ t > 0,\ X_0 = 0, \tag{2.113}$$

where $\sigma > 0$, *and consider* $\mathcal{F}_t^X = \sigma\{X_s, s \le t\}$. *Then there is a standard Brownian motion* $\overline{W}_t$ *such that*

$$dX_t = \overline{\mu}_t dt + \sigma d\overline{W}_t,\quad X_0 = 0,$$

and $\overline{\mu}_t = E\{\mu_t | \mathcal{F}_t^X\}$.

We also mention Girsanov's theorem which will be used extensively in the sequel. (See [160] and [118].)

THEOREM 2.9. *Consider a probability space $(\Omega, \mathcal{F}, P)$ under which $\{X_t, \mathcal{F}_t\}$ is an Itô process of the form $\mathrm{d}X_t = \mu_t \mathrm{d}t + \mathrm{d}W_t$; $t \leq T$, $X_0 = 0$, where $T \leq \infty$ is a constant and $\{W_t\}$ is a standard Brownian motion. Suppose that*[15]

$$E\left\{e^{\frac{1}{2}\int_0^T \mu_s^2 \mathrm{d}s}\right\} < \infty. \tag{2.114}$$

Define

$$M_t = e^{-\int_0^t \mu_s \mathrm{d}X_s + \frac{1}{2}\int_0^t \mu_s^2 \mathrm{d}s}, \quad 0 \leq t \leq T, \tag{2.115}$$[16]

and define the measure Q on the space $(\Omega, \mathcal{F}_T)$ by

$$Q_T(A) = E\{M_T(X)1_A\}, \quad A \in \mathcal{F}_T. \tag{2.116}$$

Then $\{X_t; 0 \leq t \leq T\}$ is a standard Brownian motion under Q, and the process $\{M_t; 0 \leq t \leq T\}$ is a martingale.

So far in this section we have been concerned with the question of properly defining the stochastic integral (2.103), where the integrator is a specific continuous-path martingale, that is, a Brownian motion process. We have also summarized some of the most important properties of Itô processes which are continuous-path processes defined through the stochastic integral (2.103). These properties were summarized in the above theorems.

A natural question that arises in the consideration of the stochastic integral of the form (2.103) is how to define it in the case in which the integrator is no longer a Brownian motion but a general process with càdlàg sample paths such as the Poissonian martingale of Example 2.10. Is then the progressive measurability of the integrand $\{\Phi_s\}$, a condition loosely speaking equivalent to joint measurability and adaptivity, enough to ensure that the stochastic integral with respect to a càdlàg martingale is also a càdlàg martingale? The answer to this question is in the negative as the following example demonstrates.

Example 2.11: Consider the stochastic integral $\int_0^t \Phi_s \mathrm{d}M_s$ where $M_t = N_t - \lambda t$ with N_t a homogeneous Poisson process with rate λ. Let $\Phi = I_{[0,T_1)}$ where T_1 is the time of the first jump of the process N_t. Then,

$$\begin{aligned}\int_0^t \Phi_s \mathrm{d}M_s &= \int_0^t \Phi_s \mathrm{d}N_s - \lambda \int_0^t \Phi_s \mathrm{d}s \\ &= -\lambda(t \wedge T_1),\end{aligned}$$

which is not martingale.

[15] Condition (2.114) is known as the Novikov condition.
[16] Equation (2.115) is known as the Cameron–Martin formula.

The above example indicates that we need to impose a predictability condition on the integrand $\{\Phi_s\}$ in order to preserve the martingale properties of the stochastic integral with respect to a point process.

Briefly stated, suppose that $\{X_t\}$ be a Poisson process which admits the $\mathcal{F}_t$-measurable intensity $\{\lambda_t\}$. Then the following properties hold.

- If $P\left(\int_0^t \lambda_s \mathrm{d}s < \infty\right) = 1, \forall\, t \geq 0$, then $\{X_t - \int_0^t \lambda_s \mathrm{d}s\}$ is an $\mathcal{F}_t$-local martingale.
- If $\{X_t\}$ is an $\mathcal{F}_t$-predictable process such that $E\{\int_0^t |X_s|\lambda_s \mathrm{d}s\} < \infty$, then $\{\int_0^t X_s \mathrm{d}M_s\}$ is an $\mathcal{F}_t$-martingale.
- If $\{X_t\}$ is an $\mathcal{F}_t$-predictable process such that $P(\int_0^t |X_s|\lambda_s \mathrm{d}s < \infty) = 1, \forall\, t$, then $\{\int_0^t X_s \mathrm{d}M_s\}$ is an $\mathcal{F}_t$-local martingale.

For further details on these properties please refer to [47] or [180].

3 Markov optimal stopping theory

3.1 Introduction

In this chapter, we review the theory of Markov optimal stopping. We will treat the discrete-time case in some detail, and then simply quote the analogous continuous-time results. This treatment is comprised of three main sections: Section 3.3 treats the relatively simple case of finite-horizon optimal stopping; Section 3.4 considers the infinite-horizon discrete-time case; and Section 3.5 provides a brief review of results in the continuous-time setting. Section 3.2 sets up the problem.

Most of the results contained in this set of notes can be found either in [57] or in [191], although our proofs are somewhat different in places.

3.2 Markov optimal stopping problems

We begin, then, with a discrete-time model. As a basic setting for this problem, consider a probability space $(\Omega, \mathcal{F}, P)$, equipped with a filtration $\{\mathcal{F}_k; k = 0, 1, \ldots\}$ and an adapted sequence $\{Y_k; k = 0, 1, \ldots\}$ of integrable random variables on $(\Omega, \mathcal{F})$. We call such a pair $\{Y_k, \mathcal{F}_k\}$ a *stochastic sequence.*

Suppose that, for each k, the random variable Y_k represents a reward that can be claimed at time k. For a stopping time T, $E\{Y_T\}$ is the expected reward that we will get if we stop according to T. An *optimal stopping problem* is a problem in which we try to choose a stopping time T to maximize $E\{Y_T\}$ over some interesting class of stopping times. In particular, for a class $\mathcal{S}$ of stopping times such that $E\{Y_T\}$ exists for all $T \in \mathcal{S}$, we define the *payoff*

$$V(\mathcal{S}) = \sup_{T \in \mathcal{S}} E\{Y_T\}; \tag{3.1}$$

and we are interested in computing $V(\mathcal{S})$, determining when it is achieved on $\mathcal{S}$, and, when possible, finding a stopping time in $\mathcal{S}$ that achieves it.

We will mainly be interested in so-called *Markov optimal stopping problems*, which can be described as follows. Consider a measurable space $(E, \mathcal{A})$ and a sequence $\{X_k; k = 0, 1, \ldots\}$ of measurable functions from $(\Omega, \mathcal{F})$ to $(E, \mathcal{A})$, such that X_k is $\mathcal{F}_k$-measurable for each k. (That is, $\{X_k\}$ is an adapted sequence of E-valued random variables.) The sequence $\{X_k, \mathcal{F}_k\}$ is said to be a *Markov process* if, for each $A \in \mathcal{A}$, we have

$$P(X_{k+\ell} \in A|\mathcal{F}_k) = P(X_{k+\ell} \in A|X_k)\,\forall k, \ell \geq 0. \tag{3.2}$$

Note that, for each k, ℓ and A, the quantity $P(X_{k+\ell} \in A|X_k)$ is a measurable function of X_k. The Markov process $\{X_k, \mathcal{F}_k\}$ is said to be *homogeneous* if this function does not depend on k.

A stochastic sequence $\{Y_k, \mathcal{F}_k\}$ is said to have a *Markov representation* if there is a Markov process $\{X_k, \mathcal{F}_k\}$ and a sequence of measurable functions $g_k : (E, \mathcal{A}) \to (\mathbb{R}, \mathcal{B})$, $k = 0, 1, \ldots$, such that $Y_k = g_k(X_k)$ for all k. A Markov optimal stopping problem is an optimal stopping problem whose reward sequence has a Markov representation. A Markov optimal stopping problem is said to be *stationary* if the underlying Markov process is homogeneous, and if there is a fixed function g such that $g_k = g\,\forall k$.

Note that the transition properties of a Markov process described by (3.2) do not completely specify the distribution of the process. In particular, each possible distribution for X_0 can potentially result in a different process. In the sequel, it will be convenient to think of a family $\{P_x; x \in E\}$ of distributions on the measurable space $(\Omega, \mathcal{F})$, such that $P_x(X_0 = x) = 1\,\forall x \in E$. Such a measure P_x can be described as follows. For each event F, define the measurable function $f_F : (E, \mathcal{A}) \to (\mathbb{R}, \mathcal{B})$ such that

$$f_F(X_0) = E\{1_F|X_0\}. \tag{3.3}$$

Then P_x is given by

$$P_x(F) = f_F(x). \tag{3.4}$$

In the context of Markov optimal stopping, we will usually be interested in the family of Markov optimal stopping problems

$$\sup_T E_x\{Y_T\},\ x \in E, \tag{3.5}$$

where $E_x\{\cdot\}$ denotes expectation under the measure P_x. Note that the xth problem in (3.5) is a Markov optimal stopping problem in which the initial value of the underlying Markov process is fixed at x.

3.3 The finite-horizon case: dynamic programming

3.3.1 The general case

It is of interest to consider first the above general (not necessarily Markov) optimal stopping problem in which the set of stopping times of interest is given by

$$\mathcal{S}^n = \{T \in \mathcal{T}|T \leq n\} \tag{3.6}$$

for some fixed $n > 0$, where $\mathcal{T}$ denotes the set of all stopping times with respect to the filtration $\{\mathcal{F}_k\}$. That is, we consider those stopping times that always stop by time n.

To examine this problem, let us consider for each $k = 0, 1, \ldots, n$, the set of stopping times

$$\mathcal{S}_k^n = \{T \in \mathcal{T} | k \leq T \leq n\}, \tag{3.7}$$

and the random variables $\gamma_0^n, \gamma_1^n, \ldots, \gamma_n^n$ defined by[1]

$$\gamma_k^n = \text{esssup}_{T \in \mathcal{S}_k^n} E\{Y_T | \mathcal{F}_k\}, \; k = 0, 1, \ldots, n. \tag{3.8}$$

That is, γ_k^n is the least upper bound on the possible payoff over all stopping times in $\mathcal{S}^n$ that do not stop before time k, conditioned on the information up to time k. Note that

$$|\gamma_k^n| \leq \sum_{\ell=k}^{n} E\{|Y_\ell| \,\Big|\mathcal{F}_k\}, \tag{3.9}$$

which implies that γ_k^n is integrable (and that it is a proper, not extended, random variable).

A natural stopping time to consider as a candidate for optimality is

$$T^n = \inf\{k \geq 0 | \gamma_k^n = Y_k\}. \tag{3.10}$$

That is, T^n stops the first time k such that the current reward Y_k equals the largest conditional expected reward (given current information) among all stopping times that stop from k onwards. Intuitively, one would not expect to be able to do better than this, and this intuition is borne out by the following result.

THEOREM 3.1. (BACKWARD INDUCTION). *Consider the stopping time T^n defined by (3.10). Then*

(i) *T^n is optimal over $\mathcal{S}^n$; i.e.*

$$E\{Y_{T^n}\} = V(\mathcal{S}^n) = E\{\gamma_0^n\}; \tag{3.11}$$

and

(ii) *the sequence $\gamma_0^n, \gamma_1^n, \ldots, \gamma_n^n$ satisfies the backward recursion*

$$\gamma_k^n = \max\left\{Y_k, E\{\gamma_{k+1}^n | \mathcal{F}_k\}\right\}, \; k = n-1, n-2, \ldots, 0, \tag{3.12}$$

with $\gamma_n^n = Y_n$.

Proof: Put $\gamma_n^n = Y_n$, and for $k = 0, 1, \ldots, n$, define stopping times in $\mathcal{S}_k^n$ given by

$$T_k^n = \inf\{\ell \geq k | Y_\ell = E\{\gamma_{\ell+1}^n | \mathcal{F}_\ell\}\}. \tag{3.13}$$

[1] The essential supremum (esssup) of a set $\mathcal{X}$ of random variables is any extended random variable Z having the properties

(i) $P(Z \geq X) = 1, \; \forall X \in \mathcal{X}$; and
(ii) $\{P(Y \geq X) = 1, \; \forall X \in \mathcal{X}\} \Rightarrow P(Y \geq Z) = 1, \; \forall X \in \mathcal{X}$.

The essential supremum of a set of random variables always exists.

We will show that

$$E\left\{Y_{T_k^n}\middle|\mathcal{F}_k\right\} = \gamma_k^n,\ k = 0, 1, \ldots, n. \tag{3.14}$$

This will imply that T_0^n is optimal in $\mathcal{S}^n$. Moreover, in the process of proving (3.14) we will prove (ii), which implies that $T_0^n = T^n$. The theorem will follow.

Note that (3.14) holds trivially for the case $k = n$. Suppose it also holds for the case $k = \ell$ for some nonzero $\ell < n$, and choose a stopping time $T \in \mathcal{S}_{\ell-1}^n$. Set $T' = \max\{T, \ell\}$, and note that $T' \in \mathcal{S}_\ell^n$. Since

$$E\{Y_T|\mathcal{F}_{\ell-1}\} = \begin{cases} Y_{\ell-1} & \text{on } \{T = \ell - 1\} \\ E\{Y_{T'}|\mathcal{F}_{\ell-1}\} & \text{on } \{T > \ell - 1\} \end{cases}, \tag{3.15}$$

we can write

$$\begin{aligned} E\{Y_T|\mathcal{F}_{\ell-1}\} &\le \max\{Y_{\ell-1}, E\{Y_{T'}|\mathcal{F}_{\ell-1}\}\} \\ &= \max\{Y_{\ell-1}, E\{E\{Y_{T'}|\mathcal{F}_\ell\}|\mathcal{F}_{\ell-1}\}\} \\ &\le \max\{Y_{\ell-1}, E\{\gamma_\ell^n|\mathcal{F}_{\ell-1}\}\}, \end{aligned} \tag{3.16}$$

where the equality follows from the fact that $\mathcal{F}_{\ell-1} \subseteq \mathcal{F}_\ell$, and the second inequality follows from the fact that $T' \in \mathcal{S}_\ell^n$ and the definition of γ_ℓ^n. Since T was chosen arbitrarily from $\mathcal{S}_{\ell-1}^n$ it follows from (3.16) that

$$\gamma_{\ell-1}^n \le \max\{Y_{\ell-1}, E\{\gamma_\ell^n|\mathcal{F}_{\ell-1}\}\}. \tag{3.17}$$

We now note that for $T = T_{\ell-1}^n$, we have $T' = T_\ell^n$. So, by the inductive hypothesis, the second inequality in (3.16) is an equality. This further implies that the first inequality in (3.16) is also an equality, since, by construction we have

$$T_{\ell-1}^n = \ell - 1 \iff Y_{\ell-1} = \max\{Y_{\ell-1}, E\{\gamma_\ell^n|\mathcal{F}_{\ell-1}\}\}. \tag{3.18}$$

So, we conclude that

$$E\left\{Y_{T_{\ell-1}^n}\middle|\mathcal{F}_{\ell-1}\right\} = \max\{Y_{\ell-1}, E\{\gamma_\ell^n|\mathcal{F}_{\ell-1}\}\}, \tag{3.19}$$

which implies

$$\gamma_{\ell-1}^n \ge \max\{Y_{\ell-1}, E\{\gamma_\ell^n|\mathcal{F}_{\ell-1}\}\}. \tag{3.20}$$

Combining (3.17), (3.19), and (3.20), we see that (3.14) holds for $k = \ell - 1$. Thus, induction implies the validity of (3.14) for all k. Note further that the recursion (ii) follows from (3.17) and (3.20). (That $\gamma_n^n = Y_n$ is obvious.) ■

We illustrate the above result with the following example.

Example 3.1 *(the selection problem [57,191]):* Suppose we have a box containing $n > 2$ objects that are rank ordered according to the ranks $1, 2, \ldots, n$, with 1 denoting the highest ranking. We are allowed to draw the objects at random from the box one at a time, and after drawing an object we can determine its rank relative to those objects already drawn. After drawing an object, we can either stop drawing from the box and keep that object, or we can discard the object and draw another one from the box (assuming we have not yet exhausted the n objects). Our objective is to maximize the probability that the object we stop with is the highest-ranking object in the box.

This problem can be cast within the framework of Theorem 3.1 as follows. Let Ω be the set of all permutations of the numbers $1, 2, \ldots, n$; $\mathcal{F}$ be the set of all subsets of Ω (i.e., $\mathcal{F} = 2^{\Omega}$); and P be the uniform distribution on Ω; i.e.,

$$P(F) = \frac{|F|}{n!}, \quad F \subseteq \Omega, \tag{3.21}$$

where $|F|$ denotes the number of elements in F. We observe the random variables $X_1, X_2, \ldots, X_n$, where X_k is the rank of the kth object drawn among those drawn previously to it; i.e., $X_k(\omega)$ is the number of the integers $\omega_1, \omega_2, \ldots, \omega_k$ that are not greater than ω_k. The information evolves according to the filtration $\mathcal{F}_k -\sigma(X_1, X_2, \ldots, X_k), k = 1, 2, \ldots, n$.

Ideally, we would like to stop at the value of k such that $\omega_k = 1$; or more realistically, we would like to maximize the probability that $\omega_T = 1$, over stopping times $T \leq n$. That is, we would like to solve the problem

$$\sup_{T \in \mathcal{S}^n} E\left\{1_{\{\omega_T = 1\}}\right\}. \tag{3.22}$$

Note, however, that the sequence $\{1_{\{\omega_k=1\}}\}$ is not adapted to $\{\mathcal{F}_k\}$, since, for example, we cannot say that the kth object drawn ranks first among all of the n objects just because it ranks first among those drawn up to time k. Fortunately, we can still formulate this problem as an optimal stopping problem, by considering the adapted sequence of rewards defined by

$$Y_k = E\{1_{\{\omega_k=1\}} | \mathcal{F}_k\}, \; k = 1, 2, \ldots, n. \tag{3.23}$$

For any $T \in \mathcal{S}^n$ we have

$$\begin{aligned} E\{Y_T\} &= \sum_{k=1}^{n} E\{Y_k | T = k\} P(T = k) \\ &= \sum_{k=1}^{n} E\left\{E\left\{1_{\{\omega_k=1\}} | \mathcal{F}_k\right\} | T = k\right\} P(T = k) \\ &= \sum_{k=1}^{n} E\left\{1_{\{\omega_k=1\}} | T = k\right\} P(T = k) \\ &= E\left\{1_{\{\omega_T=1\}}\right\}, \end{aligned} \tag{3.24}$$

where the third equality follows from the fact that $\{T = k\} \in \mathcal{F}_k$. So, the optimal stopping problem

$$\sup_{T \in \mathcal{S}^n} E\{Y_T\} \tag{3.25}$$

is the same as (3.22).

In order to solve (3.25), we would like to find the sequence $\{\gamma_k^n\}$ defined in Theorem 3.1. It is straightforward to show that the random variables $X_1, X_2, \ldots, X_n$ are independent with

$$P(X_k = \ell) = \frac{1}{k}, \ \ell = 1, 2, \ldots, k; \tag{3.26}$$

and that Y_k is given for each k by

$$Y_k = \frac{k}{n} 1_{\{X_k=1\}}. \tag{3.27}$$

These two properties imply that, for each $k = 2, \ldots, n$, the random variable γ_k^n is independent of $\mathcal{F}_{k-1}$ so that

$$\gamma_k^n = \max\left\{Y_k, E\{\gamma_{k+1}^n\}\right\}, k = n-1, n-2, \ldots, 1, \tag{3.28}$$

with $E\{\gamma_n^n\} = 1/n$. On applying (3.26) and (3.27), we can use (3.28) to write, for $k = n-1, n-2, \ldots, 0$,

$$E\{\gamma_k^n\} = \begin{cases} E\{\gamma_{k+1}^n\} & k < nE\{\gamma_{k+1}^n\} \\ \frac{1}{n} + \frac{k-1}{k} E\{\gamma_{k+1}^n\} & k \geq nE\{\gamma_{k+1}^n\} \end{cases}. \tag{3.29}$$

It can be seen (consider, e.g. the sequence $nE\{\gamma_k^n\}/(k-1)$) that the solution to (3.29) is given by

$$E\{\gamma_k^n\} = \begin{cases} \frac{k^*-1}{n} \sum_{\ell=k^*}^n \frac{1}{\ell-1} & k = 0, 1, \ldots, k^* \\ \frac{k-1}{n} \sum_{\ell=k}^n \frac{1}{\ell-1} & k = k^*+1, \ldots, n, \end{cases} \tag{3.30}$$

where

$$k^* = \min\left\{k \in \{2, 3, \ldots, n\} \middle| \sum_{\ell=k+1}^{n} \frac{1}{\ell-1} \leq 1\right\}. \tag{3.31}$$

It then follows from Theorem 3.1 that an optimal stopping time is given by

$$T^n = \begin{cases} \inf\{k \geq k^* | X_k = 1\} & \text{if } \min\{X_{k^*}, \ldots, X_n\} = 1 \\ n & \text{otherwise} \end{cases}; \tag{3.32}$$

that is, the optimal stopping strategy is to first rank $k^* - 1$ objects, and then to stop on the next object that ranks first among its predecessors. Theorem 3.1 also implies that the maximal expected reward is given by

$$E\{\gamma_1^n\} = \frac{k^*-1}{n} \sum_{\ell=k^*}^{n} \frac{1}{\ell-1}. \tag{3.33}$$

It can be shown that $k^* \sim n/\text{e}$, from which it follows that $E\{\gamma_1^n\} \sim 1/\text{e}$. That is, the probability that the optimal stopping time stops with the highest-ranked object is asymptotically $1/\text{e}$.

3.3.2 The Markov case

Theorem 3.1 characterizes the solution to general finite-horizon optimal stopping problems in terms of the random variables $\gamma_0^n, \gamma_1^n, \ldots, \gamma_n^n$. In the above example, the computation of this sequence is greatly facilitated by the fact that the variables $E\{\gamma_{k+1}^n|\mathcal{F}_k\}$ have a simple form; in particular, they are constants over the sample space. This example is actually a Markov optimal stopping problem in which the state space $E = \{1, 2, \ldots, n\}$; the sequence $X_1, X_2, \ldots, X_n$ of successive ranks is the Markov process; and the functions g_k are given by $g_k(x) = k1_{\{x=1\}}/n,\ k = 1, 2, \ldots, n$. The problem is particularly simple here because the Markov process is actually a sequence of independent random variables. However, it is true in general that the backward inductive procedure for obtaining the random variables γ_k^n can be simplified significantly for Markov optimal stopping problems. This idea is summarized in the following.

COROLLARY 3.2. (DYNAMIC PROGRAMMING). *Consider a Markov optimal stopping problem with Markov process* $\{X_k, \mathcal{F}_k\}$ *and reward sequence* $Y_k = g_k(X_k),\ k = 0, 1, \ldots$ *Then*

$$\gamma_k^n = f_k^n(X_k), k = 0, 1, \ldots, n, \tag{3.34}$$

where the functions f_k^n *are determined by the recursion*

$$f_k^n(x) = \max\left\{g_k(x), E\{f_{k+1}^n(X_{k+1})|X_k = x\}\right\}, x \in E,\ k = n-1, \ldots, 0, \tag{3.35}$$

with $f_n^n = g_n$.

Proof: We can prove this by induction. Clearly, since $\gamma_n^n = Y_n = g_n(X_n)$, (3.34) holds for $k = n$. Suppose it holds for $k = \ell \leq n$, and consider $\gamma_{\ell-1}^n$. We can write

$$E\{\gamma_\ell^n|\mathcal{F}_{\ell-1}\} = E\{f_\ell^n(X_\ell)|\mathcal{F}_{\ell-1}\} = E\{f_\ell^n(X_\ell)|X_{\ell-1}\}, \tag{3.36}$$

where the first equality follows from the inductive hypothesis, and the second equality follows from the Markovity of $\{X_k, \mathcal{F}_k\}$. Using (3.12) we have

$$\gamma_{\ell-1}^n = \max\left\{X_{\ell-1}, E\{f_\ell^n(X_\ell)|X_{\ell-1}\}\right\}, \tag{3.37}$$

and thus (3.34) holds for $k = \ell - 1$. Corollary 3.2 follows. ■

Theorem 3.1 and Corollary 3.2 imply that, for a finite-horizon Markov optimal stopping problem, the stopping time

$$T^n = \inf\{k \geq 0 | g_k(X_k) = f_k^n(X_k)\} \tag{3.38}$$

is optimal over $\mathcal{S}^n$. Moreover, Corollary 3.2 provides a backward recursion for computing the functions f_k^n from the reward functions g_k and the one-step transition properties of the underlying Markov process. This result is illustrated by the following example.

Example 3.2 *(American call options):* An American call option (see, e.g. [80]) is a financial instrument giving the holder the right to buy a fixed number of shares of a security at a given price (the *strike price*) at any time on or before a given date (the *expiration date*). Ignoring transaction costs and assuming that the option covers one share of the security, the time-discounted reward to the holder who exercises such an option on date k is given by

$$Y_k = \frac{(X_k - K)^+}{R^k}, k = 0, 1, \ldots, n, \tag{3.39}$$

where:

- X_k is the price of the security on date k;
- K is the strike price of the option;
- n is the expiration date of the option; and
- $R \geq 1$ is the riskless rate of return available to the option holder.

The option holder can observe the price of the security on each date, and would like to act on this information so as to maximize the expected reward returned by the option. That is, the option holder's information evolves according the filtration $\mathcal{F}_k = \sigma(X_0, X_1, \ldots, X_k),\ k = 0, 1, \ldots,$ and the goal is to solve the problem

$$\sup_{T \in \mathcal{S}^n} E\{Y_T\} \tag{3.40}$$

where $\mathcal{S}^n$ is the set of all stopping times with respect to $\{\mathcal{F}_k\}$ that are bounded by n.

The price process $X_0, X_1, \ldots$ can be written as

$$X_k = X_{k-1}\phi_k,\ k = 1, 2, \ldots, \tag{3.41}$$

where ϕ_k is the *return* on one unit of the security held during date $k - 1$. A common model for the behavior of security prices is that the returns on successive dates are independent of the past prices. We make this assumption, in which case $\{X_k, \mathcal{F}_k\}$ is a Markov process, and the above stopping problem thus becomes a Markov optimal stopping problem with

$$g_k(x) = \frac{(x - K)^+}{R^k},\ k = 0, 1, \ldots, n. \tag{3.42}$$

In order to find the optimal stopping time, we can proceed as in Corollary 3.2 to compute the sequence γ_k^n. In particular, the first step of the recursion (3.35) is given by

$$\begin{aligned} f_{n-1}^n(x) &= \max\left\{\frac{(x - K)^+}{R^{n-1}}, E\left\{\frac{(X_n - K)^+}{R^n}\middle| X_{n-1} = x\right\}\right\} \\ &= \max\left\{\frac{(x - K)^+}{R^{n-1}}, E\left\{\frac{(x\phi_n - K)^+}{R^n}\right\}\right\}. \end{aligned} \tag{3.43}$$

Straightforward calculus give

$$E\{(x\phi_n - K)^+\} = x\int_{K/x}^{\infty} P(\phi_n > v)dv,\ x > 0. \tag{3.44}$$

Setting $y = K/x$, we have

$$R^{n-1}yf_{n-1}^n(K/y)/K = \max\{(1 - y)^+, f(y)\},\ y > 0, \tag{3.45}$$

with

$$f(y) = \frac{1}{R}\int_y^\infty P(\phi_n > v)dv. \tag{3.46}$$

Of course for $y \geq 1$, we have $\max\{(1-y)^+, f(y)\} = f(y)$. For $y < 1$, there are two cases of interest. Note that

$$f(0) = \frac{E\{\phi_n\}}{R} \text{ and } f'(y) = -\frac{P(\phi_n > y)}{R} \leq 0. \tag{3.47}$$

Since $P(\phi_n > y)$ is non-increasing in y, it follows that f is convex. We also note that $f'(0) > -1$ and $\lim_{y\to\infty} f(y) \geq 0$ ($= 0$ if $E\{\phi_n\} < \infty$). From these properties we can conclude that $\max\{(1-y)^+, f(y)\} = f(y)$ for all $y > 0$ if $E\{\phi_n\} \geq R$. And, if $E\{\phi_n\} < R$, then

$$\max\{(1-y)^+, f(y)\} = \begin{cases} (1-y)^+ & 0 \leq y < y^* \\ f(y) & y \geq y^* \end{cases}, \tag{3.48}$$

where, if $P(\phi_n > 1) > 0$, $y^* \in (0,1)$ is the unique root of

$$y + f(y) = 1,\ 0 < y < 1; \tag{3.49}$$

and otherwise $y^* = \infty$.

From the above, it follows that

$$f_{n-1}^n(x) = \begin{cases} g_{n-1}(x) & \text{if } E\{\phi_n\} < R \text{ and } x > x^* \\ E\{f_n^n(X_n) \big| X_{n-1} = x\} & \text{otherwise,} \end{cases} \tag{3.50}$$

where $x^* = K/y^*$ and y^* is as above. We conclude that the optimal strategy on date $n-1$ is to exercise the option if the expected return is lower than the riskless rate *and* the current price X_{n-1} is higher than a threshold price x^*. Otherwise, we should wait until date n. The threshold price is zero if the probability that the price will increase is zero.

The above type of computation can be carried out recursively to obtain the optimal stopping time. It is straightforward to see that, as long as

$$E\left\{\prod_{j=\ell+1}^{n} \phi_j\right\} \geq R^{n-\ell},\ \forall \ell = k, \ldots, n, \tag{3.51}$$

the recursion (3.35) will yield

$$f_j^n(x) = E\{f_{j+1}(X_{j+1}) \big| X_j = x\} > g_j(x),\ x > 0,\ j = k, \ldots, n-1. \tag{3.52}$$

This means that the optimal strategy is to hold the option until expiration if it is still held at such a value of k. For the largest k such that

$$E\left\{\prod_{j=k+1}^{n} \phi_j\right\} < R^{n-k} \tag{3.53}$$

the recursion (3.35) is as in (3.50) with x^* determined as before, but with f given by

$$f(y) = \frac{1}{R^{n-k}} \int_y^\infty P(\phi_{k+1} \cdots \phi_n > v) dv, \ y > 0. \tag{3.54}$$

For j smaller than this largest k, the relationship (3.52) will again hold if $E\{\phi_{j+1}\} \geq R$; and a more complex expression will result otherwise.

A conclusion of the above is that, if (3.51) holds for $k = 0$, then the optimal strategy is to hold the option until its expiration date. An example of such a case is that in which the ϕ_k's are i.i.d. with $E\{\phi_1\} \geq R$.[2]

In order to extend the favorable properties of Markov optimal stopping seen in Corollary 3.2 to the infinite-horizon situation, it is convenient to first particularize the result of Corollary 3.2 to the stationary case.

Consider now a stationary Markov optimal stopping problem with underlying Markov process $\{X_k, \mathcal{F}_k\}$ and reward sequence $Y_k = g(X_k), \ k = 0, 1, \ldots$ For a set $\mathcal{S}$ of stopping times such that $E\{Y_T\}$ exists for all $T \in \mathcal{S}$, define the payoff

$$V_x(\mathcal{S}) = \sup_{T \in \mathcal{S}} E_x\{Y_T\}, \tag{3.55}$$

and for each non-negative integer n define the function

$$v_n(x) = V_x(\mathcal{S}^n), \ x \in E, \tag{3.56}$$

where, as before, $\mathcal{S}^n$ is the set of stopping times that are bounded by n.

It is convenient to define two operators on the set of functions $\mathcal{L}$ defined by

$$\mathcal{L} = \{f : (E, \mathcal{A}) \to (\mathbb{R}, \mathcal{B}) \mid E_x\{[f(X_1)]^-\} < \infty\}. \tag{3.57}$$

In particular, we define the operator $\mathcal{R}$ by

$$\mathcal{R}f(x) = E_x\{f(X_1)\}, \tag{3.58}$$

and the operator $\mathcal{Q}$ by

$$\mathcal{Q}f(x) = \max\{f(x), \mathcal{R}f(x)\}. \tag{3.59}$$

The following result summarizes the situation for the finite-horizon stationary Markov case.

COROLLARY 3.3. *Consider a stationary Markov optimal stopping problem with Markov process $\{X_k, \mathcal{F}_k\}$ and reward sequence $\{g(X_k)\}$. Then*

$$\gamma_k^n = v_{n-k}(X_k), \ k = 0, 1, \ldots, n, \tag{3.60}$$

and

$$v_n(x) = \max\{g(x), \mathbf{R}v_{n-1}(x)\} = \mathcal{Q}^n g(x), \forall x \in E, \ n = 1, 2, \ldots, \tag{3.61}$$

with $v_0 = g$.

[2] That a hold strategy is optimal for the i.i.d. case with $E\{\phi_1\} \geq R$ could also have been reached by noting that $\{g(X_k), \mathcal{F}_k; k = 0, 1, \ldots, n\}$ is a submartingale under this condition.

Proof: Consider the functions f_k^n defined in Corollary 3.2. By the homogeneity of $\{X_k\}$, we can write

$$f_k^n(x) = f_0^{n-k}(x). \tag{3.62}$$

Also, by Corollary 3.2, we can write $f_0^{n-k}(X_0) = E\{\gamma_0^{n-k}|X_0\}$. It follows that

$$f_0^{n-k}(x) = E_x\{\gamma_0^{n-k}\}, x \in E. \tag{3.63}$$

By Theorem 3.1 and Corollary 3.2, $E_x\{\gamma_0^{n-k}\} = v_{n-k}(x)$, and so (3.60) and the first equality in (3.61) follow.

Now consider the second equality in (3.61). This equality is proved if we can show

$$\max\{g(x), \mathbf{R}v_{n-1}(x)\} = \max\{v_{n-1}(x), \mathbf{R}v_{n-1}(x)\},\ n = 1, 2, \ldots. \tag{3.64}$$

To show this, we first note that

$$g(x) = v_0(x) \leq v_1(x) \leq \cdots,\ \forall x \in E, \tag{3.65}$$

and that

$$\{f(x) \geq h(x),\ \forall x \in E\} \Rightarrow \{\mathbf{R}f(x) \geq \mathbf{R}h(x),\ \forall x \in E\}. \tag{3.66}$$

Suppose $g(x) \geq \mathbf{R}v_{n-1}(x)$ for some $x \in E$. Then (3.65) and (3.66) imply that $g(x) \geq \mathbf{R}v_{n-2}(x)$, which implies in turn that $g(x) = v_{n-1}(x)$, via the equality $v_{n-1}(x) = \max\{g(x), \mathbf{R}v_{n-2}(x)\}$. Now suppose that $v_{n-1}(x) \geq \mathbf{R}v_{n-1}(x)$ for some $x \in E$. Then $v_{n-1}(x) \geq \mathbf{R}v_{n-2}(x)$, which again implies $g(x) = v_{n-1}(x)$ via $v_{n-1}(x) = \max\{g(x), \mathbf{R}v_{n-2}(x)\}$. So, $g(x)$ and $v_{n-1}(x)$ are equal whenever either is larger than $\mathbf{R}v_{n-1}(x)$, and the second equality in (3.61) follows. ■

For fixed n, Corollary 3.3 does not really allow for significant practical simplification of the recursions of Corollary 3.2. However, by allowing the "forward" computation of the sequence of functions $v_0, v_1, \ldots$, Corollary 3.3 will be useful in the computation of optimal solutions to the stationary Markov optimal stopping problems in the infinite-horizon case. This objective will be pursued in the following section.

3.4 The infinite-horizon case

In this section, we extend the results of the preceding section to the infinite-horizon situation. In particular, for a stochastic sequence $\{Y_k, \mathcal{F}_k\}$ we consider the problem

$$\sup_{T \in \mathcal{T}} E\{Y_T\}, \tag{3.67}$$

where $\mathcal{T}$ denotes the set of all stopping times such that $E\{(Y_T)\}$ exists. In (3.67), and in general for a stochastic sequence $\{Y_k, \mathcal{F}_k\}$, we define

$$Y_\infty = \limsup_{k \to \infty} Y_k. \tag{3.68}$$

We first re-interpret the results of the preceding section, and then we use this re-interpretation to suggest approaches to the infinite-horizon case.

3.4.1 A martingale interpretation of the finite-horizon results

The finite-horizon results of the preceding section were based on the backward induction argument used to prove Theorem 3.1. In the infinite-horizon case, this approach is not suitable since there is no maximal time at which to start such an induction.

To gain some insight into how we might proceed in the infinite-horizon case, it is instructive to examine the result of Theorem 3.1 from a different point of view. In particular, suppose we start with the recursion (3.12); i.e.

$$\gamma_k^n = \max\left\{Y_k,\, E\{\gamma_{k+1}^n|\mathcal{F}_k\}\right\},\ k = n-1, n-2, \ldots, 0, \tag{3.69}$$

with $\gamma_n^n = Y_n$. Note that this recursion defines a unique set of random variables (up to sets of zero probability), which we know from Theorem 3.1 to be the essential suprema of $E\{Y_T|\mathcal{F}_k\}$ over the sets $\mathcal{S}_k^n$. However, for the moment, let us focus only on the fact that these variables satisfy (3.69).

The relationship (3.69) implies that $\{\gamma_k^n, \mathcal{F}_k; k = 0, 1, \ldots, n\}$ is a supermartingale, and that it *dominates* the reward sequence $\{Y_k; k = 0, 1, \ldots, n\}$; i.e.,

$$\gamma_k^n \geq Y_k,\ k = 0, 1, \ldots, n. \tag{3.70}$$

These properties imply that

$$E\{Y_T\} \leq E\left\{\gamma_T^n\right\} \leq E\left\{\gamma_0^n\right\},\ \forall\, T \in \mathcal{S}^n, \tag{3.71}$$

where the first inequality follows from the domination of $\{Y_k\}$ by $\{\gamma_k^n\}$, and the second follows from optional sampling via the fact that any bounded stopping time is regular for all martingales. Note that the left-hand inequality in (3.71) becomes an equality for the stopping time T^n of (3.10):

$$T^n = \inf\{k \geq 0 | Y_k = \gamma_k^n\}. \tag{3.72}$$

Using the Doob decomposition, we can write the middle term in (3.71) as

$$E\left\{\gamma_T^n\right\} = E\{M_T\} - E\{A_T\} = E\{M_0\} - E\{A_T\} = E\left\{\gamma_0^n\right\} - E\{A_T\}, \tag{3.73}$$

where $\{M_k, \mathcal{F}_k\}$ is a martingale and $\{A_k\}$ is an increasing process with respect to $\{\mathcal{F}_k\}$. Here, we have again used the regularity of $T \in \mathcal{S}^n$. Since the process $\{A_k\}$ is predictable (i.e., A_{k+1} is $\mathcal{F}_k$-measurable for each k), it is possible to define a stopping time T such that $A_T \equiv 0$. For example, the random variable

$$\tilde{T}^n = \inf\{k \geq 0 | A_{k+1} \neq 0\} \tag{3.74}$$

is such a stopping time, as is any stopping time bounded by $\tilde{T}^n$. For any such stopping time, the right-hand inequality of (3.71) becomes an equality.

Recall that the increasing process $\{A_k\}$ is given by

$$A_k = \sum_{\ell=0}^{k-1} \left(\gamma_\ell^n - E\{\gamma_{\ell+1}^n|\mathcal{F}_\ell\}\right),\ k = 1, \ldots, n,\quad A_0 = 0. \tag{3.75}$$

So, the stopping time $\tilde{T}^n$ of (3.74) is the first time that $\gamma_k^n > E\{\gamma_{k+1}^n | \mathcal{F}_k\}$. Since $\gamma_k^n = \max\{Y_k, E\{\gamma_{k+1}^n | \mathcal{F}_k\}\}$, the stopping time T^n of (3.72) is bounded by $\tilde{T}^n$. It follows that both inequalities in (3.71) are equalities for the stopping time T^n; and thus that T^n is optimal. Note that, since

$$\gamma_{\tilde{T}^n} = Y_{\tilde{T}^n} \tag{3.76}$$

it is also the case that $\tilde{T}^n$ is optimal. And, in fact, we see that any stopping time satisfying $T \leq \tilde{T}^n$ and $Y_T = \gamma_T^n$ is optimal. T^n is the minimal stopping time that satisfies these two conditions.

Note that several of the above relationships (e.g., (3.71) and (3.73)) rely only on the fact that $\{\gamma_k^n\}$ is a supermartingale dominating $\{Y_k\}$. It follows inductively from (3.69) that any supermartingale dominating $\{Y_k\}$ also must dominate $\{\gamma_k^n\}$; i.e.

$$Z_k \geq \max\{Y_k, E\{Z_{k+1} | \mathcal{T}_k\}\}, k - 0, \ldots, n-1, \text{ and } Z_n \geq Y_n, \tag{3.77}$$

imply

$$Z_k \geq \gamma_k^n, \ k = 0, 1, \ldots, n. \tag{3.78}$$

That is, $\{\gamma_k^n\}$ is the *minimal* supermartingale dominating $\{Y_k\}$. This minimality of $\{\gamma_k^n\}$ is closely related to the optimality of stopping times T satisfying $Y_T = \gamma_T^n$. To see this, suppose $\{Z_k\}$ satisfies (3.77), and suppose T is a stopping time satisfying

$$E\{Y_T\} = E\{Z_T\} = E\{Z_0\}. \tag{3.79}$$

Then, since $\{\gamma_k^n\}$ is between $\{Y_k\}$ and $\{Z_k\}$, it must be true that

$$E\{Y_T\} = E\{\gamma_T^n\} = E\{\gamma_0^n\}. \tag{3.80}$$

But the right-hand equality of (3.80) implies that $T \leq \tilde{T}^n$ and the left-hand equality means that $Y_T = \gamma_T^n$.

3.4.2 The infinite-horizon case for bounded reward

From the above discussion, we see that the optimality of T^n follows directly from the recursion (3.69) without resorting to inductive arguments. However, this still does not give a direct approach for the infinite-horizon case, since the recursion itself requires a final time to serve as its starting point.

The above development does, however, allow us to state the following.

PROPOSITION 3.4. *Suppose* $\{\gamma_k, \mathcal{F}_k; k = 0, 1, \ldots\}$ *is a stochastic sequence satisfying the following conditions.*

(i) $\gamma_\infty = Y_\infty$; (3.81)

(ii) $\gamma_k = \max\{Y_k, E\{\gamma_{k+1} | \mathcal{F}_k\}\}, \ k = 0, 1, \ldots$; (3.82)

(iii) $\{\gamma_k\}$ *is* $\mathcal{T}$*-regular,* i.e. $E\{\gamma_T\}$ *exists and*

$$E\{\gamma_T\} \leq E\{\gamma_0\}, \ \forall \, T \in \mathcal{T}; \tag{3.83}$$

(iv) *the stopping time*

$$T_0 = \inf\{k \geq 0 | \gamma_k = Y_k\} \tag{3.84}$$

satisfies $E\left\{M_{T_0}\right\} = E\{M_0\}$ *for the martingale* $\{M_k\}$ *in the Doob decomposition of* $\{\gamma_k\}$;

Then $T_0 \in \mathcal{T}$, *and*

$$E\left\{Y_{T_0}\right\} = V(\mathcal{T}). \tag{3.85}$$

Proof: The proof is essentially the same as the argument given in the preceding section for the optimality of T^n over $\mathcal{S}^n$. In particular, we have

$$E\{Y_T\} \leq E\{\gamma_T\} \leq E\{\gamma_0\}, \ \forall T \in \mathcal{T}, \tag{3.86}$$

where the left-hand inequality follows from (i) and (ii) and the right-hand inequality is (iii). We have $E\{\gamma_{T_0}\} = E\{M_{T_0}\}$, since the increasing predictable process $A_k = M_k - \gamma_k$ satisfies $A_{T_0} \equiv 0$ via (i). Now, (iv) implies that $E\{M_{T_0}\} = E\{M_0\} \equiv E\{\gamma_0\}$, which in turn implies that $E\{Y_{T_0}\} = E\{\gamma_0\}$, and the proposition follows. ■

So, we see that (3.67) can be solved if we can find a sequence $\{\gamma_k\}$ satisfying the hypotheses of Proposition 3.4. To look for such a sequence there are two possible approaches suggested by the finite-horizon case:

- define

$$\gamma_k = \text{esssup}_{T \in \mathcal{T}_k} E\{Y_T | \mathcal{F}_k\}, \ k = 0, 1, \ldots, \tag{3.87}$$

  where $\mathcal{T}_k$ is the subset of $\mathcal{T}$ satisfying $P(T \geq k) = 1$; or
- consider limiting properties of γ_k^n as $n \to \infty$.

Since the second of these ideas is the easiest to analyze, we will consider it first. Note that, for $k = 0, 1, \ldots$, we have

$$\gamma_k^k \leq \gamma_k^{k+1} \leq \gamma_k^{k+2} \leq \cdots \tag{3.88}$$

So, the extended random variables

$$\overline{\gamma}_k = \lim_{n \to \infty} \gamma_k^n, \ k = 0, 1, \ldots, \tag{3.89}$$

are well defined as the limits of monotonically increasing sequences of random variables. Since

$$\gamma_k^n = \max\{Y_k, E\{\gamma_{k+1}^n | \mathcal{F}_k\}\}, \ n = k+1, k+2, \ldots, \tag{3.90}$$

the monotone convergence theorem implies that

$$\overline{\gamma}_k = \max\{Y_k, E\{\overline{\gamma}_{k+1} | \mathcal{F}_k\}\}, \ k = 0, 1, \ldots \tag{3.91}$$

That is, $\{\overline{\gamma}_k, \mathcal{F}_k\}$ satisfies condition (i) of Proposition 3.4.

Now, we would like to find conditions under which $\{\overline{\gamma}_k, \mathcal{F}_k\}$ satisfies the other conditions of the proposition. The following result provides such conditions.

THEOREM 3.5. *Suppose*

$$E\left\{\sup_k |Y_k|\right\} < \infty. \tag{3.92}$$

Then $\{\overline{\gamma}_k, \mathcal{F}_k\}$ satisfies the conditions of Proposition 3.4, and hence the stopping time

$$\overline{T} = \inf\{k \geq 0 | Y_k = \overline{\gamma}_k\}, \tag{3.93}$$

is optimal in $\mathcal{T}$.

Proof: Put $Z = \sup_k |Y_k|$. Note that

$$|\overline{\gamma}_k| \leq E\{Z|\mathcal{F}_k\}, k = 0, 1, \ldots, \tag{3.94}$$

from which it follows that $\overline{\gamma}_k$ is integrable. Thus, (3.90) implies that $\{\overline{\gamma}_k, \mathcal{F}_k\}$ is a supermartingale.

To show condition (i), we first note that we trivially have $\overline{\gamma}_\infty \geq Y_\infty$ from (3.90). We also can write

$$\gamma_k^n \leq E\left\{\sup_{\ell \geq m} Y_\ell \middle| \mathcal{F}_k\right\}, \; 0 \leq m \leq k \leq n < \infty, \tag{3.95}$$

from which it follows that

$$\overline{\gamma}_k \leq E\left\{\sup_{\ell \geq m} Y_\ell \middle| \mathcal{F}_k\right\}, \; 0 \leq m \leq k < \infty. \tag{3.96}$$

By taking limits in k in the above equation, we can write

$$\overline{\gamma}_\infty \leq \lim_{k\to\infty} E\left\{\sup_{\ell \geq m} Y_\ell \middle| \mathcal{F}_k\right\} = E\left\{\sup_{\ell \geq m} Y_\ell \middle| \mathcal{F}_\infty\right\} = \sup_{\ell \geq m} Y_\ell, \tag{3.97}$$

where the first equality follows from Example 2.6 of Section 2.3.1, and the second equality follows from the $\mathcal{F}_\infty$-measurability of $\sup_{\ell \geq m} Y_\ell$. Taking the limit as $m \to \infty$ of this expression implies $\overline{\gamma}_\infty \leq Y_\infty$, and condition (i) follows.

Since

$$-\overline{\gamma}_k \geq -E\left\{\sup_{\ell \geq 0}(Y_\ell)^- | \mathcal{F}_k\right\}, \tag{3.98}$$

and the hypothesis of the theorem implies

$$E\left\{\sup_{\ell \geq 0}(Y_\ell)^-\right\} < \infty, \tag{3.99}$$

the optional sampling theorem for submartingales implies that $\{\overline{\gamma}_k, \mathcal{F}_k\}$ satisfies condition (iii).

To show condition (iv), let us write, for $n = 1, 2, \ldots,$

$$E\{M_{T_0}\} = \sum_{k=0}^{n} \int_{\{T_0=k\}} M_k \mathrm{d}P + \int_{\{T_0>n\}} M_n \mathrm{d}P + \int_{\{T_0>n\}} (M_{T_0} - M_n)\mathrm{d}P. \tag{3.100}$$

By recursively working backward and using the properties of conditional expectation and the martingale property, we can write

$$\sum_{k=0}^{n} \int_{\{T_0=k\}} M_k \mathrm{d}P + \int_{T_0>n} M_n \mathrm{d}P = \int_{T_0 \geq 0} M_0 \mathrm{d}P = E\{M_0\},\ \forall n \geq 1. \tag{3.101}$$

In order to show the validity of condition (iv), it is sufficient to show that

$$\lim_{n\to\infty} \int_{\{T_0>n\}} (M_{T_0} - M_n) \mathrm{d}P = 0. \tag{3.102}$$

(This will imply that $\int_{\{T_0>n\}} (M_{T_0} - M_n)\mathrm{d}P \equiv 0$.) Note that, on $\{T_0 > n\}$, we have $M_{T_0} = \overline{\gamma}_{T_0}$ and $M_n = \overline{\gamma}_n$. So, (3.102) is equivalent to

$$\lim_{n\to\infty} \int_{\{T_0>n\}} (\overline{\gamma}_{T_0} - \overline{\gamma}_n) \mathrm{d}P = 0, \tag{3.103}$$

or, further,

$$\lim_{n\to\infty} \int 1_{\{T_0>n\}} (\overline{\gamma}_{T_0} - \overline{\gamma}_n) \mathrm{d}P = 0. \tag{3.104}$$

Note that the integrand of (3.104) converges a.s. to zero since, by the martingale convergence theorem, $\{\gamma_k\}$ is a.s. convergent. From the optional sampling theorem, we know that $\overline{\gamma}_{T_0}$ is integrable. Also, we can bound $\overline{\gamma}_n$ according to (3.94). Moreover, under integration with respect to P the bound

$$|\overline{\gamma}_n| 1_{\{T_0>n\}} \leq E\{Z|\mathcal{F}_n\} 1_{\{T_0>n\}} \tag{3.105}$$

is equivalent to the bound

$$|\overline{\gamma}_n| 1_{\{T_0>n\}} \leq Z 1_{\{T_0>n\}}. \tag{3.106}$$

Thus, the bounded convergence theorem implies (3.102). Property (iv) and the theorem follows. ■

REMARK 3.6. Theorem 3.5 remains valid if the hypothesis is relaxed to the conditions

$$E\left\{\overline{\gamma}_0\right\} < \infty \text{ and } E\left\{\sup_k Y_k^-\right\} < \infty. \tag{3.107}$$

(See [57].)

3.4.3 The general infinite-horizon case

The hypothesis of the above theorem is too strong for most applications of interest. Thus, it is of interest to relax this condition. Relaxing the lower bound condition is the most useful in this regard, and so we would like to proceed under the weaker condition

$$E\{\sup_k Y_k^+\} < \infty. \tag{3.108}$$

Unfortunately, Theorem 3.5 is no longer true in general under this condition.

However, we can turn to the first of the two approaches mentioned earlier and define a sequence $\{\gamma_k\}$ as in (3.87); i.e.

$$\gamma_k = \text{esssup}_{T \in \mathcal{T}_k} E\{Y_T | \mathcal{F}_k\}, \ k = 0, 1, \ldots, \tag{3.109}$$

where, as before, $\mathcal{T}_k$ is the subset of $\mathcal{T}$ satisfying $P(T \geq k) = 1$. Throughout the sequel, $\{\gamma_k\}$ will refer to this particular sequence.

With regard to this sequence, we have the following result.

THEOREM 3.7. *Suppose*

$$E\{\sup_k Y_k^+\} < \infty. \tag{3.110}$$

Then the sequence $\{\gamma_k\}$ from (3.109) satisfies the conditions of Proposition 3.4, and hence the stopping time

$$T_0 = \inf\{k \geq 0 | Y_k = \gamma_k\}, \tag{3.111}$$

is optimal.

Proof: We first quote the following useful lemma, a proof of which is found in the Appendix at the end of this chapter.

LEMMA 3.8. *Suppose (3.110) holds. Then there is an increasing sequence $\{T_\ell\}$ of stopping times in $\mathcal{T}_k$ such that*

$$Y_k \leq E\left\{Y_{T_\ell} \middle| \mathcal{F}_k\right\} \uparrow \gamma_k \ as \ \ell \to \infty. \tag{3.112}$$

We now can prove Theorem 3.1.

We first note that condition (i) of Proposition 3.4 can be proven here exactly as in Theorem 3.5, since only the condition (3.110) is needed there.

To prove condition (ii), we note that

$$\begin{aligned} \gamma_k &= \max\{Y_k, \text{esssup}_{T \in \mathcal{T}_{k+1}} E\{Y_T | \mathcal{F}_k\} \} \\ &= \max\{Y_k, \text{esssup}_{T \in \mathcal{T}_{k+1}} E\{E\{Y_T | \mathcal{F}_{k+1}\} | \mathcal{F}_k\} \} \\ &\leq \max\{Y_k, E\{\gamma_{k+1} | \mathcal{F}_k\} \}. \end{aligned} \tag{3.113}$$

Now choose $T_1, T_2, \ldots$ from $\mathcal{T}_{k+1}$ as in Lemma 3.8 so that

$$E\{Y_{T_\ell} | \mathcal{F}_{k+1}\} \uparrow \gamma_{k+1}. \tag{3.114}$$

Then we have

$$\gamma_k \geq E\{Y_{T_\ell} | \mathcal{F}_k\} \uparrow E\{\gamma_{k+1} | \mathcal{F}_k\}, \tag{3.115}$$

where the limit is the monotone convergence theorem. Since, $\gamma_k \geq Y_k$, condition (ii) follows.

To see condition (iii), we note that

$$\gamma_0 \geq E\{Y_T | \mathcal{F}_0\} \ \forall T \in \mathcal{T}. \tag{3.116}$$

Thus,

$$E\{\gamma_0\} \geq E\{Y_T\} \ \forall T \in \mathcal{T} \tag{3.117}$$

by iterated expectations. So, condition (iii) is trivial here.

To see condition (iv), choose increasing $T_1, T_2, \ldots$ from $\mathcal{T}_0 \equiv \mathcal{T}$ as in Lemma 3.8 so that

$$E\{Y_{T_\ell}|\mathcal{F}_0\} \quad \uparrow \quad \gamma_0, \tag{3.118}$$

and define a stopping time $T = \lim_{\ell\to\infty} T_\ell$. Then, since $\mathcal{T}_{k+1} \subset \mathcal{T}_k$, we have

$$E\{M_0\} = E\{\gamma_0\} = \lim_{\ell\to\infty} E\{Y_{T_\ell}\} \leq E\left\{\limsup_{\ell\to\infty} Y_{T_\ell}\right\} = E\{Y_T\}, \tag{3.119}$$

where the first equality is the monotone convergence theorem, the inequality is Fatou's lemma, and the second equality follows from the definition of T. Since $E\{Y_T\} \leq E\{\gamma_0\}$, we must have $E\{Y_T\} = E\{\gamma_0\}$.

Now define another stopping time $T' = \min\{T, T_0\}$. We have

$$\begin{aligned} E\{Y_{T'}\} &= \int_{\{T_0<T\}} Y_{T_0} \mathrm{d}P + \int_{\{T_0\geq T\}} Y_T \mathrm{d}P \\ &= \int_{\{T_0<T\}} \gamma_{T_0} \mathrm{d}P + \int_{\{T_0\geq T\}} Y_T \mathrm{d}P. \end{aligned} \tag{3.120}$$

The first term on the right-hand side of (3.120) satisfies

$$\begin{aligned} \int_{\{T_0<T\}} \gamma_{T_0} \mathrm{d}P &= \sum_{k=0}^{\infty} \int_{\{T_0=k<T\}} \gamma_k \mathrm{d}P \\ &\geq \sum_{k=0}^{\infty} \int_{\{T_0=k<T\}} E\{Y_T|\mathcal{F}_k\} \mathrm{d}P \\ &= \int_{\{T_0<T\}} Y_T \mathrm{d}P. \end{aligned} \tag{3.121}$$

So, we have

$$E\{Y_{T'}\} \geq E\{Y_T\}, \tag{3.122}$$

and thus that

$$E\{Y_{T'}\} = E\{\gamma_0\}. \tag{3.123}$$

If we can show that $T' = T_0$, then condition (iv) follows. To see that this is the case, assume the contrary. Then there must be a $k \geq 0$ such that the set $B_k = \{T' = k < T_0\}$ has positive probability. By definition of T_0, we must have $\gamma_k - Y_k > 0$ on B_k. Thus

$$\int_{B_k} (\gamma_k - Y_k) \mathrm{d}P = \epsilon > 0. \tag{3.124}$$

Now since $B_k \in \mathcal{F}_k$, from Lemma 3.8 we can choose a stopping time $\tilde{T}_k \in \mathcal{T}_k$ such that

$$\frac{\epsilon}{2} \geq \int_{B_k} \gamma_k \mathrm{d}P - \int_{B_k} E\{Y_{\tilde{T}_k} | \mathcal{F}_k\} \mathrm{d}P \equiv \int_{B_k} \gamma_k \mathrm{d}P - \int_{B_k} Y_{\tilde{T}_k} \mathrm{d}P. \tag{3.125}$$

So,

$$\int_{B_k} (Y_{\tilde{T}_k} - Y_k) \mathrm{d}P \geq \frac{\epsilon}{2}. \tag{3.126}$$

Now, define a stopping time $\tilde{T}'_k$ by

$$\tilde{T}'_k = \begin{cases} \tilde{T}_k & \text{on } B_k \\ T' & \text{on } B_k^c. \end{cases} \tag{3.127}$$

Then, we have

$$\begin{aligned} E\{Y_{\tilde{T}'_k}\} &= \int_{B_k} Y_{\tilde{T}_k} \mathrm{d}P + \int_{B_k^c} Y_{T'} \mathrm{d}P \\ &\geq \int_{B_k} Y_k \mathrm{d}P + \int_{B_k^c} Y_{T'} \mathrm{d}P + \frac{\epsilon}{2} \\ &= E\{Y_{T'}\} + \frac{\epsilon}{2} \\ &= E\{\gamma_0\} + \frac{\epsilon}{2}. \end{aligned} \tag{3.128}$$

But, this is a contradiction of condition (iii), and so no such k exists and $T' = T_0$.

Thus, conditions (i)–(iv) of Proposition 3.4 hold for $\{\gamma_k\}$ and the optimality of T_0 follows. ■

REMARK 3.9. Note that it was unnecessary in this case to prove conditions (i) and (ii) to prove optimality of T_0, since it follows directly from the inequalities

$$E\{Y_T\} \leq E\{\gamma_0\} = E\{Y_{T_0}\}, \ \forall \, T \in \mathcal{T}, \tag{3.129}$$

which are proved independently of conditions (i) and (ii) in the above.

REMARK 3.10. Theorem 3.7 remains valid if the hypothesis is relaxed to the conditions

$$E\{\gamma_0\} < \infty \tag{3.130}$$

and there exists an optimal stopping time [57].

We illustrate the above theorem with an example.

Example 3.3: Let $a_n = n^2 - 1$ and $b_n = n(n-1)$ and Ω be the space of sequences

$$\omega_j = (a_1, \ldots, a_j, -b_{j+1}, -b_{j+2}, \ldots),$$

for $j = 1, 2, \ldots$. Let $\mathcal{F}$ be all subsets of Ω and assign the following probability to each ω_j:

$$P(\omega_j) = \frac{1}{j} - \frac{1}{j+1},$$

for $j = 1, 2, \ldots$. For each $n = 2, 3, \ldots$, let $X_n(\omega_j)$ be the nth coordinate of ω_j and $\mathcal{F}_n = \sigma\{X_2, \ldots, X_n\}$. We have

(1) $P(X_n = a_n) = 1/n = 1 - P(X_n = b_n)$
(2) $E\{X_n\} = (n-1)\{(n+1)/n + 1 - n\}$.

It is easy to see that only stopping times of the form $T_n = \min\{T, n\}$ where

$$T = \inf\{k \geq 2;\ X_k = -b_k\}$$

need be considered in searching for an optimal stopping time.

We notice that $E\{\sup_n X_n^+\} = \infty$ and so we are not surprised to find that $\lim_{n\to\infty} X_n = -\infty$, while $\gamma_\infty = 1$. To see this, notice that $E\{X_{T_n}\} = 1 - \frac{1}{n}$. Therefore, $\gamma_\infty = 1$.

This example appears in [57; p. 85].

3.4.4 The infinite-horizon case with Markov rewards

We see from Theorems 3.5 and 3.7, that the development of an optimal stopping time depends on the derivation of the sequence $\{\overline{\gamma}_k\}$ or $\{\gamma_k\}$. This task is greatly simplified in the Markov case, as the following result suggests.

COROLLARY 3.11. *Suppose $\{Y_k, \mathcal{F}_k\}$ has a stationary Markov representation. Then*
(1) $\overline{\gamma}_k = \overline{v}(X_k), k = 0, 1, \ldots,$ *where*

$$\overline{v}(x) = \lim_{n\to\infty} \mathcal{Q}^n g(x),\ x \in E, \tag{3.131}$$

and;
(2) $\gamma_k = v(X_k),\ k = 0, 1, 2, \ldots,$ *where*

$$v(x) = \sup_{T \in \mathcal{T}} E_x\left\{g(X_T)\right\},\ x \in E. \tag{3.132}$$

So, in particular,

$$\overline{T}_0 = \inf\left\{k \geq 0 | g(X_k) \geq \overline{v}(X_k)\right\} \tag{3.133}$$

and

$$T_0 = \inf\left\{k \geq 0 | g(X_k) \geq v(X_k)\right\}. \tag{3.134}$$

Both $\overline{v}$ and v satisfy the integral equation

$$f(x) = \max\left\{g(x), \mathcal{R}f(x)\right\}, x \in E. \tag{3.135}$$

REMARK 3.12. Item (2) of Corollary 3.11 remains valid in the case

$$Y_k = g(X_k) - \sum_{\ell=0}^{k-1} c(X_k), \tag{3.136}$$

where $|g|$ is bounded and $c \geq 0$ satisfies $E\{c(X_k)\} < \infty$, $\forall k$. (See, e.g. Chapter 2 of [191].) Note that the second term on the right of (3.136) can be viewed as representing a cost of sampling.

3.5 Markov optimal stopping in continuous time

To consider the continuous-time case, we begin by the definition of a continuous-time Markov process. Just as in the discrete-time setting consider a measurable space $(E, \mathcal{A})$ and the process $\{X_t, \mathcal{F}_t; t \geq 0\}$, where each of the random variables X_t is a measurable function from $(\Omega, \mathcal{F})$ to $(E, \mathcal{A})$, such that X_t is $\mathcal{F}_t$-measurable for each t. (That is, $\{X_t\}$ is an adapted process of E-valued random variables.) The process $\{X_t, \mathcal{F}_t\}$ is said to be a *Markov process* if, for each $A \in \mathcal{A}$, we have

$$P(X_{t+s} \in A|\mathcal{F}_t) = P(X_{t+s} \in A|X_t), \ \forall\, s, t \in [0, \infty) \tag{3.137}$$

which is similar to its discrete-time equivalent (3.2). Note that, for each $s, t \in [0, \infty)$, $P(X_{t+s} \in A|X_t)$ is a measurable function of X_t. The Markov process $\{X_t, \mathcal{F}_t\}$ is said to be *homogeneous* if this function does not depend on t.

Just as in the discrete-time case the transition properties of a Markov process described by (3.137) do not completely specify the distribution of the process. In particular, each possible distribution for X_0 can potentially result in a different process. This gives rise to the family of distributions $\{P_x; x \in E\}$ on the space $(\Omega, \mathcal{F})$ such that $P_x(X_0 = x) = 1$ for all $x \in E$. Each of the measures P_x can be fully described just as in discrete time (see (3.3) and (3.4)). The quantity $P_x(X_t \in A)$ for all $x \in E$, $A \in \mathcal{A}$ and $t \in [0, \infty)$ is called the *transition function* of the Markov process. For a detailed discussion on Markov processes in continuous time please refer to [142].

Let S be the set of $\mathcal{A}$-measurable functions $f : E \to \mathbb{R}$ that are also in class $\mathcal{L}$ defined in (3.57) ($E_x\{f^-(X_t)\} < \infty$ for all $t \in [0, \infty)$ and $x \in E$) and define the mappings $\{K_t\}$ each of which (for each fixed t) maps S into itself in the following way:

$$K_t(f(x)) = \int_E f(y)P_x(X_t \in dy), \quad t \in [0, \infty).$$

We say that a given function g is *excessive* if

(1) $g \geq K_t g$, for all $t \in [0, \infty)$ and all $x \in E$; and
(2) $\lim_{t\to 0} K_t g(x) = g(x)$ for all $x \in E$.

Suppose that $f \in \mathcal{L}$ and that $E_x\{\sup_t f^-(X_t)\} < \infty$, for all $x \in E$. Then the excessive function $g \in \mathcal{L}$ is said to be an excessive majorant of f if $g(x) \geq f(x)$, for all $x \in E$. Moreover, it is said to be the smallest excessive majorant of f if it is an excessive majorant and if for any other excessive function h, $h(x) \geq g(x)$, for all $x \in E$. In connection with excessive majorants, we have the following result (see, e.g. [191]).

LEMMA 3.13. *For any function $f \in \mathcal{L}$ with $E\{\sup_t f^-(X_t)\} < \infty$, there exists a smallest excessive majorant that is a member of $\mathcal{L}$ (3.57).*

The smallest excessive majorant v of the function f in the above lemma is constructed in the following way.

$$Q_n f(x) = \max\{f(x), R_{2^{-n}} f(x)\}, \ n \in \mathcal{N}$$

and

$$v(x) = \lim_{n\to\infty} \lim_{N\to\infty} Q_n^N f(x),$$

with R the operator defined by (3.58) and Q_n^N is the Nth power of the operator Q_n. For a proof of this fact please refer to [191] page 86. We notice that $R_{2^{-n}} f(x) = K_{2^{-n}} f(x)$ in the above.

Since the state space is uncountably infinite, continuity of the functions g and v become relevant in the solution of the problem (3.55). For this reason we give a further definition concerning the continuity properties of a function.

A function g associated with a process $\{X_t\}$ is lower (upper) semicontinuous if $\liminf_{t\downarrow 0} g(X_t) \geq g(x)$ ($\limsup_{t\uparrow 0} g(X_t) \leq g(x)$) P_x a.s. for all $x \in E$. (It is known that, for a standard process, any lower semicontinuous function is continuous.)

We can now state the following main result which can be found in [191].

THEOREM 3.14. *Suppose* $\{X_t, \mathcal{F}_t\}$ *is a homogeneous Markov process. Define a process* $\{Y_t\}$ *by* $Y_t = f(X_t)$*, where* $f \in \mathcal{L}$ *satisfies* $E_x\{\sup_t f^+(X_t)\} < \infty$, $x \in E$. *Suppose further that* f *is continuous and bounded from below, and is bounded on the closed set* $\{x;\ v(x) > f(x)\}$*, where* v *is the smallest excessive majorant of* f*. Suppose also that* v *is lower semicontinuous. Then* $\overline{\gamma_t} = v(X_t)$*. Thus,* $\sup_{T\in\mathcal{T}} E_x\{Y_T\} = E_x\{f(X_{T_0})\}$ *and* $T_0 = \inf\{t > 0 | f(X_t) = v(X_t)\}$ *is an optimal stopping time.*

REMARK 3.15. Suppose that all the conditions of Theorem 3.14 are satisfied but we only have $E_{x_0}\{\sup_t f^+(X_t)\} < \infty$, for some $x_0 \in E$. Then the assertion of the theorem remains valid. (See for comparison, Chapter 3 of [191].)

3.6 Appendix: a proof of Lemma 3.8

Proof of Lemma 3.8 We first prove an intermediate result.

LEMMA 3.8′ *Suppose (3.110) holds. Then, for each* k*, there is a countable set* $C_k \subset \mathcal{T}_k$ *such that*

$$\gamma_k = \sup_{C_k} E\{Y_T | \mathcal{F}_k\}. \tag{3.138}$$

Proof of Lemma 3.8′ Let $\mathcal{C}$ denote the set of all countable subsets of $\mathcal{T}_k$. Define

$$s = \sup_{C\in\mathcal{C}} E\left\{\sup_{T\in C} E\{Y_T|\mathcal{F}_k\}\right\}, \tag{3.139}$$

and for each integer $\ell \geq 1$, choose $C_\ell \in \mathcal{C}$ such that

$$E\left\{\sup_{T \in C_\ell} E\{Y_T|\mathcal{F}_k\}\right\} \geq s - \frac{1}{\ell}. \tag{3.140}$$

Put

$$\mathcal{C}_k = \bigcup_{\ell=1}^{\infty} C_\ell, \tag{3.141}$$

and define

$$Z = \sup_{T \in \mathcal{C}_k} E\{Y_T|\mathcal{F}_k\}. \tag{3.142}$$

Note that $\mathcal{C}_k$ is countable, and that $E\{Z\} = s < \infty$. We would like to show that $Z = \gamma_k$. In order to do so, we must show two things:

(a) $P\left(Z \geq E\{Y_T|\mathcal{F}_k\}\right) = 1 \; \forall \, T \in \mathcal{T}_k$; and
(b) any other random variable Y satisfying (a) also satisfies $P(Y \geq Z) = 1$.

Now, suppose $T' \in \mathcal{T}_k$, and define

$$Z' = \max\left\{Z, E\{Y_{T'}|\mathcal{F}_k\}\right\}. \tag{3.143}$$

Since $Z \leq Z'$, we have

$$E\{Z\} \leq E\{Z'\} \leq s = E\{Z\}, \tag{3.144}$$

where the second inequality follows from the fact that $\mathcal{C}_k \cup \{T'\}$ is countable. Thus, $E\{Z' - Z\} = 0$ and $Z' - Z \geq 0$, which implies that $P(Z = Z') = 1$. But, we also have

$$Z' \geq E\{Y_{T'}|\mathcal{F}_k\}, \tag{3.145}$$

from which it follows that

$$P\left(Z \geq E\{Y_{T'}|\mathcal{F}_k\}\right) = 1. \tag{3.146}$$

Since T' was chosen arbitrarily, (a) follows.

Now suppose, Y is any other random variable satisfying

$$Y \geq E\{Y_T|\mathcal{F}_k\} \tag{3.147}$$

(almost surely) for all $T \in \mathcal{T}_k$. Then clearly $Y \geq Z$ almost surely. Lemma 3.8′ follows. ■

We now turn to the proof of the main lemma. Fix k and choose $\mathcal{C}_k$ as in Lemma 3.8′. Write the elements of $\mathcal{C}_k$ as $\{T_1', T_2', \ldots\}$. For each $\ell \geq 1$, define a stopping time $\tilde{T}_\ell \in \mathcal{T}_k$ by

$$\tilde{T}_\ell = \inf\left\{m \geq k \middle| E\left\{Y_{T_\ell'}\middle|\mathcal{F}_m\right\} \leq Y_m\right\}. \tag{3.148}$$

Since Y_∞ is $\mathcal{F}_\infty$-measurable, $\tilde{T}_\ell$ is always well defined. Now define a further sequence of stopping times by $T_1 = \tilde{T}_1$ and

$$T_\ell = \max\{T_{\ell-1}, \tilde{T}_\ell\}, \; \ell = 2, 3, \ldots \tag{3.149}$$

We will show that this sequence satisfies the property:

$$\max\left\{Y_k, \max_{1\le j\le \ell} E\left\{Y_{T'_j}\middle|\mathcal{F}_k\right\}\right\} \le E\left\{Y_{T_\ell}\middle|\mathcal{F}_k\right\} \le E\left\{Y_{T_{\ell+1}}\middle|\mathcal{F}_k\right\} \le \gamma_k. \quad (3.150)$$

Lemma 3.8 will follow since $\sup_{0\le j<\infty} E\left\{Y_{T'_j}\middle|\mathcal{F}_k\right\} = \gamma_k \ge Y_k$ by the choice of $T'_1, T'_2, \ldots$

To see that this property holds, choose $\ell \ge 1$, $j \ge k$, and $F \in \mathcal{F}_j$. We can write

$$\begin{aligned}\int_{F\cap\{\tilde{T}_\ell\ge j\}} Y_{\tilde{T}_\ell}\mathrm{d}P &= \sum_{m=j}^{\infty}\int_{F\cap\{\tilde{T}_\ell=m\}} Y_m\mathrm{d}P\\ &\ge \sum_{m=j}^{\infty}\int_{F\cap\{\tilde{T}_\ell=m\}} E\{Y_{T'_\ell}|\mathcal{F}_m\}\mathrm{d}P\\ &= \sum_{m=j}^{\infty}\int_{F\cap\{\tilde{T}_\ell=m\}} Y_{T'_\ell}\mathrm{d}P\\ &= \int_{F\cap\{\tilde{T}_\ell\ge j\}} Y_{T'_\ell}\mathrm{d}P. \end{aligned} \quad (3.151)$$

Since $\{\tilde{T}_\ell \ge j\} \in \mathcal{F}_{j-1} \subset \mathcal{F}_j$, it follows from (3.151) and the fact that F was chosen arbitrarily from $\mathcal{F}_j$ that

$$\int_{F\cap\{\tilde{T}_\ell\ge j\}} E\{Y_{\tilde{T}_\ell}|\mathcal{F}_j\}\mathrm{d}P \ge \int_{F\cap\{\tilde{T}_\ell\ge j\}} E\{Y_{T'_\ell}|\mathcal{F}_j\}\mathrm{d}P \;\; \forall\, F \in \mathcal{F}_j, \quad (3.152)$$

or, equivalently, that

$$\{\tilde{T}_\ell\} \ge j \Rightarrow E\{Y_{\tilde{T}_\ell}|\mathcal{F}_j\} \ge E\{Y_{T'_\ell}|\mathcal{F}_j\}. \quad (3.153)$$

It follows that

$$\left\{\{\tilde{T}_\ell \ge j\} \Rightarrow E\left\{Y_{\tilde{T}_\ell}\middle|\mathcal{F}_j\right\} \ge \max\{E\left\{Y_{T'_\ell}\middle|\mathcal{F}_j\right\}, Y_j\}\right\}, \; \forall\, j \ge k. \quad (3.154)$$

Now, consider $T_{\ell+1}$. Again choose $j \ge k$ and $F \in \mathcal{F}_j$. We can write

$$\begin{aligned}\int_{F\cap\{T_\ell\ge j\}} Y_{T_{\ell+1}}\mathrm{d}P &= \sum_{m=j}^{\infty}\int_{F\cap\{T_\ell=m\}} Y_{T_{\ell+1}}\mathrm{d}P\\ &= \sum_{m=j}^{\infty}\{\int_{F\cap\{T_\ell=m<T_{\ell+1}\}} E\{Y_{\tilde{T}_{\ell+1}}|\mathcal{F}_m\}\mathrm{d}P\\ &\quad + \int_{F\cap\{T_\ell=m=T_{\ell+1}\}} Y_m\mathrm{d}P\}\\ &\ge \sum_{m=j}^{\infty}\int_{F\cap\{T_\ell=m\}} Y_m\mathrm{d}P\\ &= \int_{F\cap\{T_\ell\ge j\}} Y_{T_\ell}\mathrm{d}P, \end{aligned} \quad (3.155)$$

where the second equality follows from the fact that $F \cap \{T_\ell = m < T_{\ell+1}\} \in \mathcal{F}_m$, and the inequality follows from (3.154). So, we can conclude that

$$\{T_\ell \geq j\} \Rightarrow E\{Y_{T_{\ell+1}}|\mathcal{F}_j\} \geq E\{Y_{T_\ell}|\mathcal{F}_j\}. \tag{3.156}$$

Note that (3.156) with $j = k$ implies the middle inequality in (3.150). Since the right-hand inequality in (3.150) is true by definition of γ_k, we need only show that the left-hand inequality holds in order to complete the proof. In order to show this, it is convenient to show simultaneously that the following property is satisfied for each $\ell \geq 1$.

$$\left\{\{T_\ell \geq j\} \Rightarrow E\left\{Y_{T_\ell}\middle|\mathcal{F}_j\right\} \geq Y_j\right\}, \ \forall\, j \geq k. \tag{3.157}$$

Note that (3.154) implies that both (3.157) and the left-hand inequality of (3.150) hold for $\ell = 1$. Moreover, (3.157) with $\ell = 1$ is the same as the left-hand inequality of (3.150) for $\ell = 1$. Suppose (3.157) and the left-hand inequality of (3.150) hold for $\ell = m > 1$. Since T_m satisfies (3.157), the same argument as in (3.155) leads to the conclusion

$$\{\tilde{T}_{m+1} \geq j\} \Rightarrow E\{Y_{T_{m+1}}|\mathcal{F}_j\} \geq E\{Y_{\tilde{T}_{m+1}}|\mathcal{F}_j\}. \tag{3.158}$$

Using (3.158) and (3.154) with $j = k$, we see that the left-hand inequality of (3.150) holds for $\ell = m + 1$. Moreover, since

$$\{T_{m+1} \geq j\} = \{T_m \geq j\} \cup \{\tilde{T}_{m+1} \geq j\}, \tag{3.159}$$

equations (3.156), (3.158), and (3.154) imply that (3.157) holds for $\ell = m + 1$. So, by induction, the proof is complete. ■

4 Sequential detection

4.1 Introduction

This chapter will formulate and solve the classical sequential detection problem as an optimal stopping problem. This problem deals with the optimization of decision rules for deciding between two possible statistical models for an infinite, homogeneous sequence of random observations. The optimization is carried out by penalizing, in various ways, the probabilities of error and the average amount of time required to reach a decision. By optimizing separately over the error probabilities with the decision time fixed, this problem becomes an optimal stopping problem that can be treated using the methods of the preceding chapter. As this problem is treated in many sources, the primary motivation for including it here is that it serves as a prototype for developing the tools needed in the related problem of quickest detection.

With this in mind, both Bayesian and non-Bayesian, as well as discrete- and continuous-time formulations of this problem will be treated. In the course of this treatment, a set of analytical techniques will be developed that will be useful in the solution and performance analysis of problems of quickest detection to be treated in subsequent chapters. Specific topics to be included are Bayesian optimization, the Wald–Wolfowitz theorem, the fundamental identity of sequential analysis, Wald's approximations, diffusion approximations, and Poisson approximations.

Sequential testing displays certain advantages over fixed sample testing in that it helps the user reach a decision between two hypotheses after a minimal average number of experiments. In particular, in the case of a fixed sample the sample size is fixed, and may be redundantly large for making a reasonably good inference on which of the two hypotheses is true. With sequential testing on the other hand, no observations are wasted. In fact, as soon as we can declare that one of the two hypotheses is true with reasonable certainty, we stop taking observations. For this reason, it is clear that sequential testing is a method of testing that is less costly on average than its competitor fixed sample size testing, if there is a cost associated with each observation.

4.2 Optimal detection

Consider a sequence $\{Z_k; k = 1, 2, \dots\}$ of i.i.d. real observations that obey one of two statistical hypotheses:

$$H_0: \quad Z_k \sim Q_0, \ k = 1, 2, \ldots$$

versus

$$H_1: \quad Z_k \sim Q_1, \ k = 1, 2, \ldots \tag{4.1}$$

where Q_0 and Q_1 are two distinct distributions on $(\mathbb{R}, \mathcal{B})$ satisfying $Q_0 \equiv Q_1$. We assume for now that hypothesis H_1 occurs with prior probability π and H_0 with prior $1 - \pi$. Let q_0 and q_1 denote densities of Q_0 and Q_1 with respect to some common dominating measure (e.g. $Q_0 + Q_1$).

We would like to decide between these hypotheses in a way that minimizes an appropriate measure of error probability and sampling cost.

Let us consider a probability space $(\Omega, \mathcal{F}, P) = (\mathbb{R}^\infty, \mathcal{B}^\infty, P_\pi)$ where

$$P_\pi = (1 - \pi) P_0 + \pi P_1 \tag{4.2}$$

and where P_0 and P_1 are two probability measures on $(\mathbb{R}^\infty, \mathcal{B}^\infty)$ such that, under $P_j, Z_1, Z_2, \ldots$ are i.i.d. with marginal distribution Q_j, for $j = 0, 1$.[1] The distributions P_0 and P_1 describe the hypotheses H_0 and H_1, respectively. Note that these measures are mutually singular (i.e. $P_0 \perp P_1$) since (see Example 2.3 of Section 2.3.1)

$$\Lambda_n \triangleq \prod_{k=1}^{n} \frac{q_1(Z_k)}{q_0(Z_k)} \to \begin{cases} 0 & \text{a.s. under } P_0 \\ \infty & \text{a.s. under } P_1 \end{cases} \tag{4.3}$$

$$\text{as } n \to \infty.$$

That is, we can tell these two distributions apart from the limiting value of the likelihood ratio. So, if we observe $\{Z_k; k = 1, 2, \ldots\}$ we can decide perfectly between the two hypotheses of interest.

Suppose, however, there is a cost for sampling – say, a linear cost $c > 0$ per sample taken. Suppose further that we observe $\{Z_k; k = 1, 2, \ldots\}$ sequentially, generating the filtration $\{\mathcal{F}_k; k = 1, 2, \ldots\}$, with

$$\mathcal{F}_k = \sigma(Z_1, Z_2, \ldots, Z_k), \ k = 1, 2, \ldots, \ \mathcal{F}_0 = (\Omega, \emptyset). \tag{4.4}$$

Clearly, the quality of a decision between H_0 and H_1 will improve with an increasing number of samples. However, the cost of sampling tempers the net benefit of that improvement, and thus a tradeoff develops between the cost of errors and the cost of sampling. The optimization of such tradeoffs can be examined in the context of sequential decision rules.

A *sequential decision rule* (s.d.r.) is a pair consisting of a stopping time $T \in \mathcal{T}$, where $\mathcal{T}$ denotes the set of all stopping times with respect to the filtration $\{\mathcal{F}_k\}$, and a sequence $\delta = \{\delta_k\}$ of *terminal decision rules,* where for each k, δ_k is an $\mathcal{F}_k$-measurable function taking values in the set $\{0, 1\}$. Let $\mathcal{D}$ denote the set of all such δ. To elaborate on the notion of an s.d.r., define the random variable

$$\delta_T = \sum_{k=0}^{\infty} \delta_k \mathbf{1}_{\{T=k\}}. \tag{4.5}$$

[1] Note that the observations are not independent under P_π, except in the cases $\pi = 0$ and $\pi = 1$.

Notice that δ_T is an $\mathcal{F}_T$-measurable random variable also taking values in the set $\{0, 1\}$. One can now describe an s.d.r. as the pair (T, δ), in which T declares the time to stop sampling, and once the value of T is given, δ_T takes the value 0 or 1 declaring which of the two hypotheses to accept. For an s.d.r. (T, δ) there are two performance indices of interest: the *average cost of errors*,[2]

$$c_e(T, \delta) = (1 - \pi)c_0 P_0(\delta_T = 1) + \pi c_1 P_1(\delta_T = 0), \tag{4.6}$$

where $c_j > 0$ is the cost of falsely rejecting hypothesis H_j, and the cost of sampling,

$$cE_\pi\{T\}, \tag{4.7}$$

where $E_\pi\{\cdot\}$ denotes expectation under the measure P_π.

So, we might like to choose an s.d.r. to solve the following optimization problem:

$$\inf_{T \in \mathcal{T}, \delta \in \mathcal{D}} [c_e(T, \delta) + cE_\pi\{T\}]. \tag{4.8}$$

This problem can be converted into a Markov optimal stopping problem via the following proposition.

PROPOSITION 4.1. *For any $T \in \mathcal{T}$, we have*

$$\inf_{\delta \in \mathcal{D}} c_e(T, \delta) = E_\pi \left\{ \min \left\{ c_1 \pi_T^\pi, c_0(1 - \pi_T^\pi) \right\} \right\}, \tag{4.9}$$

where the sequence $\{\pi_k^\pi\}$ is defined by the recursion

$$\pi_k^\pi = \frac{\pi_{k-1}^\pi q_1(Z_k)}{\pi_{k-1}^\pi q_1(Z_k) + (1 - \pi_{k-1}^\pi) q_0(Z_k)}, \quad k = 1, 2, \ldots, \quad \pi_0^\pi = \pi. \tag{4.10}$$

Moreover, the infimum in (4.9) is achieved by the terminal decision rule

$$\delta_k^0 = \begin{cases} 1 & \text{if } \pi_k^\pi \geq c_0/(c_0 + c_1) \\ 0 & \text{if } \pi_k^\pi < c_0/(c_0 + c_1) \end{cases} \tag{4.11}$$

REMARK 4.2. The random variable π_k^π is the posterior probability that H_1 is true, given $Z_1, \ldots, Z_k$. To see this, let p_k^π denote this posterior probability, and notice that

$$p_k^\pi = \frac{\pi \prod_{i=1}^k q_1(Z_i)}{\pi \prod_{i=1}^k q_1(Z_i) + (1 - \pi) \prod_{i=1}^k q_0(Z_i)}. \tag{4.12}$$

From the above equation it follows that

$$\begin{aligned} p_k^\pi &= \frac{\frac{\pi}{1-\pi} \prod_{i=1}^k \frac{q_1(Z_i)}{q_0(Z_i)}}{\frac{\pi}{1-\pi} \prod_{i=1}^k \frac{q_1(Z_i)}{q_0(Z_i)} + 1} \\ &= \frac{\frac{q_1(Z_k)}{q_0(Z_k)} \frac{\pi}{1-\pi} \prod_{i=1}^{k-1} \frac{q_1(Z_i)}{q_0(Z_i)}}{\frac{q_1(Z_k)}{q_0(Z_k)} \frac{\pi}{1-\pi} \prod_{i=1}^{k-1} \frac{q_1(Z_i)}{q_0(Z_i)} + 1}. \end{aligned} \tag{4.13}$$

[2] $P_0(\delta_T = 1)$ and $P_1(\delta_T = 0)$ are also known as the type I and type II errors, respectively.

Notice that

$$\frac{p_{k-1}^{\pi}}{1-p_{k-1}^{\pi}} = \frac{\pi}{1-\pi}\prod_{i=1}^{k-1}\frac{q_1(Z_i)}{q_0(Z_i)} \quad \forall\ k \geq 1.$$

Therefore the above equation (4.13) becomes

$$\begin{aligned} p_k^{\pi} &= \frac{\frac{p_{k-1}^{\pi}}{1-p_{k-1}^{\pi}}\frac{q_1(Z_k)}{q_0(Z_k)}}{\frac{p_{k-1}^{\pi}}{1-p_{k-1}^{\pi}}\frac{q_1(Z_k)}{q_0(Z_k)}+1} \\ &= \frac{p_{k-1}^{\pi}q_1(Z_k)}{p_{k-1}^{\pi}q_1(Z_k)+(1-p_{k-1}^{\pi})q_0(Z_k)}. \end{aligned} \tag{4.14}$$

On noting that $p_0^{\pi} = \pi$, and comparing (4.14) with (4.10), the assertion that $\pi_k^{\pi} = p_k^{\pi}$ follows.

Proof: Since the proposition is rather obvious if $\pi = 0$ or $\pi = 1$, we restrict attention to the case $\pi \in (0, 1)$.

Fix $T \in \mathcal{T}$. For any $\delta \in \mathcal{D}$ we can write

$$c_e(T,\delta) = \sum_{k=0}^{\infty}\left[(1-\pi)c_0\int_{\{\delta_k=1,T=k\}}\mathrm{d}P_0 + \pi c_1\int_{\{\delta_k=0,T=k\}}\mathrm{d}P_1\right]. \tag{4.15}$$

Consider the second summand in this expression. Using the Radon–Nikodym theorem and the fact that P_π dominates P_1 for $\pi > 0$, we can write

$$\int_{\{\delta_k=0,T=k\}}\mathrm{d}P_1 = \int_{\{\delta_k=0,T=k\}}\frac{\mathrm{d}P_1}{\mathrm{d}P_\pi}\mathrm{d}P_\pi. \tag{4.16}$$

Now, since the events $\{\delta_k = 0\}$ and $\{T = k\}$ are in $\mathcal{F}_k$, we can write

$$\begin{aligned} \pi\int_{\{\delta_k=0,T=k\}}\frac{\mathrm{d}P_1}{\mathrm{d}P_\pi}\mathrm{d}P_\pi &= \pi\int_{\{\delta_k=0,T=k\}}E_\pi\left\{\frac{\mathrm{d}P_1}{\mathrm{d}P_\pi}\middle|\mathcal{F}_k\right\}\mathrm{d}P_\pi \\ &= \int_{\{\delta_k=0,T=k\}}\pi_k^{\pi}\,\mathrm{d}P_\pi, \end{aligned} \tag{4.17}$$

with

$$\pi_k^{\pi} = \frac{\pi\prod_{j=1}^k q_1(Z_j)}{(1-\pi)\prod_{j=1}^k q_0(Z_j)+\pi\prod_{j=1}^k q_1(Z_j)}, \tag{4.18}$$

where we use the convention that $\prod_a^b = 1$ if $b < a$. To see why the above equality is true, notice that the $\frac{\mathrm{d}P_1}{\mathrm{d}P_\pi}$ can be described as the ratio of the probability of observing $Z_1, \ldots, Z_k$ if P_1 is the true measure divided by the probability of observing $Z_1, \ldots, Z_k$, if P_π is the true measure. In other words,

$$\frac{\mathrm{d}P_1}{\mathrm{d}P_\pi} = \frac{\prod_{i=1}^k q_1(Z_i)}{\prod_{i=1}^k q_1(Z_i)\pi+\prod_{i=1}^k q_0(Z_i)(1-\pi)}.$$

Using equation (4.12) the result follows.

Similarly,

$$(1-\pi)\int_{\{\delta_k=1,T=k\}} \mathrm{d}P_0 = \int_{\{\delta_k=1,T=k\}} (1-\pi_k^\pi)\mathrm{d}P_\pi, \tag{4.19}$$

and so

$$c_e(T,\delta) = \sum_{k=0}^{\infty}\left[c_0\int_{\{\delta_k=1,T=k\}}(1-\pi_k^\pi)\mathrm{d}P_\pi + c_1\int_{\{\delta_k=0,T=k\}}\pi_k^\pi \mathrm{d}P_\pi\right]. \tag{4.20}$$

It is clear from (4.20) that $c_e(T,\delta)$ achieves its minimum over $\delta \in \mathcal{D}$ at the terminal decision rule

$$\delta_k^0 = \begin{cases} 1 & \text{if } c_1\pi_k^\pi \geq c_0(1-\pi_k^\pi) \\ 0 & \text{if } c_1\pi_k^\pi < c_0(1-\pi_k^\pi), \end{cases} \tag{4.21}$$

which is the same as (4.11). Thus, we conclude that

$$\begin{aligned} \min_{\delta\in\mathcal{D}} c_e(T,\delta) &= \sum_{k=0}^{\infty}\int_{\{T=k\}} \min\left\{c_1\pi_k^\pi, c_0\left(1-\pi_k^\pi\right)\right\}\mathrm{d}P_\pi \\ &= E_\pi\left\{\min\left\{c_1\pi_T^\pi, c_0(1-\pi_T^\pi)\right\}\right\}. \end{aligned} \tag{4.22}$$

It is straightforward to check that $\{\pi_k^\pi\}$ satisfies the recursion (4.10), and the proposition follows. ■

Proposition 4.1 reduces the problem (4.8) to the alternative problem

$$\inf_{T\in\mathcal{T}} E_\pi\left\{\min\left\{c_1\pi_T^\pi, c_0(1-\pi_T^\pi)\right\} + cT\right\}. \tag{4.23}$$

Because of the recursivity of $\{\pi_k^\pi\}$ this new problem can be embedded in a Markov optimal stopping problem, which can be solved straightforwardly using the methods of Chapter 3. In particular, we can prove the following result.

PROPOSITION 4.3. *There are constants π_{L} and π_{U} satisfying $0 \leq \pi_{\mathrm{L}} \leq c_0/(c_0 + c_1) \leq \pi_{\mathrm{U}} \leq 1$, such that problem (4.23) is solved by the stopping time*

$$T_{\mathrm{opt}} = \inf\{k \geq 0 | \pi_k^\pi \notin (\pi_{\mathrm{L}}, \pi_{\mathrm{U}})\}. \tag{4.24}$$

Proof: Consider a homogeneous Markov process $\{X_k; k = 0, 1, \ldots\}$, with state space $E = [0,1]\times\mathcal{Z}$. Let $X_0 = \begin{pmatrix}\pi \\ m\end{pmatrix}$, and consider a family of measures $\{P_{(\pi,m)}; \begin{pmatrix}\pi \\ m\end{pmatrix} \in E\}$ such that

$$P_{(\pi,m)}\left(X_0 = \begin{pmatrix}\pi \\ m\end{pmatrix}\right) = 1, \tag{4.25}$$

and let $E_{(\pi,m)}\{\cdot\}$ denote expectation under $P_{(\pi,m)}$. Let A_1 be a Borel subset of [0, 1]. For any A Borel subset of E, we have that $A = A_1 \times Z_1$, where Z_1 is a subset of the integers. We have that

$$P_{(\pi,m)}\left(X_k = \begin{pmatrix}\pi_k^\pi \\ m+k\end{pmatrix} \in A \,\middle|\, X_0 = \begin{pmatrix}\pi \\ m\end{pmatrix}\right) = P_{(\pi,m)}\left(\pi_k^\pi \in A_1 \,\middle|\, X_0 = \begin{pmatrix}\pi \\ m\end{pmatrix}\right).$$

In other words, the first element of the vector X_k will evolve in time k according to the recursion (4.10), and the second will deterministically increase by k units.

The problem (4.23) is a special case ($m = 0$) of the Markov optimal stopping problem

$$\sup_{T \in \mathcal{T}} E_{(\pi,m)} \{g(X_T)\}, \tag{4.26}$$

where

$$g(x) = -\min\{c_1\pi, c_0(1-\pi)\} - cm, \ x = \left(\pi, m\right) \in E. \tag{4.27}$$

Since

$$\sup_k g^+(X_k) \leq c(-m)^+, \tag{4.28}$$

Theorem 3.7 implies that the optimal stopping time is

$$T_{\text{opt}} = \inf\{k \geq 0 | g(X_k) = v(X_k)\} \tag{4.29}$$

with

$$v(x) = \sup_{T \in \mathcal{T}} E_{(\pi,m)} \{g(X_T)\}, \ x = \begin{pmatrix} \pi \\ m \end{pmatrix} \in E. \tag{4.30}$$

Clearly, we have

$$v(x) = -s(\pi) - cm, \tag{4.31}$$

with

$$s(\pi) = -\sup_{T \in \mathcal{T}} E_{(\pi,0)} \{g(X_T)\}, \ \pi \in [0,1], \tag{4.32}$$

and so

$$T_{\text{opt}} = \inf\{k \geq 0 | \min\{c_1\pi_k^\pi, c_0(1-\pi_k^\pi)\} = s(\pi_k^\pi)\}. \tag{4.33}$$

Let us examine the function s. We can rewrite s as

$$\begin{aligned} s(\pi) &= \inf_{T \in \mathcal{T}} E_\pi \left\{\min\left\{c_1\pi_T^\pi, c_1(1-\pi_T^\pi)\right\} + cT\right\} \\ &= \inf_{T \in \mathcal{T}, \delta \in \mathcal{D}} [c_e(T,\delta) + cE_\pi\{T\}] \end{aligned} \tag{4.34}$$

where the second equality is from Proposition 4.1. For fixed (T, δ), the objective $c_e(T,\delta) + cE_\pi\{T\}$ of the right-most optimization in (4.34) is a linear function of π. In particular, $c_e(T,\delta)$ is seen to be linear from (4.15), and the term $E_\pi\{T\}$ is given by

$$E_\pi\{T\} = (1-\pi)E_0\{T\} + \pi E_1\{T\}. \tag{4.35}$$

It follows from this linearity that s is a concave function of π. It further follows from (4.34) that s is bounded according to

$$0 \leq s(\pi) \leq \min\{c_1\pi, c_0(1-\pi)\}, \ 0 \leq \pi \leq 1, \tag{4.36}$$

where the inequalities must be equalities at $\pi = 0$ and $\pi = 1$. These relationships are illustrated in Figure 4.1. That T_{opt} must be of the form described in the proposition is thus clear. ∎

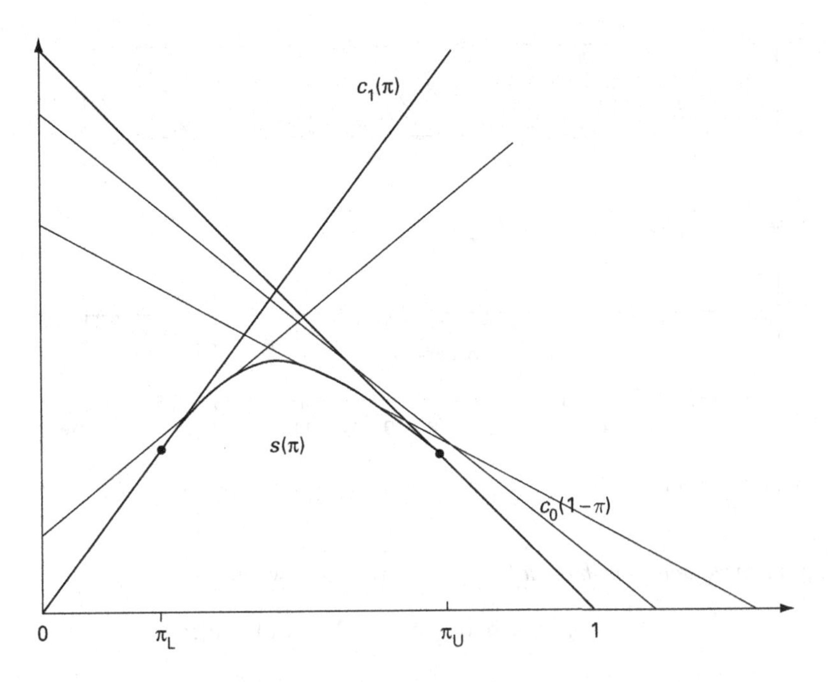

Fig. 4.1. An illustration of $s(\pi)$.

REMARK 4.4. The thresholds π_L and π_U are determined from the minimal cost function,

$$s(\pi) = \inf_{T \in \mathcal{T}, \delta \in \mathcal{D}} [c_e(T, \delta) + cE_\pi\{T\}], \ \pi \in [0, 1], \tag{4.37}$$

via

$$\pi_L = \sup\{0 \le \pi \le 1/2 \mid s(\pi) = c_1 \pi\}, \tag{4.38}$$

and

$$\pi_U = \inf\{1/2 \le \pi \le 1 \mid s(\pi) = c_0(1-\pi)\}. \tag{4.39}$$

Suppose the prior $\pi \in (0, 1)$. Then, it follows that $\pi_k^\pi \in (0, 1)$, $\forall k$. We can conclude that both π_L and π_U are also in (0, 1), since otherwise we would have $E_\pi\{T_{\text{opt}}\} = \infty$, a contradiction to the optimality of T_{opt}. From this result and Corollary 3.3 (combined with Theorem 3.7 and equation (4.28)), we have the following computational method for the optimal cost $s(\pi)$ and for the thresholds π_L and π_U.

PROPOSITION 4.5. *The minimal cost s of (4.37) is the monotone pointwise limit from above of the sequence of functions*

$$s_n(\pi) = \min\{\min\{c_1\pi, c_0(1-\pi)\}, \mathcal{Q}s_{n-1}(\pi)\}, \ n = 1, 2, \ldots, \tag{4.40}$$

with $s_0(\pi) = \min\{c_1\pi, c_0(1-\pi)\}$, *where the operator* $\mathcal{Q}$ *is defined by*

$$\mathcal{Q}s_{n-1}(\pi) = E_\pi\left\{s_{n-1}(\pi_n^\pi)\right\}. \tag{4.41}$$

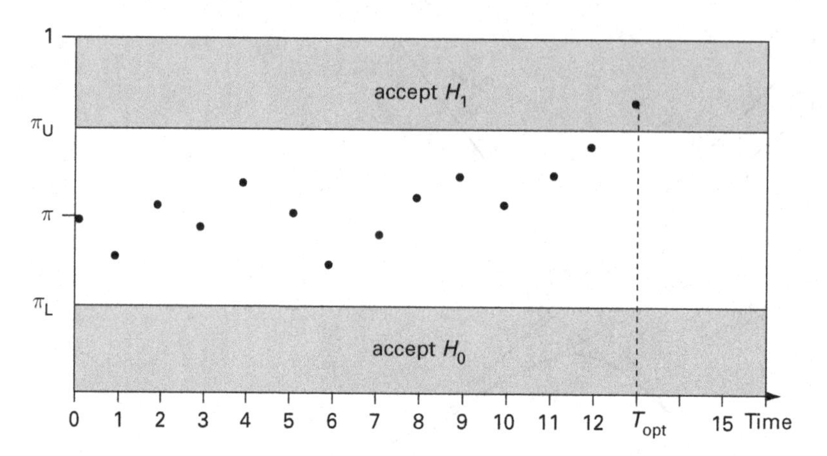

Fig. 4.2. An illustration of T_{opt}.

As a consequence, we have $\pi_L^n \downarrow \pi_L$ *and* $\pi_U^n \uparrow \pi_U$, *where*

$$\pi_L^n = \sup\{0 \leq \pi \leq 1/2 \mid s_n(\pi) = c_1\pi\}, \tag{4.42}$$

and

$$\pi_U^n = \inf\{1/2 \leq \pi \leq 1 \mid s_n(\pi) = c_0(1-\pi)\}. \tag{4.43}$$

Propositions 4.1, 4.3 and 4.5 specify the optimal sequential decision rule for (4.8), which is summarized in the following.

THEOREM 4.6. *Consider the optimization problem of (4.8). The optimal solution is given by the sequential decision rule* (T, δ) *with* T *from (4.24) and* δ *from (4.11). That is, the optimal s.d.r. continues sampling until* $\pi_k^\pi \notin (\pi_L, \pi_U)$, *at which time it chooses hypothesis* H_1 *if* $\pi_T^\pi \geq \pi_U$, *and it chooses* H_0 *otherwise.*

Figure 4.2 illustrates the operation of this s.d.r.

Note that the stopping time T of (4.24) can equivalently be written as

$$T = \inf\{k \geq 0 \mid \Lambda_k \notin (A, B)\}, \tag{4.44}$$

where $\{\Lambda_k\}$ is the sequence of likelihood ratios,

$$\Lambda_k = \prod_{j=1}^{k} \frac{q_1(Z_j)}{q_0(Z_j)} \equiv \frac{q_1(Z_k)}{q_0(Z_k)}\Lambda_{k-1},\ k = 1, 2, \ldots,\ \Lambda_0 = 1, \tag{4.45}$$

and where the thresholds A and B are given by

$$A = \frac{1-\pi}{\pi}\frac{\pi_L}{1-\pi_L} \text{ and } B = \frac{1-\pi}{\pi}\frac{\pi_U}{1-\pi_U}. \tag{4.46}$$

It follows that the Bayes optimal sequential decision rule (4.24) and (4.11) can also be viewed as a test that continues sampling as long as $\Lambda_k \in (A, B)$ and then, at the

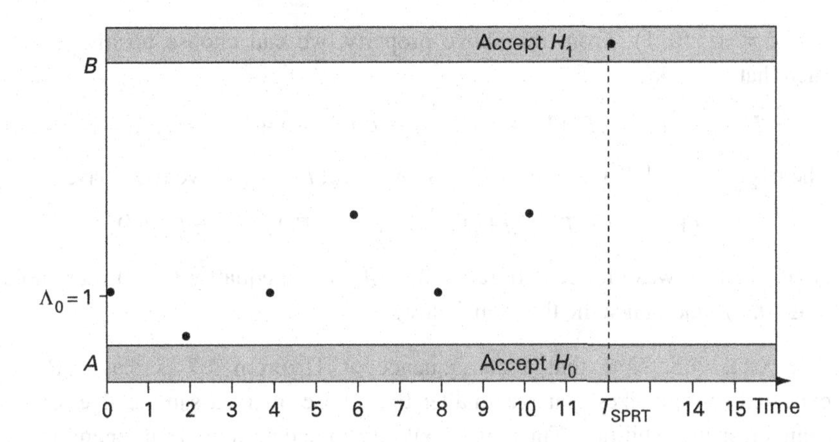

Fig. 4.3. An illustration of T_{SPRT}.

first exit of Λ_k from (A, B), decides H_1 if the exit is to the right of this interval, and decides H_0 if the exit is to the left. This test is referred to as the *sequential probability ratio test with boundaries A and B* (SPRT(A, B)), and it is of independent interest beyond its optimality in the Bayesian setting of the preceding section. The SPRT(A, B) is illustrated in Figure 4.3.

SPRT's exhibit minimal expected stopping time (i.e. minimal *runlength*) among all s.d.r.'s having given error probabilities. In particular, we have the following well-known result from Wald and Wolfowitz (1948).

THEOREM 4.7. *Suppose (T, δ) is the SPRT(A, B) with $0 < A \leq 1 \leq B < \infty$, and let (T', δ') denote any other s.d.r. with* $\max\left\{E_0\{T'\}, E_1\{T'\}\right\} < \infty$, *and satisfying*

$$P_0(\delta'_{T'} = 1) \leq P_0(\delta_T = 1) \;\; and \;\; P_1(\delta'_{T'} = 0) \leq P_1(\delta_T = 0), \tag{4.47}$$

with

$$P_0(\delta_T = 1) + P_1(\delta_T = 0) < 1.$$

Then

$$E_0\{T'\} \geq E_0\{T\} \;\; and \;\; E_1\{T'\} \geq E_1\{T\}. \tag{4.48}$$

Proof: The crux of this result is a partial converse to Theorem 4.6. In particular, it can be shown that, for any $\pi \in (0, 1)$ and any $0 < A \leq 1 \leq B < \infty$, there are positive costs c_0, c_1 and c, such that SPRT(A, B) is the Bayes test for this prior and these costs. That is, we can find positive costs such that π_{L} and π_{U} from (4.46) are the thresholds of the corresponding Bayes test from Theorem 4.6. This property follows from monotonicity and continuity properties of the thresholds and the minimal cost as functions of c, c_0 and c_1. Its proof can be found, for example, in [132], and it will not be repeated here. Note however, that in this proof we need the condition $P_1(\delta_T = 0) + P_0(\delta_T = 1) < 1$.

Fix $\pi \in (0,1)$. From the above property, we can choose positive c, c_0, and c_1 such that

$$c_e(T,\delta) + (1-\pi)E_0\{T\} + \pi E_1\{T\} \le c_e(T',\delta') + (1-\pi)E_0\{T'\} + \pi E_1\{T'\}.$$

The hypothesis (4.47) implies that $c_e(T,\delta) \ge c_e(T',\delta')$, so we must have

$$(1-\pi)E_0\{T\} + \pi E_1\{T\} \le (1-\pi)E_0\{T'\} + \pi E_1\{T'\}. \tag{4.49}$$

Now, since π was chosen arbitrarily in (0,1), the inequality (4.49) must hold for all $\pi \in (0,1)$, and hence the theorem follows. ■

REMARK 4.8. Note that a consequence of Theorem 4.7 is that SPRT's require expected sample sizes that are smaller than those of fixed-sample-size tests with the same error probabilities. Thus, the flexibility to decide quickly in unambiguous cases at the expense of taking more time to decide on ambiguous cases, pays off on the average.

REMARK 4.9. A further consequence of Theorem 4.7 is that the stopping times of SPRT's have finite expected values under both H_0 and H_1, a fact that is easily proved independently. (See for comparison, Proposition 8.21 of [198].)

4.3 Performance analysis

As a Bayes s.d.r., the performance of an SPRT can be computed via Proposition 4.5. However, a more basic understanding of the properties of SPRT's can be obtained by examining the relationships among the thresholds, the two conditional error probabilities, and the two expected runlengths. Such relationships are considered in the following paragraphs.

We first consider relationships among the thresholds and the error probabilities. Fix $0 < A \le 1 \le B < \infty$, and let (T,δ) denote the SPRT(A,B). For definiteness, we assume $A < B$. Recall that $L(Z_k) = \frac{q_1(Z_k)}{q_0(Z_k)}$, where q_0 and q_1 are densities of Q_0 and Q_1 with respect to a common dominating measure, and Q_0, Q_1 are distinct. To simplify notation, define

$$\alpha = P_0(\delta_T = 1) \text{ and } \gamma = P_1(\delta_T = 0). \tag{4.50}$$

The thresholds and error probabilities are related via the following result.

PROPOSITION 4.10.

$$B \le \frac{1-\gamma}{\alpha} \quad \textit{and} \quad A \ge \frac{\gamma}{1-\alpha}. \tag{4.51}$$

Proof: Since T must be almost surely finite under P_0 (compare Remark 4.9), we can write

$$\alpha = P_0(\Lambda_T \ge B) = \sum_{k=0}^{\infty} \int_{\{\Lambda_k \ge B\ ,\ T=k\}} \mathrm{d}P_0 \le \frac{1}{B} \sum_{k=0}^{\infty} \int_{\{\Lambda_k \ge B\ ,\ T=k\}} \Lambda_k \mathrm{d}P_0. \tag{4.52}$$

Now, since both $\{T = k\}$ and $\{\Lambda_k \geq B\}$ are in $\mathcal{F}_k$, we can write

$$\int_{\{\Lambda_k \geq B\ ,\ T=k\}} \Lambda_k \mathrm{d}P_0 = \int_{\{\Lambda_k \geq B\ ,\ T=k\}} \mathrm{d}P_1, \tag{4.53}$$

from which we have (T is also almost surely finite under P_1)

$$\alpha \leq \frac{1}{B} \sum_{k=0}^{\infty} \int_{\{\Lambda_k \geq B\ ,\ T=k\}} \mathrm{d}P_1 = \frac{P_1\,(\Lambda_T \geq B)}{B} = \frac{1-\gamma}{B}. \tag{4.54}$$

A similar argument gives $\gamma \leq A(1-\alpha)$, and thus we have the relationships in (4.51). ∎

A further pair of inequalities can also be used to relate the error probabilities of SPRT(A, B) to the expected runlengths. These are given in the following result.

PROPOSITION 4.11. *Suppose the random variable* $\log \Lambda_1$ *has finite means* d_0 *and* d_1*, respectively, under hypotheses* H_0 *and* H_1*. Then*

$$E_0\,\{T\} \geq d_0^{-1} \left[\alpha \log \left(\frac{1-\gamma}{\alpha} \right) + (1-\alpha) \log \left(\frac{\gamma}{1-\alpha} \right) \right] \tag{4.55}$$

and

$$E_1\,\{T\} \geq d_1^{-1} \left[(1-\gamma) \log \left(\frac{1-\gamma}{\alpha} \right) + \gamma \log \left(\frac{\gamma}{1-\alpha} \right) \right]. \tag{4.56}$$

REMARK 4.12. In view of the Wald–Wolfowitz theorem (Theorem 4.7), the inequalities of Proposition 4.11 must hold for the stopping times of arbitrary sequential decision rules having error probabilities α and γ.

Proof: We will prove (4.55), the proof of (4.56) being essentially the same. Similarly to the argument used in proving Proposition 4.10, we can write

$$1 - \gamma = \sum_{k=0}^{\infty} \int_{\{\Lambda_k \geq B, T=k\}} \mathrm{d}P_1 = \sum_{k=0}^{\infty} \int_{\{\Lambda_k \geq B, T=k\}} \Lambda_k \mathrm{d}P_0. \tag{4.57}$$

On defining Γ to be the event $\{\Lambda_T \geq B\}$, we then have

$$1 - \gamma = E_0\,\{\Lambda_T \times 1_\Gamma\} = E_0\,\{\Lambda_T |\, 1_\Gamma\}\, P_0(\Gamma) = \alpha\ E_0\,\{\Lambda_T |\, 1_\Gamma\}. \tag{4.58}$$

Using Jensen's inequality, we can write

$$1 - \gamma \leq \alpha\ \exp\left(E_0\,\{\log \Lambda_T |\, 1_\Gamma\}\right) = \alpha\ \exp\left(E_0\,\{\log \Lambda_T \times 1_\Gamma\}/\alpha\right), \tag{4.59}$$

or, equivalently,

$$E_0\,\{\log \Lambda_T \times 1_\Gamma\} \geq \alpha \log \left(\frac{1-\gamma}{\alpha} \right). \tag{4.60}$$

A similar sequence of steps yields a second inequality

$$E_0\,\{\log \Lambda_T \times 1_{\Gamma^c}\} \geq (1-\alpha) \log \left(\frac{\gamma}{1-\alpha} \right), \tag{4.61}$$

which, when added to (4.60), yields

$$E_0\{\log \Lambda_T\} \geq \alpha \log\left(\frac{1-\gamma}{\alpha}\right) + (1-\alpha)\log\left(\frac{\gamma}{1-\alpha}\right). \tag{4.62}$$

We now use Wald's identity (2.83) to write

$$E_0\{\log \Lambda_T\} = d_0 E_0\{T\}, \tag{4.63}$$

and (4.55) follows. (That d_0 and d_1 are non-zero follows from Jensen's inequality and the distinctness of Q_0 and Q_1.) ■

The inequalities of Propositions 4.10 and 4.11 provide universal relationships among the performance indices of SPRT's. It is interesting to consider conditions under which equality is achieved in these inequalities. Note that the inequalities in (4.52) and (4.54) (and hence the first inequality in (4.51)) would be equalities if $\Lambda_T = B$ almost surely on the event $\Gamma = \{\Lambda_T \geq B\}$; that is, if the likelihood ratio hits the upper boundary exactly when it crosses it. Similarly, the second inequality in (4.51) would be an equality if the likelihood ratio were to always hit the lower boundary upon exiting there. Examination of the proof of Proposition 4.11 reveals that these two conditions are also sufficient for equality in (4.55) and (4.56). In particular, the condition $\Lambda_T = B$ almost surely on the event Γ, is sufficient for equality in Jensen's inequality as used in (4.59), and the condition $\Lambda_T = A$ almost surely on the event Γ^c, similarly yields equality in (4.61).

An example in which these conditions are met is the following.

Example 4.1: Suppose the observations $\{Z_k\}$ are Bernoulli (0–1 valued) random variables with $Q_j(Z_k = 1) = p_j \in (0,1),\ j \in \{0,1\}$. Suppose further that $p_0 = 1 - p_1 < 1/2$. Here, we have

$$\log \Lambda_k = \log\left(\frac{p_1}{p_0}\right)\sum_{j=1}^{k}(2Z_j - 1). \tag{4.64}$$

So, the SPRT stops and decides at the first violation of the inequality

$$a' \triangleq \frac{\log A}{\log(p_1/p_0)} < \sum_{j=1}^{k}(2Z_j - 1) < \frac{\log B}{\log(p_1/p_0)} \triangleq b'. \tag{4.65}$$

Since the sum $\sum_{j=1}^{k}(2Z_j - 1)$ is integer-valued, no generality is lost in assuming that a' and b' are integers. Moreover, since $(2Z_j - 1) \in \{-1, +1\}$, the likelihood ratio will exactly touch the boundary at T. (See Figure 4.4.)

Thus, in this example, the inequalities in Propositions 4.10 and 4.11 are equalities.

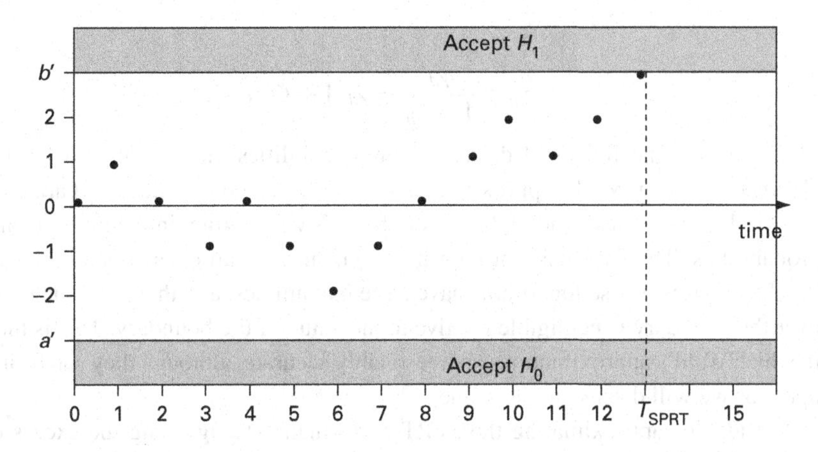

Fig. 4.4. An illustration of T_{SPRT} in the case of Bernoulli trials with $a' = -3$ and $b' = 3$.

More generally, the conditions for equality in Propositions 4.10 and 4.11 will not be met. However, these inequalities can be considered to be approximate equalities if the "excess over the boundaries" (i.e. $\Lambda_T - B$ or $A - \Lambda_T$) can be assumed to be negligible. In this situation, from Proposition 4.10 we have

$$B \cong \frac{1-\gamma}{\alpha} \text{ and } A \cong \frac{\gamma}{1-\alpha}, \tag{4.66}$$

or, equivalently,

$$\alpha \cong \frac{1-A}{B-A} \text{ and } \gamma \cong A\frac{B-1}{B-A}, \tag{4.67}$$

and from Proposition 4.11 we have

$$E_0\{T\} \cong d_0^{-1}\left[\alpha \log\left(\frac{1-\gamma}{\alpha}\right) + (1-\alpha)\log\left(\frac{\gamma}{1-\alpha}\right)\right] \tag{4.68}$$

and

$$E_1\{T\} \cong d_1^{-1}\left[(1-\gamma)\log\left(\frac{1-\gamma}{\alpha}\right) + \gamma\log\left(\frac{\gamma}{1-\alpha}\right)\right]. \tag{4.69}$$

These approximations are known as *Wald's approximations.*

Wald's approximations can be used to choose the boundaries A and B to yield a given level of error-probability performance. For example, if we wish to design a test with approximate error probabilities α_d and γ_d, then we can use (4.66) to choose boundaries

$$B = \frac{1-\gamma_d}{\alpha_d} \text{ and } A = \frac{\gamma_d}{1-\alpha_d}. \tag{4.70}$$

The inequalities (4.51) then tell us that the actual error probabilities, α_a and γ_a are bounded according to

$$\alpha_a \leq \frac{\alpha_d}{1-\gamma_d} = \alpha_d(1+\mathcal{O}(\gamma_d)) \tag{4.71}$$

and

$$\gamma_a \leq \frac{\gamma_d}{1-\alpha_d} = \gamma_d(1+\mathcal{O}(\alpha_d)). \tag{4.72}$$

Thus, we see that for small desired error probabilities, the actual error probabilities obtained by using Wald's approximations can be bounded by values that are quite close to their desired values. And, in fact, these bounds will be tight in the limit of small error probabilities. This is consistent with intuition, since small error probabilities translate into boundaries whose logarithms have large magnitudes, and thus for which the excess over the boundary is negligible relative to the value of the boundary. This is the regime in which Wald's approximations are reasonably accurate, although they can be improved upon, as we will discuss in the sequel.

The idea of approximating the SPRT performance by ignoring the excess over the boundaries can be extended to provide approximations for the exit statistics of more general random walks. In particular, let P be the probability measure under which $\{\Delta_k; k = 1, \ldots\}$ is a sequence of i.i.d. Bernoulli(p) random variables, and let $E\{\cdot\}$, denote its corresponding expectation. Define the random walk $\{W_k; k = 1, 2, \ldots\}$ by

$$W_k = \sum_{j=1}^{k} \Delta_j = W_{k-1} + \Delta_k, \; k = 1, 2, \ldots, \quad W_0 = 0. \tag{4.73}$$

For real constants a and b satisfying $-\infty < a < b < \infty$, define the stopping time $T_{a,b}$ (we consider the filtration $\sigma(W_0, W_1, \ldots, W_k)$, $k = 0, 1, \ldots$) by

$$T_{a,b} = \inf\{k \geq 0 | W_k \notin (a, b)\}. \tag{4.74}$$

We are interested in the mean value, or *expected run length,* of $T_{a,b}$:

$$E\left\{T_{a,b}\right\}, \tag{4.75}$$

and in the *upper exit probability* of the random walk:

$$P\left(W_{T_{a,b}} \geq b\right). \tag{4.76}$$

Notice that the SPRT(A, B) corresponds to the case $\Delta_k = \log L(Z_k)$, $a = \log A$, and $b = \log B$; and the expected run lengths and upper exit probabilities under P_0 and P_1 are the quantities considered in the preceding paragraphs.

The expected runlength and upper exit probability of the general random walk (and hence of the SPRT(A, B)) can be determined from the solutions to linear integral equations (see for comparison [19] pp. 202–204), which must generally be solved numerically. However, a somewhat more useful approach is to approximate them as follows.

Since $T_{a,b}$ is the first exit of the random walk from an interval, the result in the next-to-last bullet point of Section 2.3.2, implies that it is regular for the $\mathcal{F}_k$-martingale

$$e^{sW_k - k\log M(s)}, \; k = 0, 1, \ldots \tag{4.77}$$

for any $s \in \mathcal{M} \triangleq \{s | M(s) < \infty\}$, where M denotes the moment generating function of Δ_1 (see for comparison Proposition IV-4-17 of [157]). Hence, optional sampling gives us the identity

$$E\left\{e^{sW_{T_{a,b}}-T_{a,b}\log M(s)}\right\}=1,\ \forall\, s\in\mathcal{M},\tag{4.78}$$

as noted in Chapter 2. Now, we would like to find a non-zero solution, s', to the equation

$$M(s')=1.\tag{4.79}$$

In fact, there exists a unique non-zero s' that is the solution of (4.79) as long as $E\{W_1\}\neq 0$. In particular, we have that

(1) If $E\{W_1\}<0$, then $s'>0$,
(2) If $E\{W_1\}>0$, then $s'<0$.

To see the first implication above we just have to notice that $M'(0)=E\{W_1\}<0$. Moreover, $M'(s)=E\{W_1 e^{sW_1-\log M(s)}\}$ is a strictly increasing function of s. Therefore, there exists a unique $s_0>0$ such that $M'(s)<0$ for $s<s_0$ and $M'(s)>0$ for all $s>s_0$. But this implies that $M(s)$ is strictly decreasing for all $s<s_0$, while it is strictly increasing for all $s>s_0$, and $s_0>0$. Moreover, $M(0)=0$ and therefore there exists an $s'>s_0>0$ such that $M(s')=0$. A similar argument implies that $s'<0$ if $E\{W_1\}>0$.

For such s', (4.78) becomes

$$E\left\{e^{s'W_{T_{a,b}}}\right\}=1.\tag{4.80}$$

Now, if we ignore the excess over the boundaries, then $W_{T_{a,b}}$ is approximately equal to either a or b, and (4.80) becomes

$$e^{s'b}P\left(W_{T_{a,b}}\geq b\right)+e^{s'a}\left[1-P\left(W_{T_{a,b}}\geq b\right)\right]\cong 1.\tag{4.81}$$

On rearranging, we have

$$P\left(W_{T_{a,b}}\geq b\right)\cong\frac{1-e^{s'a}}{e^{s'b}-e^{s'a}}.\tag{4.82}$$

In the case of the SPRT(A,B), it is easily seen that (4.79) is solved under P_0 and P_1 by $s'=1$ and $s'=-1$, respectively. On substituting these values into (4.82), Wald's approximations (4.67) are obtained.

The expected value of the stopping time $T_{a,b}$ can also be approximated for the general random walk. In this case, if we assume $E\{\Delta_1\}\neq 0$, we can use Wald's identity (2.83) to write

$$E\left\{T_{a,b}\right\}=\frac{E\left\{W_{T_{a,b}}\right\}}{E\left\{\Delta_1\right\}}.\tag{4.83}$$

Again ignoring the excess over the boundaries, we thus have

$$E\left\{T_{a,b}\right\}\cong\frac{bP\left(W_{T_{a,b}}\geq b\right)+a\left[1-P\left(W_{T_{a,b}}\geq b\right)\right]}{E\left\{\Delta_1\right\}},\tag{4.84}$$

which, together with (4.82) reduces to (4.68) and (4.69) for the SPRT(A,B) under P_0 and P_1, respectively. Alternatively, if $E\{\Delta_1\}=0$ and $E\{\Delta_1^2\}<\infty$, then the second Wald identity (2.84) yields the approximation

$$E\left\{T_{a,b}\right\}\cong\frac{b^2P\left(W_{T_{a,b}}\geq b\right)+a^2\left[1-P\left(W_{T_{a,b}}\geq b\right)\right]}{E\left\{\Delta_1^2\right\}}.\tag{4.85}$$

In the context of sequential detection, the interest in the general approximations (4.82), (4.84) and (4.85) lies in the fact that they provide simple estimates of error probabilities and average runlengths for certain sequential decision rules other than the SPRT(A, B). This is of particular interest when the exact distribution of the observations is unknown under one or both of the two hypotheses.

Suppose, for example, that we have a parametric family $\{f_\theta;\ \theta \geq 0\}$ of possible marginal densities for the i.i.d. observations $\{Z_k\}$. For a given s.d.r. (T, δ) we can consider the so-called *operating characteristic* (*oc*)

$$oc(\theta) = P_\theta(\delta_T = 1),\ \theta \geq 0, \tag{4.86}$$

and the *average sample number* (*asn*)

$$asn(\theta) = E_\theta\{T\},\ \theta \geq 0 \tag{4.87}$$

where P_θ and $E_\theta\{\cdot\}$ denote probability and expectation, respectively, under the model that $\{Z_k\}$ are i.i.d. with marginal density f_θ. Consider the following pair of hypotheses:

$$H_0: \quad Z_k \sim f_0,\ k = 1, 2, \ldots$$

versus

$$H_1: \quad Z_k \sim f_\theta,\ k = 1, 2, \ldots,\ \theta > 0. \tag{4.88}$$

Since H_1 is composite, we cannot, in general, design a single SPRT that will optimize performance in testing between these hypotheses. We might, however, consider SPRT's between H_0 and the particular alternative marginal density f_θ, for some specific θ, in which case the test is as described in Section 4.2., with $f_\theta = q_1$, and $f_0 = q_0$ (Here, without loss of generality, we consider the real parameter θ to be translated and scaled so that the marginal under H_0 is f_0 and a nominal marginal under H_1 is f_θ.) The performance of this test, as measured by oc and asn, are computable by solving appropriate integral equations, as noted above. However, (4.82), (4.84) and (4.85), give the simpler, and more revealing, approximations

$$oc(\theta) = \frac{1 - e^{a\theta}}{e^{b\theta} - e^{a\theta}},\ \theta > 0,$$

$$oc(\theta) = \frac{e^{a\theta} - 1}{e^{b\theta} - e^{a\theta}},\ \theta < 0$$

and

$$asn(\theta) = \frac{(b - a) \cdot oc(\theta) + a}{d_\theta}, \tag{4.89}$$

where s_θ solves

$$E_\theta\left\{[\Lambda_1]^{s_\theta}\right\} = 1,$$

and

$$d_\theta = E_\theta\{\log \Lambda_1\}.$$

Whereas, in the case for which $d_\theta = 0$, we obtain

$$asn(\theta) = \frac{b^2 oc(\theta) + a^2(1 - oc(\theta))}{E_\theta\{(\log \Lambda_1)^2\}}.$$

The performance approximations made in this section can be improved upon by a more careful modeling of the excess over the boundary. (See for comparison, Chapter X of [198].)

4.4 The continuous-time case

The results of the preceding sections can be extended to certain continuous-time problems in which the observations have stationary independent increments under either hypothesis. Here, we consider this problem in the important cases in which the observations are either Brownian motions or homogeneous Poisson counting processes. We also give a result on the extended optimality of the SPRT to convex cost functions, which applies to stochastic processes in discrete time, continuous time, as well as in the case of processes with jumps.

4.4.1 The Brownian case

Consider the following measurable space:

(1) $\Omega = C[0, \infty]$: the space of continuous functions, and
(2) $\mathcal{F} = \mathcal{B}(C[0, \infty])$: the σ-algebra of Borel sets on the space of continuous functions.

Let us consider first the case of Brownian observations, about which we wish to test hypotheses concerning the rate of drift. In particular, suppose we observe a continuous-time random process $\{Z_t; t \geq 0\}$ which behaves statistically according to one of two possible models:

$$H_0: \quad Z_t = \sigma W_t + \mu_0 t \;, \; t \geq 0$$

versus

$$H_1: \quad Z_t = \sigma W_t + \mu_1 t \;, \; t \geq 0 \tag{4.90}$$

where $\{W_t; t \geq 0\}$ is a standard Brownian motion with respect to the filtration $\{\mathcal{F}_t;\ t \geq 0\}$, with

$$\mathcal{F}_t = \sigma\,(Z_u; 0 \leq u \leq t)\,.$$

As before, we can consider a family of probability distributions $\{P_\pi; \pi \in [0, 1]\}$ where

$$P_\pi = (1 - \pi)P_0 + \pi P_1, \tag{4.91}$$

with P_0 and P_1 denoting the distributions of the observed process under H_0 and H_1, respectively.

We again define a sequential decision rule as a pair $(T, \delta) \in \mathcal{T} \times \mathcal{D}$ where $\mathcal{T}$ denotes the set of all $\{\mathcal{F}_t; t \geq 0\}$ stopping times, and where $\mathcal{D}$ now denotes the set of terminal decision rules, δ_T, which are $\mathcal{F}_T$-measurable random variables for every $T \in \mathcal{T}$.[3]

As in the discrete-time case, for positive constants c_0, c_1, and c, we wish to solve the following problem:

$$\inf_{T \in \mathcal{T}, \delta \in \mathcal{D}} [c_e(T, \delta) + cE_\pi\{T\}] \tag{4.92}$$

with

$$c_e(T, \delta) = (1 - \pi)c_0 P_0(\delta_T = 1) + \pi c_1 P_1(\delta_T = 0). \tag{4.93}$$

We have the following result, analogous to Theorem 4.6.

THEOREM 4.13. *There are constants π_{L} and π_{U} satisfying $0 \leq \pi_{\mathrm{L}} \leq c_0/(c_0 + c_1) \leq \pi_{\mathrm{U}} \leq 1$ such that the infimum in (4.92) is achieved by the s.d.r. (T, δ) consisting of the stopping time*

$$T = \inf\left\{t \geq 0 | \pi_t^\pi \notin (\pi_{\mathrm{L}}, \pi_{\mathrm{U}})\right\} \tag{4.94}$$

and the terminal decision rule

$$\delta_T = \begin{cases} 1 & \text{if } \pi_T^\pi \geq c_0/(c_0 + c_1) \\ 0 & \text{if } \pi_T^\pi < c_0/(c_0 + c_1), \end{cases} \tag{4.95}$$

where the random process $\{\pi_t^\pi; t \geq 0\}$ is given by

$$\pi_t^\pi = \frac{\pi \Lambda_0^t}{\pi \Lambda_0^t + (1 - \pi)}, \quad \pi_0^\pi = \pi, \tag{4.96}$$

with

$$\Lambda_0^t = \exp\left\{\frac{\mu_1 - \mu_0}{\sigma^2}\left(Z_t - \frac{\mu_1 + \mu_0}{2}\right)\right\}, \quad t \geq 0. \tag{4.97}$$

REMARK 4.14. Equation (4.97) gives the likelihood ratio between the hypotheses of (4.90) over the interval $[0, t]$, a result known as the Cameron–Martin formula (see, for example, Theorem 2.9 or [176]). From this result and the Radon–Nikodym theorem (compare (2.116) and (2.19)) it follows that

$$\pi_t^\pi = \pi E_\pi\left\{\frac{\mathrm{d}P_1}{\mathrm{d}P_\pi}\middle| \mathcal{F}_t\right\}. \tag{4.98}$$

From this it can be shown that, as in the discrete-time case, the random variable π_t^π is the posterior probability that H_1 is true, given $\{Z_u; 0 \leq u \leq t\}$. A straightforward application of Itô's rule (compare Theorem 2.6) shows that π_t^π evolves via the Itô stochastic differential equation

$$\mathrm{d}\pi_t^\pi = \frac{\mu_1 - \mu_0}{\sigma^2}\pi_t^\pi(1 - \pi_t^\pi)\mathrm{d}Z_t - \frac{(\mu_1 - \mu_0)^2}{\sigma^2}(\pi_t^\pi)^2(1 - \pi_t^\pi)\mathrm{d}t. \tag{4.99}$$

[3] Recall that $\delta_T = \sum_{k=0}^\infty \delta_k \mathbf{1}_{\{T=k\}}$ is an $\mathcal{F}_T$-measurable random variable in the discrete-time setting as well.

Proof: Aside from technical details, the proof of this result is essentially the same as the analogous discrete-time result given in Theorem 4.6 (i.e. Propositions 4.1 and 4.3) We outline the proof.

First, analogously with Proposition 4.1, we can show that, for fixed $\delta \in \mathcal{D}$,

$$\inf_{\delta\in\mathcal{D}} c_e(T,\delta) = E_\pi\left\{\min\left\{c_1\pi_T^\pi, c_0(1-\pi_T^\pi)\right\}\right\}. \tag{4.100}$$

To see this, fix $(T,\delta) \in \mathcal{T} \times \mathcal{D}$ and write

$$P_1(\delta_T = 0) = 1 - E_1\{\delta_T\}. \tag{4.101}$$

Using the Radon–Nikodym theorem and (2.116), we can write

$$\pi E_1\{\delta_T\} = \pi E_0\left\{\Lambda_0^T \delta_T\right\} = E_\pi\left\{\pi_T^\pi \delta_T\right\}, \tag{4.102}$$

where $\left\{\Lambda_0^t; t \geq 0\right\}$ is from (4.97). Similarly,

$$(1-\pi)E_0\{\delta_T\} = E_\pi\left\{(1-\pi_T^\pi)\delta_T\right\}. \tag{4.103}$$

So, we see that

$$c_e(T,\delta) = \pi c_1 + E_\pi\left\{\left[(1-\pi_T^\pi)c_0 - \pi_T^\pi c_1\right]\delta_T\right\}, \tag{4.104}$$

and (4.100) follows.

Equation (4.100) reduces the problem (4.92) to a Markov optimal stopping problem:

$$\sup_{T\in\mathcal{T}} E_{\pi,0}\{g(X_T)\}, \tag{4.105}$$

where g is defined as in (4.27); $\{X_t; t \geq 0\}$ is a continuous-time homogeneous Markov process having state space $E = [0,1] \times \mathbb{R}$. Let $X_0 = \begin{pmatrix}\pi\\ u\end{pmatrix}$, and consider the family of measures $\left\{P_{(\pi,u)}; \begin{pmatrix}\pi\\ u\end{pmatrix} \in E\right\}$ such that

$$P_{(\pi,u)}\left(X_0 = \begin{pmatrix}\pi\\ u\end{pmatrix}\right) = 1, \tag{4.106}$$

and let $E_{(\pi,u)}\{\cdot\}$ denote expectation under $P_{(\pi,m)}$. Let A_1 be a Borel subset of [0, 1]. For any Borel subset of A of E, we have that $A = A_1 \times \mathcal{B}_1$, where $\mathcal{B}_1$ is a Borel subset of $\mathbb{R}$. We have that

$$P_{(\pi,u)}\left(X_t = \begin{pmatrix}\pi_t^\pi\\ u+t\end{pmatrix} \in A \,\middle|\, X_0 = \begin{pmatrix}\pi\\ u\end{pmatrix}\right) = P_{(\pi,u)}\left(\pi_t^\pi \in A_1 \,\middle|\, X_0 = \begin{pmatrix}\pi\\ u\end{pmatrix}\right).$$

In other words, the first element of the vector X_t will evolve in time t according to (4.96), and the second will deterministically increase by t units.

To show that T of (4.94) is optimal is then essentially the same as for the discrete-time case, where we use Theorem 3.14 along with Remark 3.15 in place of Theorem 3.7. ■

The optimal s.d.r. specified by Theorem 4.13 is a continuous-time SPRT, stopping when the inequalities

$$\frac{1-\pi}{\pi}\frac{\pi_L}{1-\pi_L} \triangleq A < \Lambda_t < B \triangleq \frac{1-\pi}{\pi}\frac{\pi_U}{1-\pi_U} \tag{4.107}$$

are violated, and deciding H_1 if $\Lambda_T \geq B$ and H_0 otherwise. In this case, since Brownian motion almost surely has continuous sample paths, the likelihood ratio Λ_0^t hits the boundary exactly; i.e., with probability 1,

$$\Lambda_0^T \in \{A, B\}. \tag{4.108}$$

A result analogous to the Wald–Wolfowitz theorem (Theorem 4.7) can be proved here in a very direct manner via a stronger version of Proposition 4.11. In particular, we have the following.

THEOREM 4.15. *Consider the hypothesis pair (4.90), and suppose (T, δ) is a sequential decision rule with error probabilities*

$$P_0(\delta_T = 1) = \alpha \text{ and } P_1(\delta_T = 0) = \gamma, \tag{4.109}$$

where $\alpha, \gamma \in (0, 1)$. Then

$$E_0\{T\} \geq -\frac{2\sigma^2}{(\mu_1-\mu_0)^2}\left[\alpha \log\left(\frac{1-\gamma}{\alpha}\right) + (1-\alpha)\log\left(\frac{\gamma}{1-\alpha}\right)\right] \tag{4.110}$$

and

$$E_1\{T\} \geq \frac{2\sigma^2}{(\mu_1-\mu_0)^2}\left[(1-\gamma)\log\left(\frac{1-\gamma}{\alpha}\right) + \gamma\log\left(\frac{\gamma}{1-\alpha}\right)\right], \tag{4.111}$$

with equality if (T, δ) is an SPRT.

The proof of this result is essentially the same as that of Proposition 4.11 with appropriate modification to account for the continuous-time case (as in the proof of Theorem 4.13). The fact that the SPRT achieves the bounds in (4.51) is due to the fact that Λ_0^T is almost surely constant conditioned on δ_T. (See the discussion following Proposition 4.11.)

A Wald–Wolfowitz type result then follows as a simple corollary:

COROLLARY 4.16. *Consider the hypotheses (4.90) and suppose (T, δ) is the SPRT(A, B) with $0 < A \leq 1 \leq B < \infty$, and $A < B$. Let (T', δ') denote any other s.d.r. satisfying*

$$P_0(\delta'_{T'} = 1) \leq P_0(\delta_T = 1) \textit{ and } P_1(\delta'_{T'} = 0) \leq P_1(\delta_T = 0), \tag{4.112}$$

with

$$P_0(\delta_T = 1) + P_1(\delta_T = 0) < 1.$$

Then

$$E_0\{T'\} \geq E_0\{T\} \textit{ and } E_1\{T'\} \geq E_1\{T\}. \tag{4.113}$$

Proof: Let

$$\alpha' = P_1(\delta'_{T'} = 0) \tag{4.114}$$

and

$$\gamma' = P_0(\delta'_{T'} = 1). \tag{4.115}$$

From Theorem 4.15, it follows that

$$E_0\left\{T'\right\} \geq -\frac{2\sigma^2}{(\mu_1-\mu_0)^2}\left[\alpha' \log\left(\frac{1-\gamma'}{\alpha'}\right) + (1-\alpha')\log\left(\frac{\gamma'}{1-\alpha'}\right)\right], \tag{4.116}$$

and

$$E_1\left\{T'\right\} \geq \frac{2\sigma^2}{(\mu_1-\mu_0)^2}\left[(1-\gamma')\log\left(\frac{1-\gamma'}{\alpha'}\right) + \gamma'\log\left(\frac{\gamma'}{1-\alpha'}\right)\right]. \tag{4.117}$$

But since $\alpha + \gamma < 1$ (which implies that $\alpha' + \gamma' < 1$), we have that

$$\begin{aligned} -\tfrac{2\sigma^2}{(\mu_1-\mu_0)^2}\left[\alpha'\log\left(\tfrac{1-\gamma'}{\alpha'}\right) + (1-\alpha')\log\left(\tfrac{\gamma'}{1-\alpha'}\right)\right] \geq \\ -\tfrac{2\sigma^2}{(\mu_1-\mu_0)^2}\left[\alpha\log\left(\tfrac{1-\gamma}{\alpha}\right) + (1-\alpha)\log\left(\tfrac{\gamma}{1-\alpha}\right)\right], \end{aligned} \tag{4.118}$$

and

$$\begin{aligned} \tfrac{2\sigma^2}{(\mu_1-\mu_0)^2}\left[(1-\gamma')\log\left(\tfrac{1-\gamma'}{\alpha'}\right) + \gamma'\log\left(\tfrac{\gamma'}{1-\alpha'}\right)\right] \geq \\ \tfrac{2\sigma^2}{(\mu_1-\mu_0)^2}\left[(1-\gamma)\log\left(\tfrac{1-\gamma}{\alpha}\right) + \gamma\log\left(\tfrac{\gamma}{1-\alpha}\right)\right]. \end{aligned} \tag{4.119}$$

Combining the pair of inequalities (4.116) and (4.117) with (4.118) and with (4.119), the result follows. ■

We see from Theorem 4.15 that the bounds and approximations relating runlengths of SPRT's to their error probabilities made in the discrete-time case are exact in the case of Brownian observations. The same is true for the relationship between error probabilities and thresholds, as we see from the following result.

THEOREM 4.17. *Consider the hypotheses (4.90), and denote by (T, δ) the SPRT(A, B) with $0 < A \leq 1 \leq B < \infty$, and $A < B$. Then*

$$P_0(\delta_T = 1) = \frac{1-A}{B-A} \tag{4.120}$$

and

$$P_1(\delta_T = 0) = A\frac{B-1}{B-A}. \tag{4.121}$$

Proof: Wald's identity for stopped Brownian motion (i.e. (2.98)) implies that

$$E_0\left\{\Lambda_0^T\right\} = 1, \tag{4.122}$$

and

$$E_1\left\{(\Lambda_0^T)^{-1}\right\} = 1. \tag{4.123}$$

From (4.122), it follows that

$$B \cdot P_0(\delta_T = 1) + A \cdot P_0(\delta_T = 0) \quad = \quad 1, \tag{4.124}$$

while from (4.123), it follows that

$$B^{-1} \cdot P_1(\delta_T = 1) + A^{-1} \cdot P_1(\delta_T = 0) \quad = \quad 1. \tag{4.125}$$

From (4.124), and (4.125) and the facts that $P_0(\delta_T = 1) + P_0(\delta_T = 0) = 1$, and $P_1(\delta_T = 1) + P_1(\delta_T = 0) = 1$, the result follows. ∎

REMARK 4.18. In examining sequential tests between two Brownian motion models, we have considered the case in which the two models differ only in their drift parameters. If we also allow the two models to differ in their variance parameters, then it is straightforward to see that the two models are mutually singular over any time interval. In particular, the quadratic variation over an interval of length t, of the sample paths of a Brownian motion with variance parameter σ^2, is almost surely equal to $\sigma^2 t$. (See for example, [118], p. 141.) Thus, Brownian motion models with distinct variance parameters can be perfectly discriminated on any arbitrarily small observation interval.

REMARK 4.19. The similarity between Wald's approximations and the exact performance expressions given in Theorem 4.15 and Theorem 4.17 is not accidental. In particular, the general approximations of (4.82), (4.84), and (4.85) (of which Wald's approximations are special cases), can be obtained through the alternate route of first approximating the random walk $\{W_k\}$ with a Brownian motion (as discussed in Section 4.3), and then using these exact results for Brownian motion to approximate the exit statistics of the random walk. Such approximations, which essentially are the same as (4.82), (4.84) and (4.85) are sometimes called *diffusion approximations,* since Brownian motion is the protypical diffusion. Better approximations can be obtained by more careful consideration of the asymptotics involved in diffusion approximation. (See for example Chapter X of [198].) In general, these types of approximations are accurate when the distributions arising under the two hypotheses are sufficiently close together in a sense that will be described in Section 7.4.

4.4.2 The Brownian case – an alternative proof

Theorem 4.13 was proved using the traditional optimal stopping theory. However, this result can be proved alternatively without resorting to this theory. In this section, we provide such a proof, which will serve as a model for several proofs in the subsequent chapters.

We restrict attention to the case of (4.92) in which $c_0 = c_1 = 1$. Moreover, without loss of generality, we set $\mu_1 = \mu$, $\mu_0 = 0$, and $\sigma = 1$. Also, for stopping times T, we define

$$R_\pi(T) = \inf_\delta [c_e(T, \delta) + cE_\pi\{T\}] = E_\pi\{\min\{\pi_T, 1 - \pi_T\} + cT\}. \tag{4.126}$$

We begin with the following result.

PROPOSITION 4.20. *Define the function $f : (0, 1) \to \mathbb{R}$ by*

$$f(x) = \frac{2}{\mu^2}\left[(2x-1)\log\left(\frac{x}{1-x}\right) - (2\pi - 1)\log\left(\frac{\pi}{1-\pi}\right)\right], \quad x \in (0, 1).$$

Then, for all bounded stopping times T we have

$$E_\pi\{T\} = E_\pi\left\{f\left(\pi_T^\pi\right)\right\}, \tag{4.127}$$

where π_t^π is defined as in Theorem 4.13.

Proof: Under the measure P_π the observations can be written as

$$Z_t = \mu\,\theta\,t + W_t,\ t \geq 0, \tag{4.128}$$

where θ is a random variable indicating which hypothesis is true; i.e. $\theta = 1$ with probability π and $\theta = 0$ with probability $1 - \pi$. Since

$$\pi_t^\pi = E_\pi\left\{\theta\,|\mathcal{F}_t\right\}, \tag{4.129}$$

it follows that π_t^π obeys the Itô stochastic differential equation (as in (4.99))

$$d\pi_t^\pi = \mu\,\pi_t^\pi(1 - \pi_t^\pi)dI_t,\ \ \pi_0^\pi = \pi, \tag{4.130}$$

where $\{I_t; t \geq 0\}$ denotes the innovation process

$$Z_t - \mu\pi_t^\pi,\ t \geq 0. \tag{4.131}$$

Note that $\{I_t\}$ is a standard Brownian motion with respect to $\{\mathcal{F}_t; t \geq 0\}$ under P_π. (See, Theorem 2.8.)

Now, we note that f is twice continuously differentiable in $(0, 1)$, so that we can apply Itô's rule (compare Theorem 2.6) to yield

$$df\left(\pi_t^\pi\right) = \frac{\mu^2}{2}f''\left(\pi_t^\pi\right)\left[\pi_t^\pi(1-\pi_t^\pi)\right]^2 dt + \mu f'\left(\pi_t^\pi\right)\pi_t^\pi(1-\pi_t^\pi)dI_t,\ t \geq 0,$$

with initial condition $f\left(\pi_0^\pi\right) = f(\pi) = 0$, and

$$f''(x) = \frac{2}{\mu^2}\frac{1}{[x(1-x)]^2}, \tag{4.132}$$

so that we can write

$$f(\pi_t^\pi) = t + \mu\int_0^t f'\left(\pi_s^\pi\right)\pi_s^\pi(1-\pi_s^\pi)dI_s,\ t \geq 0. \tag{4.133}$$

Moreover, since

$$f'(x) = \frac{2}{\mu^2}\left[2\log x - 2\log(-x) + \frac{2x-1}{x(1-x)}\right], \tag{4.134}$$

the product $f'(x)x(1-x)$ is bounded on $(0, 1)$. From this it follows that

$$E_\pi\left\{\int_0^t\left[f'\left(\pi_s^\pi\right)\pi_s^\pi(1-\pi_s^\pi)\right]^2 ds\right\} < \infty, \tag{4.135}$$

and thus the expectation of the stochastic integral in (4.133) under P_π is 0, (compare Theorem 2.5). The continuous-time optional sampling theorem (2.88) given at the end of Section 2.3.3, yields

$$E_\pi\left\{\int_0^T f'\left(\pi_s^\pi\right)\pi_s^\pi(1-\pi_s^\pi)\mathrm{d}I_s\right\} = 0, \tag{4.136}$$

for all bounded stopping times T.

The proposition thus follows from (4.133). ■

We see that for bounded stopping times we can write

$$R_\pi(T) = E\left\{g\left(\pi_T^\pi\right)\right\}, \tag{4.137}$$

where

$$g(x) = \min\{x, 1-x\} + cf(x),\ x \in (0,1). \tag{4.138}$$

Straightforward analysis of the function g shows that it has exactly two global minima on $(0, 1)$, at the points $x_0 \in (0, 1/2)$ and $x_1 = 1 - x_0$ satisfying

$$cf'(x_0) = -1 \text{ and } cf'(x_1) = 1. \tag{4.139}$$

This behavior is illustrated in Figures 4.5 and 4.6. It follows that

$$R_\pi(T) \geq g(x_0) \tag{4.140}$$

for all bounded times T.

The inequality (4.140) extends to all stopping times T. To see this, we first note that it trivially holds whenever $E_\pi\{T\} = \infty$. Now, if $E_\pi\{T\} < \infty$, we consider the sequence

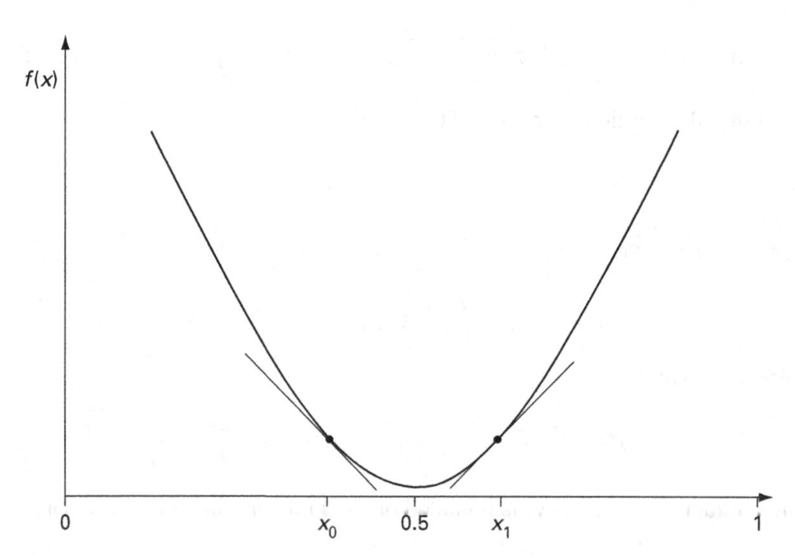

Fig. 4.5. An illustration of $f(x)$.

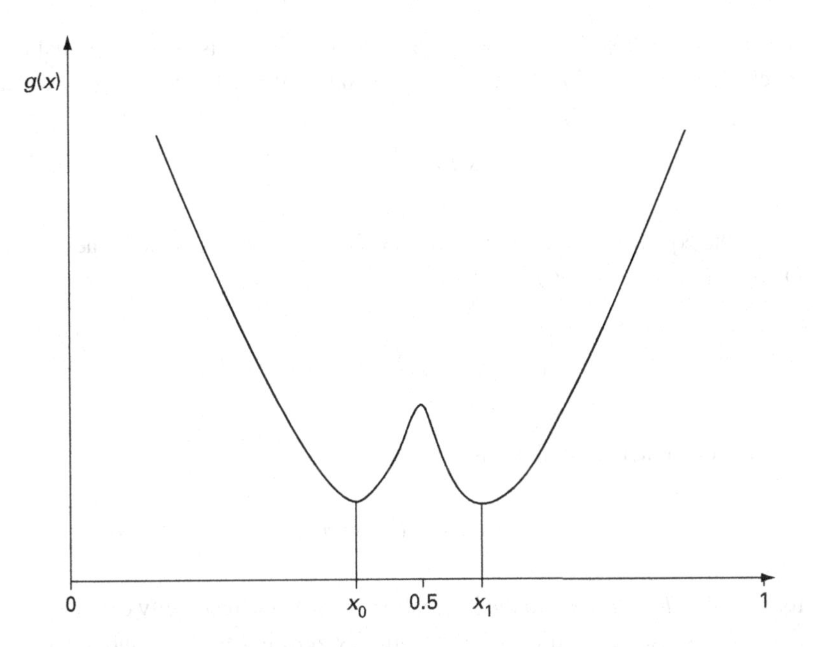

Fig. 4.6. An illustration of $g(x)$.

of stopping times $T \wedge n,\ n = 1, 2, \ldots$ Since $T \wedge n \uparrow T$ as $n \uparrow \infty$ almost surely under P_π, the monotone convergence theorem implies

$$E_\pi\{T \wedge n\} \uparrow E_\pi\{T\}. \tag{4.141}$$

Also, the path continuity of Brownian motion implies that

$$\min\{\pi_{T\wedge n}, 1 - \pi_{T\wedge n}\} \to \min\{\pi_T, 1 - \pi_T\} \text{ a.s. } P_\pi, \tag{4.142}$$

and so the bounded convergence theorem implies

$$R_\pi(T) = \lim_{n\to\infty} R_\pi(T \wedge n) \geq g(x_0). \tag{4.143}$$

Now suppose $\pi \in [x_0, x_1]$, and consider the stopping time

$$T^* = \inf\left\{t \geq 0 | \pi_t^\pi \notin (x_0, x_1)\right\}. \tag{4.144}$$

We have

$$R_\pi(T^*) = \lim_{n\to\infty} R_\pi(T^* \wedge n) = E\left\{g\left(\pi_{T^*}^\pi\right)\right\} = g(x_0), \tag{4.145}$$

where the first equality is from (4.143), the second equality follows from the bounded convergence theorem (note that g is continuous, and hence bounded, on $[x_0, x_1]$), and the third equality follows since $\pi_{T^*}^\pi \in \{x_0, x_1\}$. In view of the inequality (4.143), we see that T^* is optimal.

Now consider the case $\pi \in (0, x_0)$. Let $\mathcal{T}^*$ denote the set of bounded stopping times with respect to $\{\mathcal{F}_t\}$. Note that

$$\inf_{T\in\mathcal{T}^*} R_\pi(T) = \inf_{T\in\mathcal{T}^*} R_\pi(T \wedge T_0) \tag{4.146}$$

where T_0 is the first time $\pi_t^\pi = x_0$. This follows since it is clearly optimal to stop if π_t^π touches x_0. For $t \leq T \wedge T_0$, $\pi_t^\pi \in [0, x_0]$ and $\min\{\pi_t^\pi, 1 - \pi_t^\pi\} = \pi_t^\pi$, so that

$$R_\pi(T \wedge T_0) \geq E_\pi \left\{ \pi_{T \wedge T_0}^\pi \right\} = \pi, \tag{4.147}$$

where the equality on the right-hand side follows from the boundedness of $T \wedge T_o$ and the representation (4.130):

$$\pi_t^\pi = \pi + \mu \int_0^t \pi_s^\pi (1 - \pi_s^\pi) \mathrm{d}I_s. \tag{4.148}$$

So, we conclude that, in this case,

$$R_\pi(T) \geq \pi \tag{4.149}$$

for bounded T. Similarly to the inequality (4.143), this inequality extends to all stopping times T. The stopping time T^* is identically zero in this case, and achieves the lower bound (4.149). Thus, it is optimal.

A similar argument shows that T^* is also optimal for $\pi > x_1$, and thus the optimality of T^* is established for all priors $\pi \in (0, 1)$. Since T^* is the same as the stopping time (4.94) with $\pi_{\mathrm{L}} = x_0$ and $\pi_{\mathrm{U}} = x_1$, Theorem 4.13 is thus proven in this ($c_0 = c_1$) case. Note that this proof shows how the thresholds can be computed via (4.139) and the minimal Bayes cost is determined by

$$\min_{T \in \mathcal{T}} R_\pi(T) = \begin{cases} g(x_0) & \text{if } \pi \in (x_0, x_1) \\ \min\{\pi, 1 - \pi\} & \text{otherwise.} \end{cases} \tag{4.150}$$

4.4.3 An interesting extension of Wald–Wolfowitz

In this section we give an extension of the continuous-time Wald–Wolfowitz result of Corollary 4.16. Let $(\Omega, \mathcal{F})$ be a measurable space, $\mathcal{K} = \{0, 1, 2, \ldots\}$ or $\mathcal{K} = [0, \infty)$ the time set, and $\{\mathcal{F}_t; t \in \mathcal{K}\}$ an increasing sequence of sub-σ-algebras of $\mathcal{F}$. Let P_0 and P_1 be probability measures on $(\Omega, \mathcal{F}.)$ A sequential decision rule of P_0 versus P_1, (T, δ) is given by a stopping time

$$T : \Omega \to \mathcal{K}$$

and a terminal decision function δ that is $\mathcal{F}_T$-measurable. Let

$$\Lambda_t = \left. \frac{\mathrm{d}P_1}{\mathrm{d}P_0} \right|_{\mathcal{F}_t},$$

where for each t, $\left.\frac{dP_1}{dP_0}\right|_{\mathcal{F}_t}$ is the Radon–Nikodym derivative of the measure P_1 with respect to P_0 restricted to the σ-algebra $\mathcal{F}_t$.[4] To assert that Λ_t is a process with well-behaved paths in the continuous-time case we make the assumption that the filtration $\{\mathcal{F}_t\}$ is right-continuous and complete with respect to $(P_0 + P_1)/2$, and according to [112], we may assume that Λ_t has càdlàg paths. Furthermore Λ_∞ exists, $P_0(\sup_t \Lambda_t = \infty) = 0$ and

$$\Lambda_T = \left.\frac{dP_1}{dP_0}\right|_{\mathcal{F}_T},$$

for all stopping times T.

DEFINITION 4.21. *For* $0 < a < b$, *define* $\mathcal{G}(a, b)$ *to be the set of all convex functions* $g : [0, \infty) \to \mathbb{R}$, *with* $\lim_{x\to\infty} g(x)/x = \infty$, *such that* $g'(a) < g'(b)$.

Naturally, we define the SPRT$(A, B) = \inf\{t \geq 0 | \Lambda_t \notin (A, B)\}$. In this context we have the following theorem:

THEOREM 4.22. *Let* (T, δ) *be an SPRT with stopping boundaries* $0 < A \leq 1 \leq B$, *and error probabilities* $\gamma = P_1(\delta_T = 0)$, *and* $\alpha = P_0(\delta_T = 1)$, *such that* $\Lambda_T \in \{A, B\}$ *a.s.* P_0 *and a.s.* P_1. *Then for any sequential test* (T', δ') *with error probabilities* $\alpha' \leq \alpha$ *and* $\gamma' \leq \gamma$, *the following hold:*

(1) $P_0(T' < \infty) = 1$ *implies*

$$E_0\{g(\Lambda_{T'})\} \geq E_0\{g(\Lambda_T)\} \;\forall g \in \mathcal{G}(A, B); \text{ and}$$

(2) $P_1(T' < \infty) = 1$ *implies*

$$E_1\{g(\Lambda_{T'})\} \geq E_1\{g(\Lambda_T)\} \;\forall g \in \mathcal{G}(A, B).$$

Thus, the Wald–Wolfowitz property of the SPRT extends to all convex functions of the stopped likelihood ratio. For a proof of this result including relaxation of the requirements on g, please see [109].

4.4.4 The case of Itô processes

As we have noted, Brownian motion is the prototypical diffusion process. In this section we discuss the continuous-time sequential detection problem for more general diffusions, namely for Itô processes. On the same measurable space as in the Brownian case,

[4] That is, $\left.\frac{dP_1}{dP_0}\right|_{\mathcal{F}_t}$ for a fixed t is a random variable that satisfies

$$P_1(A) = \int_A \left.\frac{dP_1}{dP_0}\right|_{\mathcal{F}_t} dP_0$$

for all $A \in \mathcal{F}_t$. (See (2.19).)

consider the situation in which we observe $\{Z_t; t \geq 0\}$ with either one of the following dynamics:

$$H_0 : \mathrm{d}Z_t = \mathrm{d}W_t, \quad Z_0 = 0, \; t \geq 0,$$

versus

$$H_1 : \mathrm{d}Z_t = \theta_t \mathrm{d}t + \mathrm{d}W_t, \quad Z_0 = 0, \; t \geq 0,$$

where $\{W_t; t \geq 0\}$ is a standard Brownian motion and $\{\theta_t; t \geq 0\}$ is a stochastic process independent of $\{W_t; t \geq 0\}$. Here the process $\{\theta_t; t \geq 0\}$ can be interpreted as a signal and the Brownian motion process as noise, and so the problem involves testing two hypotheses concerning the presence (hypothesis H_1) or the absence (hypothesis H_0) of the signal $\{\theta_t; t \geq 0\}$. As before, the information available at each time instant t is summarized in $\mathcal{F}_t = \sigma\{Z_s; \; s \leq t\}$. We assume that $\{\theta_t; t \geq 0\}$ is $\mathcal{F}_t$-adapted.

Let us first make the following technical assumptions:

(1) $E_1\{|\theta_t|\} < \infty, \;\; \forall\, t \geq 0,$
(2) $E_0\{|\theta_t|\} < \infty, \;\; \forall\, t \geq 0,$
(3) $P_1\left(\int_0^\infty [E_1\{\theta_t \mid \mathcal{F}_t\}]^2 \, \mathrm{d}t = \infty\right) = 1$, and
(4) $P_0\left(\int_0^\infty [E_0\{\theta_t \mid \mathcal{F}_t\}]^2 \, \mathrm{d}t = \infty\right) = 1.$

Note that conditions (3) and (4) have the physical interpretation that the signal has sufficient energy for detection.

As before we define an s.d.r. as the pair (T, δ), with T a stopping time with respect to the filtration $\{\mathcal{F}_t; t \geq 0\}$, and the terminal decision rule δ_T that is an $\mathcal{F}_T$-measurable r.v. taking values in the set $\{0, 1\}$. We also assume that

(5) $E_0\{\int_0^T [E_0\{\theta_t \mid \mathcal{F}_t\}]^2 \, \mathrm{d}t\} < \infty$, and
(6) $E_1\{\int_0^T [E_1\{\theta_t \mid \mathcal{F}_t\}]^2 \, \mathrm{d}t\} < \infty.$

Naturally, we define the SPRT$(A, B) = \inf\{t \geq 0 | \Lambda_t \notin (A, B)\}$, where

$$\Lambda_t = \mathrm{e}^{\int_0^t E_0\{\theta_s \mid \mathcal{F}_s\}\mathrm{d}Z_s - \frac{1}{2}\int_0^t E_0\{\theta_s \mid \mathcal{F}_s\}^2 \mathrm{d}s}.$$

We then have the following result for this model.

THEOREM 4.23. *Let (T, δ) be an SPRT with stopping boundaries $0 < A \leq 1 \leq B$, and error probabilities $\gamma = P_1(\delta_T = 0)$, and $\alpha = P_0(\delta_T = 1)$, with $\alpha + \gamma < 1$, satisfying conditions (1)–(6). Then for any sequential test (T', δ') with error probabilities $\alpha' \leq \alpha$ and $\gamma' \leq \gamma$, the following hold:*

$$E_0\left\{\int_0^{T'} [E_0\{\theta_t \mid \mathcal{F}_t\}]^2 \, \mathrm{d}t\right\} \geq E_0\left\{\int_0^T [E_0\{\theta_t \mid \mathcal{F}_t\}]^2 \, \mathrm{d}t\right\},$$

and

$$E_1\left\{\int_0^{T'} [E_1\{\theta_t \mid \mathcal{F}_t\}]^2 \, \mathrm{d}t\right\} \geq E_1\left\{\int_0^T [E_1\{\theta_t \mid \mathcal{F}_t\}]^2 \, \mathrm{d}t\right\}.$$

Moreover $\delta_T = 1$ when $\Lambda_T = B$, and $\delta_T = 0$ when $\Lambda_T = A$.

Aside from technical details, which can be found in Section 17.6 of [135], the proof is based on the use of Jensen's inequality, and the following lemma.

LEMMA 4.24. *In the above model, we have*

$$E_0\left\{\int_0^T E_0\{\theta_t \mid \mathcal{F}_t\}^2 \mathrm{d}t\right\} = 2\omega(\alpha, \gamma), \tag{4.151}$$

and

$$E_1\left\{\int_0^T E_1\{\theta_t \mid \mathcal{F}_t\}^2 \mathrm{d}t\right\} = 2\omega(\gamma, \alpha), \tag{4.152}$$

where

$$\omega(x, y) = (1 - x)\log\frac{1 - x}{y} + x\log\frac{x}{1 - y}.$$

REMARK 4.25. Note that Theorems 4.7, 4.15 and 4.23 follow from the more general Theorem 4.22 with $g(x) = 2x\log(x)$.

4.4.5 The Poisson case

Another continuous-time observation model of considerable interest is the homogeneous Poisson counting process (HPCP) described in Section 2.4.2. In particular, we may assume that our observation process $\{Z_t; t \geq 0\}$ is an HPCP with rate $\lambda > 0$, and we wish to test the hypotheses

$$H_0: \quad \lambda = \lambda_0$$

versus

$$H_1: \quad \lambda = \lambda_1, \tag{4.153}$$

with $\lambda_0 \neq \lambda_1$.

In the Bayesian formulation of the problem, we observe the point process $\{Z_t; t \geq 0\}$, with compensator λt, such that the random intensity λ is a random variable taking two values λ_0 and λ_1 with probabilities π and $1 - \pi$ respectively.

Consider the following probability space:

(1) Ω is the space of càdlàg functions (right-continuous left limits),
(2) $\mathcal{F} = \cup_{t>0}\mathcal{F}_t$, where $\mathcal{F}_t = \sigma\{Z_s; s \leq t\}$, and
(3) $\{P_\pi; \pi \in [0, 1]\}$: the family of probability measures describing the above Poisson model indexed by π, where $P_\pi = \pi P_1 + (1 - \pi)P_0$.

We assume that the random variable λ is $\mathcal{F}_0$-measurable. That is to say, nature selects the value of the intensity at time $t = 0$ when we begin observing the process $\{Z_t; t \geq 0\}$. And we have $P_\pi(\lambda = \lambda_0) = \pi$, and $P_\pi(\lambda = \lambda_1) = 1 - \pi$.

As before, we consider sequential decision rules $(T, \delta) \in \mathcal{T} \times \mathcal{D}$ where $\mathcal{T}$ denotes the set of all $\{\mathcal{F}_t; t \geq 0\}$ stopping times, and where $\mathcal{D}$ is the set of terminal decision rules.

As in the previous cases for positive constants c_0, c_1, and c, we wish to solve the following problem:

$$\inf_{T\in\mathcal{T},\delta\in\mathcal{D}} [c_e(T, \delta) + cE_\pi\{T\}] \tag{4.154}$$

with

$$c_e(T,\delta) = c_0 P_0(\delta_T = 1) + c_1 P_1(\delta_T = 0). \tag{4.155}$$

We have the following result, analogous to Theorem 4.6.

THEOREM 4.26. *There are constants π_L and π_U satisfying $0 \le \pi_L \le c_0/(c_0+c_1) \le \pi_U \le 1$ such that the infimum in (4.92) is achieved by the s.d.r. (T,δ) consisting of the stopping time*

$$T = \inf\left\{t \ge 0 | \pi_t^\pi \notin (\pi_L, \pi_U)\right\} \tag{4.156}$$

and the terminal decision rule

$$\delta_T = \begin{cases} 1 & \text{if } \pi_T^\pi \ge c_0/(c_0+c_1) \\ 0 & \text{if } \pi_T^\pi < c_0/(c_0+c_1), \end{cases} \tag{4.157}$$

where the random process $\{\pi_t^\pi; t \ge 0\}$ is given by

$$\pi_t^\pi = P_\pi(\lambda = \lambda_1 | \mathcal{F}_t). \tag{4.158}$$

REMARK 4.27. It follows from the Cameron–Martin formula and the Radon–Nikodym theorem ([47] pp. 165–167 and 2.19) that

$$\Lambda_t = \left(\frac{\lambda_1}{\lambda_0}\right)^{Z_t} e^{(\lambda_0 - \lambda_1)t}, \ t \ge 0. \tag{4.159}$$

Moreover, we have

$$\pi_t^\pi = \pi \left.\frac{dP_1}{dP_\pi}\right|_{\mathcal{F}_t}. \tag{4.160}$$

Combining (4.159) and (4.160), we have

$$\pi_t^\pi = \frac{\pi \Lambda_t}{1 - \pi + \pi \Lambda_t}. \tag{4.161}$$

Notice the similarity of (4.161) to (4.13).

In what follows we will briefly outline the proof of Theorem 4.26.

Outline of proof: First, analogously with Proposition 4.1, we can show that, for fixed $\delta \in \mathcal{D}$,

$$\inf_{\delta \in \mathcal{D}} c_e(T,\delta) = E_\pi \left\{\min\left\{c_1 \pi_T^\pi, c_0(1-\pi_T^\pi)\right\}\right\}. \tag{4.162}$$

This follows by arguments similar to those in the proof of Theorem 4.13.

The problem is therefore reduced to solving

$$V(\pi) = \inf_T E_\pi \left[T + g(\pi_T^\pi)\right], \tag{4.163}$$

where $g(\pi) = \min\{c_0(1-\pi), c_1\pi\}$. The solution of this problem involves the solution of a Stefan problem that appears in [168].

The determination of π_L and π_U in Theorem 4.26 is given in terms of the solution of a system of equations that involve the function V in (4.163). However, in what follows we will give an alternative derivation of π_L and π_U.

Let us first restate the Wald–Wolfowitz Theorem 4.7, which remains valid in this case.

THEOREM 4.28. *Suppose* (T, δ) *is the SPRT*(A, B) *with* $0 < A \leq 1 \leq B < \infty$, *and let* (T', δ') *denote any other s.d.r. with* $\max\{E_0\{T'\}, E_1\{T'\}\} < \infty$, *and satisfying*

$$P_0(\delta'_{T'} = 1) \leq P_0(\delta_T = 1) \;\; and \;\; P_1(\delta'_{T'} = 0) \leq P_1(\delta_T = 0), \tag{4.164}$$

with

$$P_0(\delta_T = 1) + P_1(\delta_T = 0) < 1.$$

Then

$$E_0\{T'\} \geq E_0\{T\} \;\; and \;\; E_1\{T'\} \geq E_1\{T\}. \tag{4.165}$$

Let (T, δ) be the SPRT(A, B). The SPRT(A, B) for the hypotheses (4.153) stops at the first exit from (A, B) of the likelihood ratio

$$\Lambda_t = \left(\frac{\lambda_1}{\lambda_0}\right)^{Z_t} e^{(\lambda_0-\lambda_1)t}, \; t \geq 0. \tag{4.166}$$

It is important to point out the difference in the character of Λ_t in the each of the following two cases:

(1) $\lambda_1 > \lambda_0$, and
(2) $\lambda_0 > \lambda_1$.

We notice that in the former case, $\{\Lambda_t; t \geq 0\}$ increases only through jumps of the Poisson process $\{Z_t; t \geq 0\}$, while continuously decreasing otherwise (see Figure 4.7). In the latter case, however, $\{\Lambda_t; t \geq 0\}$ decreases through the jumps of the Poisson process $\{Z_t; t \geq 0\}$, while continuously increasing otherwise (see Figure 4.8).

As a result of the discussion in Section 4.3, it follows that in the former case, when Λ_t exits through B there will be overshooting of the barrier B (i.e. $\Lambda_T \geq B$), whereas when it exits through A, we have that $\Lambda_T = A$. The contrary is true in the latter case. For an illustration of this phenomenon please refer to Figures 4.7 and 4.8. In what follows we will consider the former case, as the latter can be treated similarly. Thus, henceforth, let us suppose that $\lambda_1 > \lambda_0$. Moreover, let $b = \log B$ and $a = \log A$ and

$$\alpha = P_0(\delta_T = 1) \;\text{ and }\; \gamma = P_1(\delta_T = 0).$$

In view of the above discussion and Proposition 4.10, we have that

$$b \leq \log \frac{1-\gamma}{\alpha} \;\text{ and }\; a = \log \frac{\gamma}{1-\alpha}.$$

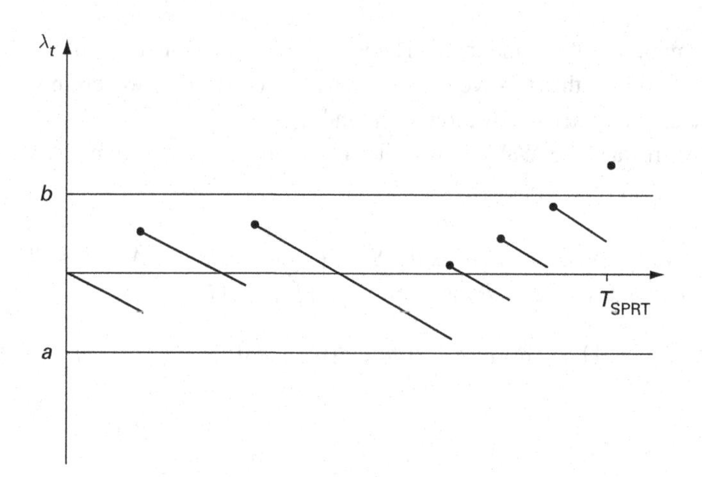

Fig. 4.7. An illustration of $\lambda_t = \log \Lambda_t$, when $\lambda_1 > \lambda_0$.

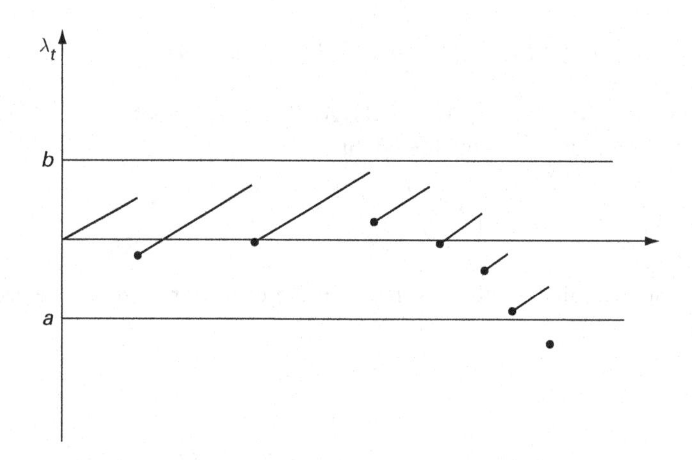

Fig. 4.8. An illustration of $\lambda_t = \log \Lambda_t$, when $\lambda_1 < \lambda_0$.

The problem now becomes to determine b, through the $oc(\text{SPRT}(A, B))$ and to calculate $asn(\text{SPRT}(A, B))$, where

(1) $oc(\text{SPRT}(A, B)) = P_\lambda(\log \Lambda_T \geq B)$, and
(2) $asn(\text{SPRT}(A, B)) = E_\lambda\{T\}$,

in which P_λ is the same as P_i when $\lambda = \lambda_i, i = 1, 2$.

REMARK 4.29. Notice that

$$b \geq -\log \frac{\lambda_1}{\lambda_0} + \log \frac{1-\gamma}{\alpha}.$$

This comes as a result of the fact that when $\log \Lambda_t$ increases it does so in increments of $\log(\lambda_1/\lambda_0)$. (Recall that $\lambda_1 > \lambda_0$.)

Before proceeding to the calculation of $oc(\text{SPRT}(A, B))$ and $asn(\text{SPRT}(A, B))$, let us define a few useful quantities. In particular, define

$$K = \frac{a}{\log(\lambda_1/\lambda_0)}, \tag{4.167}$$

$$J = \frac{b}{\log(\lambda_1/\lambda_0)}, \tag{4.168}$$

$$c = \frac{\lambda_1 - \lambda_0}{\log(\lambda_1/\lambda_0)}, \tag{4.169}$$

$$d = \frac{\lambda}{c}, \tag{4.170}$$

and

$$x = \frac{e^d}{d}. \tag{4.171}$$

PROPOSITION 4.30. *Suppose (T, δ) is the SPRT(A, B). We have*

$$\begin{aligned} oc(\text{SPRT}(A, B)) &= P_\lambda(\log \Lambda_T \geq b) \\ &= 1 - \frac{\sum_{l=0}^{[J]} \frac{(-1)^l}{l!} \left(\frac{(J-l)}{x}\right)^l}{\sum_{l=0}^{[J-K]} \frac{(-1)^l}{l!} \left(\frac{(J-K-l)}{x}\right)^l}. \end{aligned} \tag{4.172}$$

Proof: Let us first consider the process

$$X_t^r = (Z_t + r) \log \frac{\lambda_1}{\lambda_0} - (\lambda_1 - \lambda_0)t, \; t \geq 0.$$

The idea is that for every path that leads to overshooting the barrier b, if we were to start the process at a specific $r \neq 0$, the upper boundary b would have been hit exactly. As an illustration of this, please refer to Figure 4.9.

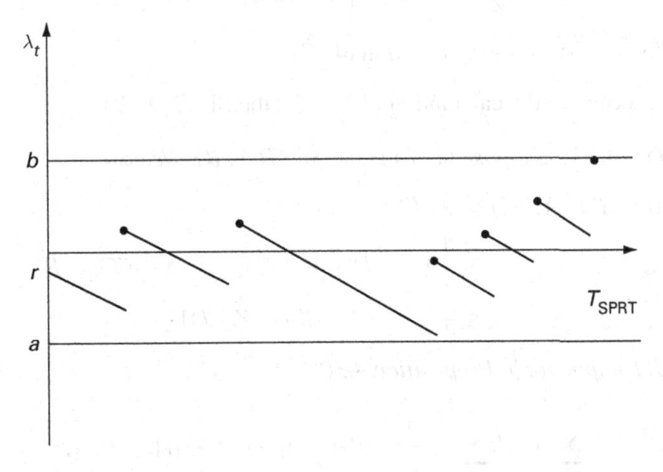

Fig. 4.9. The path of $\lambda_t = \log \Lambda_t$ of Figure 4.7 started at r.

Define $V_\lambda(r) = P_\lambda(X_t^r = b)$. It follows that the *oc* of (4.172) is $V_\lambda(0)$. Using the analysis that appears in [81], it is seen that $V_\lambda(r)$ satisfies the following delay differential equation (DDE):

$$c\frac{\mathrm{d}}{\mathrm{d}r}V_\lambda(r) + \lambda V_\lambda(r) = \lambda V_\lambda(r+1), \quad K < r < J,\ r \neq J-1, \tag{4.173}$$

with boundary conditions

$$V_\lambda(K) = 0 \text{ and } V_\lambda(r) = 1, \ \forall\, r \geq J.$$

By making the transformation

$$t = J - r$$

(4.173) is simplified. More specifically, we can now write

$$V_\lambda(r) = 1 - \frac{U_d(J-r)}{U_d(J-K)}, \quad K \leq r \leq J, \tag{4.174}$$

where U_d satisfies

$$\frac{\mathrm{d}}{\mathrm{d}t}U_d(t) - \mathrm{d}U_d(t) + \mathrm{d}U_d(t-1) \ = \ 0, \ t > 1, \tag{4.175}$$

subject to the initial condition

$$U_d(t) = \mathrm{e}^{dt}, \quad 0 \leq t \leq 1.$$

Following the steps outlined in [73], we obtain the series representation

$$U_d(t) = \mathrm{e}^{dt}\sum_{l=0}^{[t]}\frac{(-1)^l}{l!}\left[\frac{(t-l)}{x}\right]^l, \quad t \geq 0, \tag{4.176}$$

with x defined in (4.171). Substituting (4.176) into (4.174), we obtain

$$V_\lambda(r) = 1 - \mathrm{e}^{dr}\cdot\frac{\sum_{l=0}^{[J-r]}\frac{(-1)^l}{l!}\left(\frac{(J-r-l)}{x}\right)^l}{\sum_{l=0}^{[J-K]}\frac{(-1)^l}{l!}\left(\frac{(J-K-l)}{x}\right)^l}, \quad K \leq r < J. \tag{4.177}$$

The result follows by substituting $r = 0$ in (4.177). ■

Let us now proceed to the calculation of *asn* for the SPRT(A, B).

PROPOSITION 4.31. *Suppose (T, δ) is the SPRT(A, B). We have*

$$\begin{aligned} asn(SPRT(A,B)) &= E_\lambda(T) \\ &= \frac{1}{\lambda}\left[1 - dY_d(J-K)\right]\cdot oc\left(SPRT(A,B)\right), \\ &\quad + d\left[Y_d(J-K) - Y_d(J)\right] \end{aligned} \tag{4.178}$$

where $oc(SPRT)$ appears in Proposition 4.30 and

$$Y_d(t) = \frac{1}{d}\mathrm{e}^{d(t-1)}\sum_{l=0}^{[t-1]}\mathrm{e}^{-dl}\sum_{j=0}^{l}\frac{(-1)^j}{j!}(d(t-1-l))^j - [t], \quad t > 1. \tag{4.179}$$

Proof: Consider the process X_t^r as in the proof of Proposition 4.30 and denote by T the first time that X_t^r exits the interval (a, b). Let $Z_\lambda(r) = E_\lambda\{T\}$. Notice that $asn(\text{SPRT}(A, B)) = Z_\lambda(0)$. It can be seen (see [81]) that $Z_\lambda(r)$ satisfies the following DDE:

$$c\frac{\mathrm{d}}{\mathrm{d}r}Z_\lambda(r) + \lambda Z_\lambda(r) = 1 + \lambda Z_\lambda(r+1), \ K < r < J, \ r \neq J-1, \tag{4.180}$$

with boundary conditions

$$Z_\lambda(K) = 0 \text{ and } Z_\lambda(r) = 0, \ \forall r \geq J.$$

Again the transformation $t = J - r$, allows us to reduce the problem to

$$\lambda Z_\lambda(r) = [1 - \mathrm{d}Y_d(J-K)]\, V_\lambda(r) + \mathrm{d}\,[Y_d(J-K) - Y_d(J-r)], \ K \leq r \leq J, \tag{4.181}$$

where Y_d solves the DDE

$$\frac{\mathrm{d}}{\mathrm{d}t}Y_d(t) - \mathrm{d}Y_d(t) + \mathrm{d}Y_d(t-1) = 1, \ t > 1, \tag{4.182}$$

and

$$Y_d(t) = 0, \ 0 \leq t \leq 1.$$

The solution to the above DDE is given by (4.179). See [73].

The result then follows by substituting $r = 0$ in (4.181). ■

REMARK 4.32. Using Proposition 4.30, we can determine the exact value of $b = \log B$. We have that $J = \frac{b}{\log(\lambda_1/\lambda_0)}$ and J is to be chosen so that

$$1 - \frac{\sum_{l=0}^{[J]} \frac{(-1)^l}{l!}\left(\frac{(J-l)}{x}\right)^l}{\sum_{l=0}^{[J-K]} \frac{(-1)^l}{l!}\left(\frac{(J-K-l)}{x}\right)^l} = \alpha, \tag{4.183}$$

where x is given by (4.171) and $a = (\lambda_0/c)$, with c as in (4.169). Notice that equivalently J can be chosen to satisfy

$$1 - \frac{\sum_{l=0}^{[J]} \frac{(-1)^l}{l!}\left(\frac{(J-l)}{x}\right)^l}{\sum_{l=0}^{[J-K]} \frac{(-1)^l}{l!}\left(\frac{(J-K-l)}{x}\right)^l} = 1 - \gamma, \tag{4.184}$$

with x as in (4.171) and $a = (\lambda_1/c)$. Notice that K is given explicitly by

$$K = \frac{a}{\log(\lambda_1/\lambda_0)},$$

and

$$a = \log\frac{\gamma}{1-\alpha}.$$

REMARK 4.33. The limits π_L and π_U in (4.156) are given by the following equations:

$$\frac{1-\pi}{\pi}\frac{\pi_{\mathrm{L}}}{1-\pi_{\mathrm{L}}} = A = \mathrm{e}^{a}, \tag{4.185}$$

and

$$\frac{1-\pi}{\pi}\frac{\pi_{\mathrm{U}}}{1-\pi_{\mathrm{U}}} = B = \mathrm{e}^{b}. \tag{4.186}$$

4.4.6 The compound Poisson case

We now generalize the results of the previous section to the compound Poisson case. To do so let $\{N_t; t \geq 0\}$ be a standard homogeneous Poisson counting process with arrival rate λ on some probability space $(\Omega, \mathcal{F}, P_\pi)$ as described in the previous subsection. Independent of the process $\{N_t; t \geq 0\}$, let $Y_1, Y_2, \ldots,$ be i.i.d. $\mathbb{R}$-valued random variables with marginal distribution ν. The pair (λ, ν) is the unknown characteristic of the compound Poisson process $Z_t = \sum_{i=1}^{N_t} Y_i,\ \ t \geq 0$. We wish to test the hypotheses

$$H_0: \quad (\lambda, \nu) = (\lambda_0, \nu_0)$$

versus

$$H_1: \quad (\lambda, \nu) = (\lambda_1, \nu_1), \tag{4.187}$$

where it is assumed that both ν_1 and ν_0 are known and that $\nu_1 \equiv \nu_0$. At time $t = 0$, we know only that the hypotheses H_0 and H_1 are correct with prior probabilities $1 - \pi$ and π, respectively. As in all previous cases, for positive constants c_0, c_1 and c, we wish to solve the following problem:

$$\inf_{T \in \mathcal{T}, \delta \in \mathcal{D}} [c_e(T, \delta) + cE_\pi\{T\}] \tag{4.188}$$

with

$$c_e(T, \delta) = c_0 P_0(\delta_T = 1) + c_1 P_1(\delta_T = 0). \tag{4.189}$$

The likelihood ratio process, in this case, takes the form

$$\Lambda_t = \mathrm{e}^{-(\lambda_1 - \lambda_0)t} \prod_{i=1}^{N_t} \left[\frac{\lambda_1}{\lambda_0} f(Y_i)\right],\ t \geq 0, \tag{4.190}$$

where f is the likelihood ratio of the distribution of ν_1 with respect ν_0.

THEOREM 4.34. *There are constants A and B such that* $0 < A < \frac{c_0(1-\pi)}{c_1\pi} < B < \infty$, *for which the s.d.r. that optimally solves (4.188) is given by the stopping time*

$$T = \inf\{t \geq 0 | \Lambda_t \notin (A, B)\} \tag{4.191}$$

and the terminal decision rule

$$\delta_T = \begin{cases} 1 & \text{if } \Lambda_T > \frac{c_0(1-\pi)}{c_1\pi} \\ 0 & \text{if } \Lambda_T \leq \frac{c_0(1-\pi)}{c_1\pi}. \end{cases} \tag{4.192}$$

Proof: A proof of this result as well as a method for computing the upper and lower limits A and B can be found in [72]. The crux of the result however is based on expressing $[c_e(T, \delta) + cE_\pi\{T\}]$ as a function of the process $\{\Lambda_t\}$. ■

The problem of extending two alternatives into many in this setting has been studied in [70].

4.5 Discussion

A great deal has been written about sequential testing, and here we have treated only the most fundamental of results as needed in the sequel. More comprehensive treatments of this problem, including practical issues of tuning, etc., can be found, for example, in [94] and [198].

5 Bayesian quickest detection

5.1 Introduction

In Chapter 4, we considered the problem of optimally deciding, with a cost on sampling, between two statistical models for a set of sequentially observed data. Within each of these two models the data are homogeneous; that is, the data obey only one of the two alternative statistical models during the entire period of observation. In most of the remainder of this book, we turn to a generalization of this problem in which it is possible for the statistical behavior of observed data to change from one model to another at some unknown time during the period of observation. The objective of the observer is to detect such a change, if one occurs, as quickly as possible. This objective must be balanced with a desire to minimize false alarms. Such problems are known as *quickest detection* problems. In this and subsequent chapters, we analyze several useful formulations of this type of problem. Again, our focus is on the development of optimal procedures, although the issue of performance analysis will also be considered to a degree.

A useful framework for quickest detection problems is to consider a sequence $Z_1, Z_2, \ldots$ of random observations, and to suppose that there is a change point $t \geq 1$ (possibly $t = \infty$) such that, given t, $Z_1, Z_2, \ldots, Z_{t-1}$ are drawn from one distribution and $Z_t, Z_{t+1}, \ldots,$ are drawn from another distribution. The set of detection strategies of interest corresponds to the set of (extended) stopping times with respect to the observed sequence, with the interpretation that the stopping time T decides that the change point t has occurred at or before time k when $T = k$. We will be more specific about this model in subsequent sections.

The design of quickest detection procedures typically involves the optimization of a trade-off between two types of performance indices, one being a measure of the delay between the time a change occurs and it is detected (i.e., $(T - t + 1)^+$, where $x^+ = \max\{0, x\}$), and the other being a measure of the frequency of false alarms (i.e. events of the type $\{T < t\}$). In this chapter, we will adopt a Bayesian viewpoint of this problem, in which the unknown change point is assumed to be a random variable with a known prior distribution.

We will consider several versions of this problem, in discrete time and in continuous time with Brownian and Poisson observations.

5.2 Shiryaev's problem

To examine this problem we begin with the discrete-time case, considering a random sequence $\{Z_k; k = 1, 2, \ldots\}$ with a random change point t. We further assume that conditioned on t, $\{Z_k; k = 1, 2, \ldots\}$ is an independent sequence with $Z_1, \ldots, Z_{t-1}$, being i.i.d. with marginal distribution Q_0, and with $Z_t, Z_{t+1}, \ldots$ being i.i.d. with marginal distribution Q_1. For simplicity of exposition we assume $Q_0 \equiv Q_1$. Thus we consider a probability distribution P that describes both the (prior) distribution of t and the distribution of $\{Z_k; k = 1, 2, \ldots\}$ induced by this prior and the above conditional behavior. The observations $\{Z_k; k = 1, 2, \ldots\}$ generate the filtration $\{\mathcal{F}_k; k = 1, 2, \ldots\}$, with

$$\mathcal{F}_k = \sigma(Z_1, \ldots, Z_k, \{t = 0\}), \; k = 1, 2, \ldots$$

and $\mathcal{F}_0$ contains not only Ω but also the set $\{t = 0\}$.

In this situation, for a stopping time T, as a measure of delay (see Section 5.1) we can adopt the expected delay:

$$E\left\{(T - t + 1)^+\right\}, \tag{5.1}$$

where $E\{\cdot\}$ denotes expectation under the probability measure P. Similarly, as a measure of false alarm rate we can adopt the false alarm probability:

$$P(T < t). \tag{5.2}$$

We would like to determine stopping times T that effect optimal trade-offs between the two objectives of small detection delay and small false-alarm rate.

A convenient way of implementing such a trade-off is to seek $T \in \mathcal{T}$ to solve the optimization problem

$$\inf_{T \in \mathcal{T}} \left[P(T < t) + cE\left\{(T - t + 1)^+\right\}\right], \tag{5.3}$$

where $c > 0$ is a constant controlling the relative importance of the two performance indices. This problem was first posed by Kolmogorov and Shiryaev. (See [194].)[1] Notice that it suffices to consider stopping times T such that $E\{T\} < \infty$. This readily follows from the fact that $E\left\{(T - t + 1)^+\right\} = E\{T\} - E\{(T \wedge (t - 1))\}$, so that if $E\{T\} = \infty$ we have

$$\left[P(T < t) + cE\left\{(T - t + 1)^+\right\}\right] = \infty.$$

We follow Shiryaev [191].

We begin by transforming (5.3) into an optimal stopping problem of the type treated in Chapter 3. This is accomplished via the following result.

PROPOSITION 5.1. *Suppose $E\{t\} < \infty$, $E\{T\} < \infty$ and define the sequence $\{\pi_k\}$ by*

$$\pi_k = P(t \le k | \mathcal{F}_k), \; k = 0, 1, \ldots \tag{5.4}$$

[1] [194] uses $(T - t)^+$ in place of $(T - t + 1)^+$ in this formulation. With an appropriate modification of the constant c, these two problems are equivalent.

Then, for each $T \in \mathcal{T}$, we can write

$$P(T < t) + cE\left\{(T - t + 1)^+\right\} = E\left\{1 - \pi_T + c\sum_{m=0}^{T}\pi_m\right\}. \tag{5.5}$$

REMARK 5.2. Notice that

$$\begin{aligned} E\{\pi_{n+1}|\mathcal{F}_n\} &= E\{P(t \le n+1|\mathcal{F}_{n+1})|\mathcal{F}_n\} \\ &= P(t \le n|\mathcal{F}_n) + P(t = n|\mathcal{F}_n) \\ &\ge P(t \le n|\mathcal{F}_n) = \pi_n. \end{aligned} \tag{5.6}$$

Therefore, $\{\pi_n\}$ is an $\mathcal{F}_k$-submartingale. Since $|\pi_n| \le 1 \ \forall\ n$, by the bounded convergence theorem we have

$$1 = \lim_{n\to\infty} E\{\pi_n\} = E\{\lim_{n\to\infty} \pi_n\}. \tag{5.7}$$

Proof of Proposition 5.1: For every stopping time T, we have that

$$P(T < t) = E\left\{1_{\{T<t\}}\right\} = E\{1 - \pi_T\}. \tag{5.8}$$

Since T can take only countably many values, the above equality holds for every T, even if it can take the value ∞, since then, in view of Remark 5.2 the right-hand side would equal 0, and trivially, so would the left-hand side. Now, it remains to show that

$$E\left\{(T - t + 1)^+\right\} = E\left\{\sum_{m=0}^{T}\pi_m\right\}. \tag{5.9}$$

We can write

$$E\left\{(T - t + 1)^+\right\} = E\{D_T\}, \tag{5.10}$$

where

$$D_k = E\left\{(k - t + 1)^+ \middle| \mathcal{F}_k\right\}. \tag{5.11}$$

We have

$$\begin{aligned} D_k &= \sum_{m=0}^{k}(k - m + 1)P(t = m|\mathcal{F}_k) \\ &= \sum_{m=0}^{k} P(t \le m|\mathcal{F}_k) \\ &= \sum_{m=0}^{k}\pi_m + M_k, \end{aligned} \tag{5.12}$$

where

$$M_k = \sum_{m=0}^{k-1}[P(t \le m|\mathcal{F}_k) - \pi_m],\ k = 0, 1, \ldots \tag{5.13}$$

Consider the sequence $\{M_k\}$. Since $\mathcal{F}_{k-1} \subset \mathcal{F}_k$, the iteration property of conditional expectation (2.31) implies that

$$\begin{aligned} E\{P(t \leq m|\mathcal{F}_k)|\mathcal{F}_{k-1}\} &= E\left\{E\left\{1_{\{t \leq m\}}\middle|\mathcal{F}_k\right\}\middle|\mathcal{F}_{k-1}\right\} \\ &= E\left\{1_{\{t \leq m\}}\middle|\mathcal{F}_{k-1}\right\} \\ &= P(t \leq m|\mathcal{F}_{k-1}). \end{aligned} \tag{5.14}$$

Similarly, since $\mathcal{F}_m \subset \mathcal{F}_{k-1},\ m = 1, \ldots, k-1$, we have

$$E\{P(t \leq m|\mathcal{F}_k)|\mathcal{F}_m\} = \pi_m, \forall\, m < k. \tag{5.15}$$

Thus,

$$E\{M_k|\mathcal{F}_{k-1}\} = M_{k-1}, \tag{5.16}$$

and so $\{M_k\}$ is an $\mathcal{F}_k$-martingale.

Now consider the quantity

$$M = \sum_{m=0}^{\infty}(1 - \pi_m) - t. \tag{5.17}$$

Since,

$$E\left\{\sum_{m=0}^{k}(1 - \pi_m)\right\} = \sum_{m=0}^{k} P(t > m) \uparrow E\{t\} < \infty, \tag{5.18}$$

it follows from Fatou's lemma that $E\{|M|\} < 2E\{t\} < \infty$, and thus that M is an integrable random variable. On writing,

$$t = \sum_{m=0}^{\infty} 1_{\{t>m\}}, \tag{5.19}$$

we have

$$M = \sum_{m=0}^{\infty}\left[1 - \pi_m - 1_{\{t>m\}}\right], \tag{5.20}$$

which by dominated convergence has conditional expectation

$$\begin{aligned} E\{M|\mathcal{F}_k\} &= \sum_{m=0}^{\infty}[E\{P(t > m|\mathcal{F}_m)|\mathcal{F}_k\} - P(t > m|\mathcal{F}_k)] \\ &= \sum_{m=0}^{k-1}[P(t > m|\mathcal{F}_m) - P(t > m|\mathcal{F}_k)] = M_k, \end{aligned} \tag{5.21}$$

where the second equality follows from the fact that $\mathcal{F}_k \subset \mathcal{F}_m,\ m > k$, and the iteration property of expectation. Thus, $\{M_k\}$ is also regular, and so optional sampling (2.78) implies

$$E\{M_T\} = E\{M_0\} = 0. \tag{5.22}$$

Since $D_k = \sum_{m=0}^{k} \pi_m + M_k$, the proposition follows. ■

In order to find optimal stopping times for (5.3), it is necessary to assume a specific prior distribution for the change point t. A useful prior model arises from the assumption that the r.v. t is distributed according to the following rules:

$$P(t=0)=\pi \text{ and } P(t=k|t\geq k)=\rho, \tag{5.23}$$

where π and ρ are two constants lying in the interval (0, 1). That is, there is a probability π that a change has already occurred when we start observing the sequence; and there is a conditional probability ρ that the sequence will transition to the post-change state at any time, given that it has not done so prior to that time. This behavior is a good model for the description of many phenomena, such as failure times. This model gives rise to a geometric prior distribution

$$P(t=k)=\begin{cases}\pi & \text{if } k=0\\ (1-\pi)\rho(1-\rho)^{k-1} & \text{if } k=1,2,\ldots\end{cases} \tag{5.24}$$

which we henceforth assume.

The solution to problem (5.3) with the geometric prior (5.24) is summarized in the following result.

THEOREM 5.3. *For appropriately chosen threshold $\pi^*\in[0,1]$, the stopping time*

$$T_{\mathrm{B}}=\inf\{k\geq 0|\pi_k\geq\pi^*\} \tag{5.25}$$

is Bayes optimal (i.e., it solves (5.3) with the geometric prior (5.24)). Moreover, if $c\geq 1$, then $\pi^*=0$.

REMARK 5.4. Under the geometric prior (5.24), the sequence $\{\pi_k\}$ evolves via the recursion

$$\pi_k=\frac{L(Z_k)\left[\pi_{k-1}+\rho(1-\pi_{k-1})\right]}{L(Z_k)\left[\pi_{k-1}+\rho(1-\pi_{k-1})\right]+(1-\rho)(1-\pi_{k-1})},\quad k=1,2,\ldots, \tag{5.26}$$

where $L=\frac{\mathrm{d}Q_1}{\mathrm{d}Q_0}$ and $\pi_0=\pi$.

Proof of Theorem 5.3: Here $E\{t\}<\infty$, and so Proposition 5.1 can be applied to write (5.3) as

$$\inf_{T\in\mathcal{T}} E\left\{1-\pi_T+c\sum_{m=0}^{T}\pi_m\right\}. \tag{5.27}$$

In view of Remark 5.4, we see that the sequence $\{\pi_k\}$ is a homogeneous Markov process. Moreover, in view of Remark 5.2, $\pi_k\stackrel{\text{a.s.}}{\to}1$, and in turn

$$\sum_{m=0}^{k}\pi_m\to\infty \text{ a.s.} \tag{5.28}$$

We thus see that (5.27) is in the form mentioned in Remark 3.12, and so the optimal stopping time is

$$T_{\mathrm{opt}}=\inf\{k\geq 0\mid 1+(c-1)\pi_k=s(\pi_k)\} \tag{5.29}$$

with

$$s(\pi) = \inf_{T\in\mathcal{T}} E_\pi\left\{1 - \pi_T + c\sum_{m=0}^{T}\pi_m\right\} \tag{5.30}$$

where E_π denotes expectation assuming the prior (5.24).

For fixed $T \in \mathcal{T}$, we have (via Proposition 5.1 and (5.24)) that

$$E_\pi\left\{1 - \pi_T + c\sum_{m=0}^{T}\pi_m\right\} = (1-\pi)\left[P_0(T < t) + cE_0\left\{(T - t + 1)^+\right\}\right]$$
$$+ \pi c E_1\left\{(T+1)^+\right\}, \tag{5.31}$$

where subscripts refer to values of π under which the corresponding quantities are computed. Since the objective is thus linear in π, the infimum s must be concave in π. It further follows from (5.31) that

$$c\pi \le s(\pi) \le 1 + (c-1)\pi,\ 0 \le \pi \le 1, \tag{5.32}$$

with equality in both inequalities when $\pi = 1$, where the upper bound is the cost incurred by the stopping time $T \equiv 0$.

We conclude from these properties (see Figure 5.1) that

$$T_{\text{opt}} = \inf\left\{k \ge 0\,|\pi_k \ge \pi^*\right\}, \tag{5.33}$$

where

$$\pi^* = \inf\{\pi \in [0,1]|s(\pi) = 1 + (c-1)\pi\}. \tag{5.34}$$

To examine the case $c \ge 1$, let us rewrite the objective of (5.27) as

$$E\left\{1 + (c-1)\pi_T + \sum_{m=0}^{T-1}\pi_m\right\} = 1 + (c-1)P(T \ge t) + cE\left\{\sum_{m=0}^{T-1}\pi_m\right\}. \tag{5.35}$$

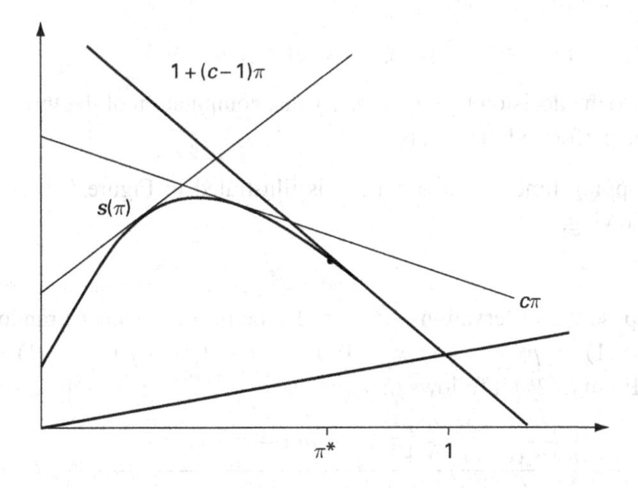

Fig. 5.1. An illustration of $s(\pi)$.

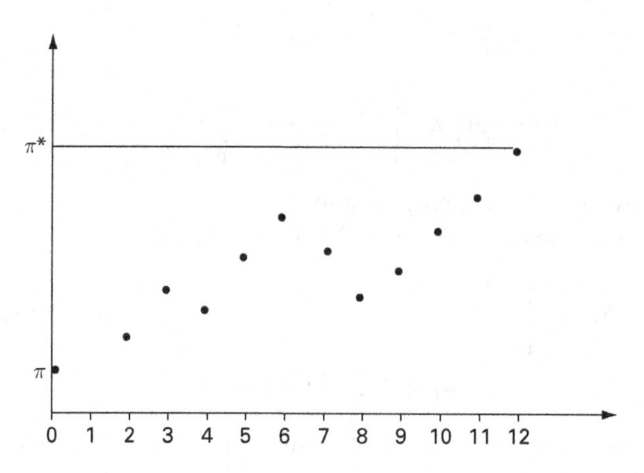

Fig. 5.2. An illustration of π_k and T of Theorem 5.3.

For $c \geq 1$, this quantity is clearly minimized by the stopping time $T = 0$, or, equivalently, by (5.33) with $\pi^* = 0$.

This completes the proof of the theorem. ■

REMARK 5.5. The payoff s is the pointwise monotone limit from above of the sequence of functions $\{\mathcal{Q}^n g\ ; n=0, 1, \ldots\}$, (compare Corollary 3.3), where g is the line

$$g(\pi) = 1 + (c-1)\pi, \ \pi \in [0, 1], \tag{5.36}$$

and where $\mathcal{Q}$ is the operator

$$\mathcal{Q}^1 h(\pi) = \min\left\{g(\pi)\,,\ \int h\left[\frac{x\,[\pi + \rho(1-\pi)]}{x\,[\pi + \rho(1-\pi)] + (1-\pi)(1-\rho)}\right] \mathrm{d}Q_1(x)\right\},$$
$$\pi \in [0, 1]. \tag{5.37}$$

Since $\{\mathcal{Q}^n g\ ; n = 0, 1, \ldots\}$ is a monotone non-increasing sequence of continuous functions, the sequence $\{\pi_n^*\}$ defined by

$$\pi_n^* = \inf\{\pi \in [0, 1] | \mathcal{Q}^n g(\pi) = g(\pi)\}, \ n = 0, 1, \ldots \tag{5.38}$$

converges upward to the decision threshold π^*. Thus, computation of the threshold and optimal cost can be performed iteratively.

The optimal stopping time of Theorem 5.3 is illustrated in Figure 5.2. A specific example is the following:

Example 5.1: Suppose the observations $\{Z_k\}$ are Bernoulli (0–1 valued) random variables with $P(Z_k = 1) = p_0 \in (0, 1), k = 0, 1, \ldots, t-1$, and $P(Z_k = 1) = p_1 \in (0, 1), k = t, \ldots$. From (5.26) it follows that

$$\pi_k = \frac{\frac{(p_1)^{Z_k}}{(p_0)^{Z_k}} \frac{(1-p_1)^{1-Z_k}}{(1-p_0)^{1-Z_k}} \left[\pi_{k-1} + \rho(1-\pi_{k-1})\right]}{\frac{(p_1)^{Z_k}}{(p_0)^{Z_k}} \frac{(1-p_1)^{1-Z_k}}{(1-p_0)^{1-Z_k}} \left[\pi_{k-1} + \rho(1-\pi_{k-1})\right] + (1-\rho)(1-\pi_{k-1})}, \ k = 1, \ldots$$

with $\pi_0 = \pi$. The optimal stopping time is then $T = \inf\{k \geq 0; \pi_k \geq \pi^*\}$, where $\pi^* = \lim_{n\to\infty} \pi_n^*$, with π_n^* defined in (5.38), and $Q^1 g = \min\{g(\pi), g(r)\}$, with

$$r = \frac{\frac{p_1}{p_0}[\pi + \rho(1-\pi)]}{\frac{p_1}{p_0}[\pi + \rho(1-\pi)] + (1-\rho)(1-\pi)} \cdot p_1$$
$$+ \frac{\frac{1-p_1}{1-p_0}[\pi + \rho(1-\pi)]}{\frac{1-p_1}{1-p_0}[\pi + \rho(1-\pi)] + (1-\rho)(1-\pi)} \cdot (1-p_1).$$

Thus, the above results completely characterize the solution to (5.3) for the geometric prior (5.24). And, in particular, we see that the optimal stopping time is the first upcrossing of a suitable threshold by the sequence $\{\pi_k\}$ of posterior probabilities of a change.

5.3 The continuous-time case

In this section, we treat Shiryaev's problem in continuous-time. In particular, as in the problem of classical sequential detection, we consider the cases in which the observations form a Brownian motion and that in which they form a homogeneous Poisson counting process.

5.3.1 Brownian observations

Here we consider the observation model

$$Z_u = \sigma W_u + \mu_0 u + (\mu_1 - \mu_0)(u-t)^+, \; u \geq 0, \tag{5.39}$$

where $\{W_u; u \geq 0\}$ is a standard Brownian motion; σ, μ_0, and μ_1 are real constants with $\sigma \neq 0$ and $\mu_1 \neq \mu_0$; and t is a random change point, independent of $\{W_u; u \geq 0\}$. Thus, (5.39) represents a Brownian motion whose rate of drift changes from μ_0 to μ_1 at the change point t. Without loss of generality, we can by simple shifting and re-scaling, replace (5.39) with the simpler model

$$Z_u = W_u + \mu(u-t)^+, \; u \geq 0, \tag{5.40}$$

where $\mu > 0$.

To analyze this problem, we consider the measurable space $(\Omega, \mathcal{F}) = \big(C[0,\infty) \times [0,\infty), \mathcal{B}(C[0,\infty)) \times \mathcal{B}([0,\infty))\big)$, where $\mathcal{B}([0,\infty))$ denotes the restriction of the real Borel sets $\mathcal{B}$ to $[0,\infty)$, and we consider the filtrations:

(1) $\mathcal{F}_s = \sigma\{Z_u, u \leq s\}$, and
(2) $\mathcal{G}_s = \sigma\{t, Z_u, u \leq s\}$.

Recall that both of the above filtrations are right-continuous (see for example [118] and Section 2.3.3). Notice that $\mathcal{F}_\infty = \sigma\{\cup_{0\leq t}\mathcal{F}_t\} \subset \mathcal{B}(C[0,\infty))$. Moreover, $\mathcal{F}_t \subset \mathcal{G}_t\ \forall\, t \in [0,\infty)$, and $\mathcal{G}_\infty = \sigma\{\cup_{0\leq t}\mathcal{G}_t\} \subset \mathcal{F}$.

We equip the above measurable space, with the probability measure $P_0 = \mu_W \times \phi$, where μ_W is Wiener measure and ϕ denotes the prior distribution on the change point t. We further construct a new probability measure P characterized by the process

$$\left.\frac{\mathrm{d}P}{\mathrm{d}P_0}\right|_{\mathcal{G}_u} = \exp\left\{\mu \int_0^u 1_{\{t \le s\}} \mathrm{d}Z_s - \frac{\mu^2}{2}\int_0^u 1_{\{t \le s\}} \mathrm{d}s\right\}. \tag{5.41}$$

Under this measure P, the process $\{Z_u; u \ge 0\}$ is a Brownian motion with drift zero in the interval $[0, t)$ and with known drift $\mu > 0$ in the interval $[t, \infty)$. Notice that under P

$$W_u = Z_u - \mu(u-t)^+$$

is a standard Brownian motion adapted to the filtration $\{\mathcal{G}_u\}$.

REMARK 5.6. The existence of the probability measure P under which $\{W_u; u \ge 0\}$ is a standard Brownian motion is guaranteed by the Novikov condition

$$E_0\left\{\exp\left\{\frac{\mu^2}{2}(u-t)^+\right\}\right\} < \infty.$$

(see for comparison (2.114) in Theorem 2.9.)

REMARK 5.7. The random variable t is $\mathcal{G}_0$-measurable and independent of the $\mathcal{G}$-Brownian motion W_t under P. Moreover, t has the same distribution under both P and P_0:

$$P(t < u) = E_0\left(\left.\frac{\mathrm{d}P}{\mathrm{d}P_0}\right|_{\mathcal{G}_0} 1_{\{t<u\}}\right) = P_0(t < u).$$

Consider now the family $\mathcal{T}$ of stopping times with respect to the filtration $\{\mathcal{F}_s\}$. Analogously with the discrete-time problem of the preceding section, we wish to choose a stopping time $T \in \mathcal{T}$ to solve the problem

$$\inf_{T \in \mathcal{T}} \left[P(T < t) + c\, E\left\{(T-t)^+\right\}\right], \tag{5.42}$$

with $c > 0$.

Notice that

$$E\{(T-t)^+\} = E\{T\} - E\{T \wedge t\}.$$

Therefore to solve (5.42) it suffices to consider T with $E\{T\} < \infty$.

We have the following result, analogous to Proposition 5.1.

PROPOSITION 5.8. *Suppose $E\{t\}$ and $E\{T\}$ are finite. Then*

$$P(T < t) + c\, E\left\{(T-t)^+\right\} = E\left\{1 - \pi_T + c\int_0^T \pi_s \mathrm{d}s\right\}, \tag{5.43}$$

where

$$\pi_s = P\,(t \le s | \mathcal{F}_s), s \ge 0. \tag{5.44}$$

Proof: For any stopping time T we can write

$$P(T < t) = E\{1 - \pi_T\}. \tag{5.45}$$

To see this let us first suppose that T is any stopping time that can take only countably many values in a set that also contains ∞. Then we have that

$$P(t > T) = \sum_{j=1}^{\infty} 1_{\{T=s_j\}} P(t > s_j) = E\left\{1 - \sum_{j=1}^{\infty} 1_{\{T=s_j\}} P(t \le s_j | \mathcal{F}_{s_j})\right\}$$
$$= E\{1 - \pi_T\}. \tag{5.46}$$

Notice that an equivalent to Remark 5.2 holds here so that $E\{\lim_{t\to\infty} \pi_t\} = \lim_{t\to\infty} E\{\pi_t\} = 1$.

Now, an arbitrary stopping time T is the a.s limit of a decreasing sequence of T_n that can take only finitely many values. This is summarized in the following remark.

REMARK 5.9. Any stopping time T is the limit of a decreasing sequence of stopping times $\{T_n\}$ which take countably many values. To see this consider the sequence of functions

$$h_n(x) = \begin{cases} 0, & x = 0, \\ \frac{(k+1)n}{2^n}, & \frac{kn}{2^n} < x \le \frac{(k+1)n}{2^n},\ k = 0, \ldots, (2^n - 1), \\ \infty, & x > n. \end{cases}$$

Notice that $\lim_{n\to\infty} h_n(x) = x$ and that the sequence of functions converges downwards to x. Now, let $T_n = h_n(T)$. To see that T_n is a stopping time, notice that

$$\left\{T_n = \frac{kn}{2^n}\right\} = \left\{\frac{(k-1)n}{2^n} < T \le \frac{kn}{2^n}\right\} \in \mathcal{F}_{\frac{kn}{2^n}}. \tag{5.47}$$

Also, for any $m \ge n$, we have that

$$\{T_n \le m\} = \{T_n \le n\} \in \mathcal{F}_n \subset \mathcal{F}_m. \tag{5.48}$$

Therefore, for each n, T_n is a stopping time. Moreover, we have that $\lim_{n\to\infty} T_n = T$.

Using the monotone convergence theorem the result follows for the left-hand side of equation (5.45).

For the right-hand side of (5.45) notice that it suffices to show that for a decreasing sequence of stopping times $\{T_n\}$ taking finitely many values and converging a.s. to T, we have that $\lim_{n\to\infty} \pi_{T_n} = \pi_T$. Consider the sequence $S_n = E\left\{1_{\{t\le T\}} | \mathcal{F}_n\right\}$. This sequence is a martingale with a last element S_∞ to which it converges a.s. (since $E\left\{1_{\{t\le T\}}\right\} < \infty$) by Example 2.6 of Section 2.3.1. Now,

$$P(S_n > c) \le \frac{E\{S_n\}}{c} = \frac{E\left\{1_{\{t\le T\}}\right\}}{c} \to 0,$$

as $c \to \infty$, while,

$$\int_{\{S_n>c\}} S_n \mathrm{d}P = \int_{\{S_n>c\}} E\left\{1_{\{t\le T\}} | \mathcal{F}_n\right\} \mathrm{d}P = \int_{\{S_n>c\}} 1_{\{t\le T\}} \mathrm{d}P \to 0, \tag{5.49}$$

as $c \to \infty$ and the second equality follows from the fact that $\{S_n > c\} \in \mathcal{F}_n$ and the definition of conditional expectation in Section 2.2.5.

Equation (5.49) shows that the sequence S_n is u.i. and therefore $S_n \to S_\infty$ a.s. implies that $E\{S_n\} \to E\{S\}$ by the comments made in Section 2.2.6.

Since this martingale has a last element, the optional sampling theorem applies (compare (2.88) in Section 2.3.3) and we have that for $m > n$, $T_m < T_n$ and $E\left\{S_{T_n}|\mathcal{F}_{T_m}\right\} = S_{T_m}$. Moreover, $\mathcal{F}_{T_n} \downarrow \mathcal{F}_T$. Let $A \in \mathcal{F}_T$. Then

$$\begin{aligned}\int_A S_\infty \mathrm{d}P &= \lim_{n\to\infty} \int_A E\left\{1_{\{t\le T\}}|\mathcal{F}_{T_n}\right\} \mathrm{d}P \\ &= \lim_{n\to\infty} \int_A E\left\{E\left\{1_{\{t\le T\}}|\mathcal{F}_{T_n}\right\}|\mathcal{F}_T\right\} \mathrm{d}P \\ &= \lim_{n\to\infty} \int_A E\left\{1_{\{t\le T\}}|\mathcal{F}_T\right\} \mathrm{d}P \\ &= \int_A E\left\{1_{\{t\le T\}}|\mathcal{F}_T\right\} \mathrm{d}P.\end{aligned}$$

Thus S_∞ is a version of $E\left\{1_{\{t\le T\}}|\mathcal{F}_T\right\}$ and $\lim_{n\to\infty} E\left\{1_{\{t\le T\}}|\mathcal{F}_{T_n}\right\} = E\left\{1_{\{t\le T\}}|\mathcal{F}_T\right\}$ a.s.

Now let $Y_m = \sup_{n\ge m}\left|1_{\{t\le T_n\}} - 1_{\{t\le T\}}\right|$. We have that $|Y_m| \le 1$. Moreover,

$$\begin{aligned}\left|E\left\{1_{\{t\le T_n\}}|\mathcal{F}_{T_n}\right\} - E\left\{1_{\{t\le T\}}|\mathcal{F}_T\right\}\right| &\le \left|E\left\{1_{\{t\le T_n\}}|\mathcal{F}_{T_n}\right\} - E\left\{1_{\{t\le T\}}|\mathcal{F}_{T_n}\right\}\right| \\ &\quad + \left|E\left\{1_{\{t\le T\}}|\mathcal{F}_{T_n}\right\} - E\left\{1_{\{t\le T\}}|\mathcal{F}_T\right\}\right| \\ &\le E\left\{Y_m|\mathcal{F}_{T_n}\right\} \\ &\quad + \left|E\left\{1_{\{t\le T\}}|\mathcal{F}_{T_n}\right\} - E\left\{1_{\{t\le T\}}|\mathcal{F}_T\right\}\right| \qquad (5.50)\end{aligned}$$

Since $|Y_m| < 1$, $\{Y_m\}$ is another martingale with an integrable last element, and therefore the optional sampling theorem applies and as before, $\lim_{n\to\infty} E\left\{Y_m|\mathcal{F}_{T_n}\right\} = E\{Y_m|\mathcal{F}_T\}$. Moreover $\lim_{m\to\infty} Y_m = 0$ a.s. and since $|Y_m| \le 1$, by the bounded convergence theorem we have that

$$\lim_{m\to\infty} E\{Y_m|\mathcal{F}_T\} = 0 \text{ a.s.} \qquad (5.51)$$

Taking the limit of both sides of (5.50) as $n \to \infty$ and as $m \to \infty$, we have that $\lim_{n\to\infty} E\{1 - \pi_{T_n}\} = E\{1 - \pi_T\}$. Note that this result holds even if T is unbounded.

Now, consider the remaining term in (5.43). From (5.40) we can write

$$\mu\,(u-t)^+ = \mu \int_0^u \pi_s \mathrm{d}s + Z_u - \mu \int_0^u \pi_s \mathrm{d}s - W_u,\ u \ge 0. \qquad (5.52)$$

Since

$$Z_u = \mu \int_0^u 1_{\{t\le s\}} \mathrm{d}s + W_u, \qquad (5.53)$$

the innovation theorem (cf., Theorem 2.8) implies that

$$I_u = Z_u - \mu \int_0^u \pi_s \mathrm{d}s \qquad (5.54)$$

is a Brownian motion with respect to $\{\mathcal{F}_u\}$.

Since $E\{T\} < \infty$, Wald's identity for Brownian motion (2.98) implies that

$$E\left\{Z_T - \mu \int_0^T \pi_s \mathrm{d}s\right\} = E\{W_T\} = 0, \tag{5.55}$$

and the proposition follows. ■

In order to solve (5.42), we can proceed rather straightforwardly from Proposition 5.8 to use optimal stopping theory to determine the optimal procedure for change detection. This development parallels that of the preceding section, and can be found, for example, in [191].

However, it is also possible to use the sample-path continuity of Brownian motion (and hence of $\{\pi_t; t \geq 0\}$), to solve the problem (5.42) in a more direct fashion. This approach is analogous to that of Section 4.2.1, in which a direct proof of the optimality of the SPRT for Brownian observations is given. Here we adopt this latter approach (which was considered in [26] in this framework), for the purposes of illustration.

As in the classical sequential detection problem, the crux of this approach is to rewrite (5.43) in the form

$$E\{g(\pi_T)\} \tag{5.56}$$

for bounded stopping times T, where g is a continuous function with a unique minimum, which lies at a point $\pi^* \in [\pi, 1]$. In order to achieve this we will begin by assuming a specific prior for the change point t. In particular, we assume that the probability measure ϕ, which describes the prior distribution of the change point t is of the form

$$\phi(0) = \pi$$
$$\phi((s, \infty)) = (1 - \pi)\mathrm{e}^{-\alpha s}, \; \alpha > 0; \tag{5.57}$$

that is, analogously with the discrete-time case, the change point t has an atom of probability π at 0 and is exponential otherwise. From (5.56) we will then be able to conclude that the optimal strategy is to stop the first time π_u equals π^*, and that the value of the stopping problem is thus $g(\pi^*)$. The first step in this development is the following result, analogous to Proposition 4.20.

PROPOSITION 5.10. *Suppose t obeys the prior distribution (5.57), and define the function $f : (0, 1) \to \mathbb{R}$ by*

$$f(x) = \frac{2}{\mu^2} \int_\pi^x \mathrm{e}^{-\Lambda H(y)} \int_\pi^y \mathrm{e}^{\Lambda H(z)} \frac{1}{z(1-z)^2} \mathrm{d}z\mathrm{d}y \tag{5.58}$$

where $\Lambda = 2\alpha/\mu^2$ and

$$H(y) = \log \frac{y}{1-y} - \frac{1}{y}. \tag{5.59}$$

Then, for all bounded stopping times T we have

$$E\left\{\int_0^T \pi_u \mathrm{d}u\right\} = E\{f(\pi_T)\}. \tag{5.60}$$

Proof: We note first that, with the exponential prior (5.57) on t, $\{\pi_u\}$ satisfies the stochastic differential equation

$$d\pi_u = \alpha\,(1-\pi_u)du + \mu\,\pi_u(1-\pi_u)dI_u,\ \pi_0 = \pi, \tag{5.61}$$

where $\{I_u\}$ is the innovation process of (5.54). So, since f is twice continuously differentiable, Itô's rule implies

$$df(\pi_u) = \mathcal{D}f(\pi_u)du + \mu\,f'(\pi_u)\pi_u(1-\pi_u)dI_u,\ u \geq 0, \tag{5.62}$$

with initial condition $f(\pi_0) = f(\pi)$, where

$$\mathcal{D}f(x) = \alpha\,(1-x)\,f'(x) + \frac{\mu^2}{2}\,[x(1-x)]^2\,f''(x). \tag{5.63}$$

Since f satisfies the differential equation

$$\mathcal{D}f(x) = x,\ \ f(\pi) = 0, \tag{5.64}$$

the proposition follows analogously with the proof of Proposition 4.20. ■

The result of Proposition 5.10, can be generalized to any arbitrary stopping time T by considering the sequence $T_n = T \wedge n$. Then, using the dominated convergence theorem on the left-hand side of (5.60), and the bounded convergence theorem on the right-hand side, which can be used because f is continuous and $\lim_{t\to\infty}\pi_t \to 1\ P\ a.s.$, we can deduce that (5.60) holds for any arbitrary stopping time T.

It follows from Proposition 5.10 that we can write the objective (5.43) in the form (5.56) with g given by

$$g(x) = 1 - x + c\,f(x),\ 0 < x < 1. \tag{5.65}$$

The function g is convex [26] and has a unique minimum at the point x_0 satisfying

$$c\,f'(x_0) = 1. \tag{5.66}$$

This allows us to prove the following result.

THEOREM 5.11. *The problem (5.42) with prior (5.57) is solved by the stopping time*

$$T^* = \inf\{u \geq 0 | \pi_u \geq x_0\}, \tag{5.67}$$

where x_0 is given by (5.66).

Proof: The proof is very similar to that of Theorem 4.13 given in Section 4.4.1. Thus, we give an outline only. From (5.56) and the fact that g achieves its minimum at x_0, we can conclude that

$$P(T < t) + c\,E\left\{(T-t)^+\right\} \geq g(x_0) \tag{5.68}$$

for all bounded stopping times T. Similarly to the argument used in Section 4.4.2, this inequality extends to all stopping times T.

The process $\{\pi_u\}$ has continuous sample paths with $\pi_0 = \pi$ and unity asymptote, almost surely P. (This result follows in exactly the same way as (5.7).) Moreover, since

g is continuous and bounded on $[0, x_0]$, it follows similarly to the argument used in (4.145) that

$$P(T^* < t) + c\, E\left\{(T^* - t)^+\right\} = g(x_0) \tag{5.69}$$

if $\pi \in [0, x_0]$. Thus, the optimality of T^* is established when $\pi \in [0, x_0]$.

Now suppose $\pi \in (x_0, 1)$. Since g is convex, Jensen's inequality implies

$$E\{g(\pi_T)\} \geq g\left(E\{\pi_T\}\right) = g\left(\, P(t \leq T)\,\right), \tag{5.70}$$

for all stopping times T. Now, since $P(t \leq T) \geq \pi$ for any T and since g is increasing on $(x_0, 1)$, we see that in this $\pi > x_0$ case

$$E\{g(\pi_T)\} \geq g(\pi) = 1 - \pi. \tag{5.71}$$

We thus have the inequality,

$$P(T < t) + c\, E\left\{(T - t)^+\right\} \geq 1 - \pi \tag{5.72}$$

for all bounded stopping times T, which extends to all stopping times. Since $T^* \equiv 0$ in this case, we see that

$$P(T^* < t) + c\, E\left\{(T^* - t)^+\right\} = 1 - \pi, \tag{5.73}$$

and so the optimality of T^* extends to $\pi \in (x_0, 1)$.

The case $\pi = 1$ is trivial, and so the theorem follows. ■

Thus, as in the discrete-time case, optimal (Bayesian) quickest detection of a change in drift of a Brownian motion is achieved by announcing a change at the first upcrossing of a threshold by the posterior probability of a change.

5.3.2 Poisson observations

We now turn to quickest detection in the case of point-process observations. In particular, we assume the observation process $\{Z_s; s \geq 0\}$ is a homogeneous Poisson counting process with rate $\lambda_0 > 0$, for $s < t$ and rate $\lambda_1 \neq \lambda_0$ for $s \geq t$. The change-point t is assumed to be a random variable independent of the observation process. This problem was initially treated in [92]. In this section, however, we will follow [22] and [169].

In the Bayesian formulation of this problem, we begin by observing the point process $\{Z_s; s \geq 0\}$, with its compensator λs, and the random intensity λ is a random variable taking two values, λ_0 for $s < t$ (t is also a random variable) and λ_1 for $s \geq t$.

To examine this problem, we consider the following measurable space:

(1) $\Omega :\ D[0, \infty) \times [0, \infty)$, and
(2) $\mathcal{F} = \mathcal{B}(D[0, \infty)) \times \mathcal{B}([0, \infty))$.

We also consider the filtrations:

(1) $\mathcal{F}_s = \sigma\{Z_u, u \leq s\}$, and
(2) $\mathcal{G}_s = \sigma\{t, Z_u, u \leq s\}$.

We point out here that both of the above filtrations are right-continuous. (See for comparison page 91 of [118]). Notice that $\mathcal{F}_\infty = \sigma\{\cup_{0\le t}\mathcal{F}_t\} \subset \mathcal{B}(D[0,\infty))$. Moreover, $\mathcal{F}_s \subset \mathcal{G}_s \ \forall\ s \in [0,\infty)$, and $\mathcal{G}_\infty = \sigma\{\cup_{0\le s}\mathcal{G}_s\} \subset \mathcal{F}$. The process $\{Z_s; s \ge 0\}$ and the r.v. t are defined by the projections on the first and second component of $(\Omega, \mathcal{F})$, respectively.

We equip the above measurable space with the initial probability measure

$$P_0 = P^0 \times \phi, \tag{5.74}$$

where P^0 is the probability measure generated on the space of càdlàg functions, $D[0,\infty)$ by a Poisson process with parameter λ_0, and ϕ is the probability measure describing the prior distribution of the change point t. Moreover, consider the intensity process which also lies in Ω, $h(s) = \lambda_0 \mathbf{1}_{\{s<t\}} + \lambda_1 \mathbf{1}_{\{s\ge t\}}$, $s \ge 0$. We now construct a new probability measure P characterized by the process

$$\Lambda_u = \left.\frac{\mathrm{d}P}{\mathrm{d}P_0}\right|_{\mathcal{G}_u} = \exp\left\{\int_0^u \log\frac{h(s)}{\lambda_0}\mathrm{d}Z_s - \int_0^u (h(s)-\lambda_0)\mathrm{d}s\right\}. \tag{5.75}$$

This process is a martingale (see [47]) and under this measure P, the process $\{Z_s; s \ge 0\}$ is a homogeneous Poisson counting process with intensity λ_0 in the interval $[0,t)$ and λ_1 in the interval $[t,\infty)$. In other words, under the measure P the process $\{Z_s - h(s); s \ge 0\}$ is a $\mathcal{G}_s$-martingale.

Moreover, for reasons that will become apparent later, we also equip $(\Omega, \mathcal{F})$ with the family of probability measures $\{P_\pi\}$ indexed by $\pi = \phi(\{0\})$. We notice that P_π for any $\pi \in [0,1)$ is the same as the probability measure P above and that $P_\pi|_{\{\pi=1\}}$ is the probability measure generated on Ω by a homogeneous Poisson process with intensity λ_1. We remark here that the reference measure P_0 of (5.74) is not the same as the probability measure $P_\pi|_{\{\pi=0\}}$.

Notice that we can rewrite (5.75) in the form

$$\left.\frac{\mathrm{d}P}{\mathrm{d}P_0}\right|_{\mathcal{G}_u} = 1_{\{t>u\}} + 1_{\{t\le u\}}\frac{L_u}{L_t}, \quad u \ge 0,$$

where

$$L_u = \left(\frac{\lambda_1}{\lambda_0}\right)^{Z_u} \mathrm{e}^{-(\lambda_1-\lambda_0)u}, \quad u \ge 0.$$

Consider now the family $\mathcal{T}$ of stopping times with respect to the filtration $\{\mathcal{F}_s\}$. Analogously with the previous sections, we wish to choose a stopping time $T \in \mathcal{T}$ to solve the problem

$$\inf_{T\in\mathcal{T}} \left[P(T<t) + c\,E\left\{(T-t)^+\right\}\right], \tag{5.76}$$

with $c > 0$.

We have the following result, analogous to Proposition 5.1. Recall again that

$$E\{(T-t)^+\} = E\{T\} - E\{T \wedge t\},$$

and thus to solve (5.76) it suffices to consider T with $E\{T\} < \infty$.

We will first give an equivalent to Propositions 5.1 and 5.8.

PROPOSITION 5.12. *Suppose that $E\{t\}$ and $E\{T\}$ are finite, and define the càdlàg process*

$$\pi_s = P(t \leq s|\mathcal{F}_s), \ s \geq 0. \tag{5.77}$$

Then, for each $T \in \mathcal{T}$, we can write

$$\left[P(T < t) + c\, E\left\{(T-t)^+\right\}\right] = E\left\{1 - \pi_T + c\int_0^T \pi_s \mathrm{d}s\right\}.$$

Proof: The proof of this result follows in a way similar to the proofs of Propositions 5.1 and 5.8. More specifically, let T be a stopping time that takes on only countably many values. Then we have that

$$\begin{aligned} P(t > T) = \sum_{j=1}^{\infty} 1_{\{T=s_j\}} P(t > s_j) &= E\left\{1 - \sum_{j=1}^{\infty} 1_{\{T=s_j\}} P(t \leq s_j|\mathcal{F}_{s_j})\right\} \\ &= E\left\{1 - \pi_T\right\} \end{aligned} \tag{5.78}$$

Now, in view of Remark 5.9 an arbitrary stopping time T is the a.s. limit of a decreasing sequence of T_n that can take only finitely many values. Hence using an argument similar to the one used in Proposition 5.8 along with the right-continuity of the filtration $\{\mathcal{F}_s\}$ the result follows. Note that the result holds even if T is unbounded.

Moreover,

$$\begin{aligned} \lambda_1(u-t)^+ = &\int_0^u E\left\{h(s)1_{\{t\leq s\}}|\mathcal{F}_s\right\} \mathrm{d}s + \lambda_1 \int_0^u 1_{\{t\leq s\}} \mathrm{d}Z_s \\ &- \int_0^u E\left\{h(s)1_{\{t\leq s\}}|\mathcal{F}_s\right\} \mathrm{d}s - \left(\int_0^u 1_{\{t\leq s\}} \mathrm{d}Z_s - \lambda_1 \int_0^u 1_{\{t\leq s\}} \mathrm{d}s\right). \end{aligned}$$

The intensity of the Poisson counting process $\{Z_s\}$ is λ_1 for $t \geq s$. Let

$$M_u = \left(\int_0^u 1_{\{t\leq s\}} \mathrm{d}Z_s - \lambda_1 \int_0^u 1_{\{t\leq s\}} \mathrm{d}s\right).$$

We notice that $1_{\{t\leq s\}}$ is a $\mathcal{G}_0$-measurable random variable and hence $\{M_u;\, u \geq 0\}$ is a $\mathcal{G}_u$-martingale.

Thus, $M_{u\wedge T}$ is also a $\mathcal{G}_u$-martingale and $E\{Z_{u\wedge T}\} = \lambda_1 E\{u \wedge T\}$. Letting $u \to \infty$ and using the monotone convergence theorem on both sides of the above equation we get $E\{Z_T\} = \lambda_1 E\{T\} < \infty$, or $E\{M_T\} = 0$.

Hence it follows, using $E\{T\} < \infty$, that

$$E\left\{\int_0^T 1_{\{t\leq s\}} \mathrm{d}Z_s - \lambda_1 \int_0^T 1_{\{t\leq s\}} \mathrm{d}s\right\} = 0.$$

Also,

$$M_u = \lambda_1 \int_0^u 1_{\{t\leq s\}} \mathrm{d}Z_s - \int_0^u E\left\{h(s)1_{\{t\leq s\}}|\mathcal{F}_s\right\} \mathrm{d}s$$

is an $\mathcal{F}_u$-martingale, since

$$E\left\{M_u|\mathcal{F}_r\right\} = M_r,$$

for all $r < u$ can be easily be verified. Therefore, using an argument similar to that for $E\{T\} < \infty$, it follows that

$$\lambda_1 \int_0^T 1_{\{t\leq s\}} \mathrm{d}Z_s - \int_0^T E\left\{h(s)1_{\{t\leq s\}}|\mathcal{F}_s\right\} \mathrm{d}s = 0.$$

Substituting for $h(s)$, obtain

$$\lambda_1 E\left\{(T-t)^+\right\} = E\left\{\int_0^T \lambda_1 \pi_s \mathrm{d}s\right\}.$$

Hence the result follows. ■

We now give an alternative representation of $P(T < t) + c\,E\left\{(T-t)^+\right\}$ that involves the odds ratio process $\{\Phi_s\}$, with $\Phi_s = \pi_s/(1-\pi_s)$, $s \geq 0$. In order to give this representation we will make a specific choice for ϕ, the prior distribution of t. Namely, we assume (as in the Brownian case) that

$$\begin{aligned} \phi\{0\} &= \pi \\ \phi\{(s,\infty)\} &= (1-\pi)\mathrm{e}^{-\alpha s}, \; s > 0, \; \alpha > 0. \end{aligned} \tag{5.79}$$

Through this choice of prior, it becomes easier to derive the form of the optimal stopping time T.

PROPOSITION 5.13. *We have*

$$P(T < t) + c\,E\left\{(T-t)^+\right\} = (1-\pi) + c(1-\pi)E_0\left\{\int_0^T \mathrm{e}^{-\alpha s}\left\{\Phi_s - \frac{\alpha}{c}\right\}\mathrm{d}s\right\}, \tag{5.80}$$

where

$$\pi_s = P\left(t \leq s|\,\mathcal{F}_s\right), \; s \geq 0, \tag{5.81}$$

and

$$\Phi_s = \frac{\pi_s}{1-\pi_s}, \quad s \geq 0. \tag{5.82}$$

Before proving Proposition 5.13 we derive the dynamics of the process Φ_s. Using Bayes' formula we have that (see [135])

$$\pi_s = \frac{E_0\left\{\Lambda_s 1_{\{t\leq s\}}|\mathcal{F}_s\right\}}{E_0\left\{\Lambda_s|\mathcal{F}_s\right\}} \tag{5.83}$$

from which it follows that

$$1-\pi_s = \frac{E_0\left\{1_{\{t>s\}}|\mathcal{F}_s\right\}}{E_0\left\{\Lambda_s|\mathcal{F}_s\right\}} = \frac{(1-\pi)\mathrm{e}^{-\alpha s}}{E_0\left\{\Lambda_s|\mathcal{F}_s\right\}}. \tag{5.84}$$

Hence

$$\Phi_s = \frac{\mathrm{e}^{\alpha s}}{1-\pi} E_0\left\{\Lambda_t 1_{\{t\le s\}}|\mathcal{F}_s\right\} \tag{5.85}$$

$$= \frac{\mathrm{e}^{\alpha s}}{1-\pi}\left[\pi L_s + (1-\pi)\int_0^s \frac{L_t}{L_u}\alpha \mathrm{e}^{-\alpha u}\mathrm{d}u\right]. \tag{5.86}$$

The process $\{L_s\}$ is the solution of the stochastic differential equation (compare Theorem 2.7)

$$\mathrm{d}L_t = \left[\frac{\lambda_1}{\lambda_0} - 1\right] L_{t-}(\mathrm{d}Z_t^{-}\lambda_0 \mathrm{d}t),\; L_0 = 1. \tag{5.87}$$

From the chain rule it follows that

$$\mathrm{d}\Phi_t = (\alpha + (\alpha - \lambda_1 + \lambda_0)\Phi_t)\mathrm{d}t + \left[\frac{\lambda_1}{\lambda_0} - 1\right]\Phi_{t-}\mathrm{d}Z_t, \tag{5.88}$$

with $\Phi_0 = \pi/(1-\pi)$.

Also,

$$\mathrm{d}\pi_t = [\alpha - (\lambda_1 - \lambda_0)\pi_t](1-\pi_t)\mathrm{d}t + \frac{(\lambda_1 - \lambda_0)\pi_{t-}(1-\pi_{t-})}{\lambda_0(1-\pi_{t-}) + \lambda_1\pi_{t-}}\mathrm{d}Z_t, \tag{5.89}$$

with $\pi_0 = \pi$.

From (5.89) and (5.88) it is evident that both processes are strongly Markovian.

We now prove Proposition 5.13.

Proof of Proposition 5.13: We have that

$$\begin{aligned}
E\left\{(T-t)^+\right\} &= E\left\{1_{\{T>t\}}\int_0^T \mathrm{d}t\right\} \\
&= E\left\{\int_0^\infty 1_{\{T>u\}}1_{\{t\le u\}}\right\}\mathrm{d}u \\
&= \int_0^\infty E_0\left\{\Lambda_t 1_{\{T>u\}}1_{\{t\le u\}}\right\}\mathrm{d}u \\
&= \int_0^\infty E_0\left\{1_{\{T>u\}}E_0\left\{\Lambda_t 1_{\{t\le u\}}|\mathcal{F}_u\right\}\right\}\mathrm{d}u \\
&= (1-\pi)E_0\left\{\int_0^T \mathrm{e}^{-\alpha u}\Phi_u \mathrm{d}u\right\}.
\end{aligned}$$

Moreover, for any T that takes countably many values in $\{s_n, n \in \mathcal{N} \cup \{\infty\}\}$ we have

$$
\begin{aligned}
P(T < t) &= \sum_{n=1}^{\infty} P(s_n < t, T = s_n) = \sum_{n=1}^{\infty} E_0 \left\{ \Lambda_{s_n} 1_{\{s_n < t\}} 1_{\{T = s_n\}} \right\} \\
&= \sum_{n=1}^{\infty} E_0 \left\{ 1_{\{t > s_n\}} 1_{\{T = s_n\}} \right\} = (1 - \pi) \sum_{n=1}^{\infty} \mathrm{e}^{-\alpha s_n} E_0 \left\{ 1_{\{T = s_n\}} \right\} \\
&= (1 - \pi) E_0 \left\{ \sum_{n=1}^{\infty} \left\{ 1 - \int_0^{s_n} \alpha \mathrm{e}^{-\alpha s} \mathrm{d}s \right\} 1_{\{T = s_n\}} \right\} \\
&= (1 - \pi) - (1 - \pi) \alpha E_0 \left\{ \int_0^T \mathrm{e}^{-\alpha s} \mathrm{d}s \right\}.
\end{aligned}
$$

An arbitrary stopping time T can be constructed as the a.s. limit of a sequence $\{T_n\}$ of stopping times that take countably many values. Now, both the function $1_{\{u<t\}}$ and the function $\int_0^u \mathrm{e}^{-\alpha s} \mathrm{d}s$ are bounded, so by the bounded convergence theorem, we can extend to all stopping times T. ■

In what follows we will use the representations given in Propositions 5.12 and 5.13 to deduce that the optimal stopping time to (5.76) is of the threshold type. The representation in Proposition 5.13 will prove more useful in characterizing the exact threshold in special cases of parameter values.

We provide the first solution according to the representation given in Proposition 5.12. Using the fact that $\{\pi_s; s \geq 0\}$ is strongly Markovian, we can use an equivalent to Remark 3.12 for the continuous-time Markov optimal stopping problem (see for comparison Chapter 3 of [191]), to deduce that the optimal stopping time is

$$
T_{\mathrm{opt}} = \inf\{s > 0 | s(\pi_s) = 1 - \pi_s\}, \tag{5.90}
$$

where

$$
s(\pi) = \inf_{T \in \mathcal{T}} \left[P(T < t) + c\, E\left\{(T - t)^+\right\}\right].
$$

Moreover, for fixed T we have that

$$
P(T < t) + c\, E\left\{(T - t)^+\right\} = (1 - \pi)\left[P_0(T < t) + c E_0\{(T - t)^+\}\right] + c\pi E_1\{T\},
$$

where P_0 and P_1 above are used to denote $P_\pi|_{\{\pi=0\}}$ and $P_\pi|_{\{\pi=1\}}$, respectively. From the above equation it follows that $s(\pi)$ is a concave function satisfying

$$
0 \leq s(\pi) \leq 1 - \pi.
$$

Therefore (5.90) can be written as

$$
T_{\mathrm{opt}} = \inf\{s > 0 | \pi_s \geq \pi^*\}, \tag{5.91}
$$

where $\pi^* = \inf\{\pi | s(\pi) = 1 - \pi\}$. That is to say, the form of the optimal solution is a threshold stopping time. The determination of the threshold is a complex problem, and we will discuss it in what follows by using the representation of the cost function in terms of the process $\{\Phi_s\}$ that appears in Proposition 5.13.

We now proceed to the solution of the problem (5.76) by using the representation of Proposition 5.13. To this effect we define the following constants:

(1) $a = \alpha - \lambda_1 + \lambda_0$,
(2) $b = \alpha + \lambda_0$,
(5) $r = \lambda_1/\lambda_0$,
(6) $\phi_d = -\alpha/a$, and
(7) $k = \alpha/c$.

We also distinguish the following cases in each of which the solution takes a different form.

(1) $r > 1$, and either $\phi_d < 0$ or $0 < k \leq \phi_d$. In this case $\phi_d < 0$ so the paths of $\{\Phi_s\}$ always increase between jumps, and since $r > 1$, the jumps are also positive. This means that once the process $\{\Phi_s\}$ leaves the interval $[0, k]$ it will never return there (see Figure 5.3). This is also the case when $\phi_d \geq k > 0$. To see this, one can inspect the dynamics of $\{\Phi_s\}$ by closely examining (5.88). It is easily seen that if $\phi_d > 0$ then the process $\{\Phi_s\}$ mean-reverts to the level ϕ_d. Now since $r > 1$ it also follows that the process $\{\Phi_s\}$ has increases between jumps (see Figure 5.3). Thus, it can readily be seen that if $\phi_d > k$, then the process $\{\Phi_s\}$ will not return to the interval $[0, k]$ once it leaves it. Therefore, from Proposition 5.13 it follows that the optimal stopping time in this case is

$$T^* = \inf\{s \geq 0 | \Phi_s \geq k\}. \tag{5.92}$$

(2) $r > 1$, and $k > \phi_d$. In this case the process $\{\Phi_s\}$ can return to the interval $[0, k]$ with positive probability even after it leaves it. The optimal stopping time is

$$T^* = \inf\{s \geq 0 | \Phi_s \geq \phi^*\}, \tag{5.93}$$

where $\phi^* > k$. The threshold ϕ^* is characterized further in [22].

(3) $r < 1$. In this case the process $\{\Phi_s\}$ can return to the interval $[0, k]$ even after it leaves it with positive probability (see Figure 5.3). The optimal stopping time is

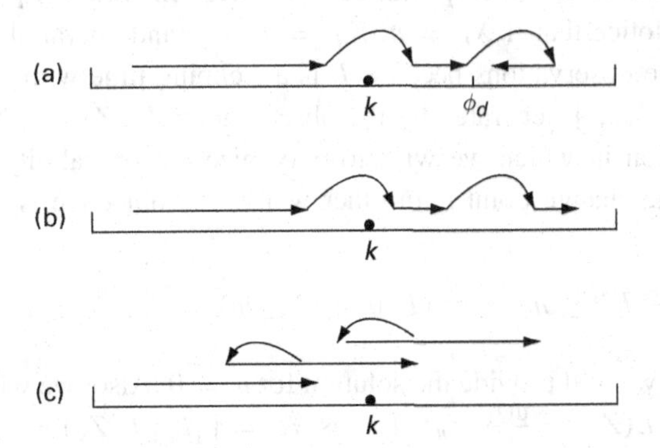

Fig. 5.3. An illustration of a path of the process $\{\Phi_s\}$ in the cases: (a) $r > 1, \phi_d > 0$; (b) $r > 1, \phi_d < 0$; and (c) $r < 1$.

$$T^* = \inf\{s \geq 0 | \Phi_s \geq \phi^*\}, \tag{5.94}$$

where $\phi^* > k$. Again, please see [22] for discussion of the threshold ϕ^*.

Note that the optimal threshold ϕ^* in both cases (2) and (3) can be determined numerically using a method described in [22].

REMARK 5.14. The stopping time of the type in (5.92), (5.93), and (5.94) can equivalently be expressed in terms of $\{\pi_s\}$. However, working with the process $\{\Phi_s\}$ proves to be easier since the dynamics of the process $\{\pi_s\}$ involve a non-linear stochastic differential equation. For the solution involving $\{\pi_s\}$ please refer to [169].

For a more detailed description of the solution as well as the minimal cost achieved by the optimal stopping time T^* in each of the above three cases please refer to [22]. For other approaches, please refer to [67,92,93,105].

5.4 A probability maximizing approach

In this section we examine a different approach to detecting the change point t within the Bayesian framework. This approach is based on maximizing the probability of having selected the right estimator for t based, of course, only on the observations. This problem was first examined by Bojdecki in its simplest form in [40].

Let us return to the discrete-time setting of Section 5.2 and consider the following problem:

Find $T^* \in \mathcal{T}$ such that

$$P(T^* < \infty, X_{T^*} \in B) = \sup_{T \in \mathcal{T}} P(T < \infty, X_T \in B), \tag{5.95}$$

where B is an appropriately measurable set and X_T is a random variable whose value depends on the observations $Z_1, Z_2, \ldots$. The dependence of the above expression on the change point t in the problem that we will consider in this section enters through the set B, as we will see by example below. Such a T^*, if it exists, will be called optimal.

In this section we are interested in a special case of (5.95) in which $X_n = n$ and $B = [t - m, t + m]$. [2] Notice that if $X_n = n$, $X_T = T$ is a random variable whose value is determined by the observations because T is a stopping time with respect to the filtration $\{\mathcal{F}_k; k = 1, 2, \ldots\}$ generated by the observations $Z_1, Z_2, \ldots$. This case corresponds to the situation in which we wish to maximize the probability of stopping within m units of the change point t. In other words, the objective is to find T such that

$$P(|t - T^*| \leq m) = \sup_T P(|t - T| \leq m). \tag{5.96}$$

For the sake of simplicity we will provide the solution for $m = 0$. Also, we will denote by L_n the likelihood ratio $L(Z_n) = \frac{dQ_1}{dQ_0}(Z_n)$. That is, $\Lambda_n = \prod_{k=1}^n L(Z_k) = \prod_{k=1}^n L_k$.

[2] Another interesting case is $B = \{\sup_n Z_n\}$, which is considered in [40].

THEOREM 5.15. *Suppose that the prior distribution of t is given by (5.24). A solution to (5.96) with $m = 0$ exists and is given by*

$$T^* = \inf\{n | L_n \geq r^*\}, \tag{5.97}$$

where $r^* = \lim_{n\to\infty} r_n$, with $r_n = \int \left[\max\{x, r_{n-1}\}\right] \mathrm{d}Q_1(x)$ and $r_0 = 1$.

Proof: It follows from Bayes' formula that

$$P(t = k|\mathcal{F}_n) = \frac{P(t = k)L_n \cdots L_k}{\sum_{i=1}^{n} P(t = i)L_n \cdots L_i + \sum_{i=n+1}^{\infty} P(t = i)}, \quad \forall\, k \leq n, \tag{5.98}$$

and

$$P(t = k|\mathcal{F}_n) = \frac{P(t = k)}{\sum_{i=1}^{n} P(t = i)L_n \cdots L_i + \sum_{i=n+1}^{\infty} P(t = i)}, \quad \forall\, k > n. \tag{5.99}$$

Using the definition of π_n from (5.4) in Section 5.1, it follows that

$$P(t = n|\mathcal{F}_n) = \frac{\rho}{1-\rho}(1 - \pi_n)L_n. \tag{5.100}$$

Therefore,

$$Y_n = h(\pi_n, L_n), \tag{5.101}$$

where $h(x, y) = \frac{\rho}{1-\rho}(1 - x)y$.

Moreover,

$$1 - \pi_{n+1} = \frac{\rho(1 - \pi_n)}{L_{n+1}(1 - (1 - \rho)(1 - \pi_n)) + \rho(1 - \pi_n)},$$

and

$$\begin{aligned} P(t > n + 1|\mathcal{F}_n) &= 1 - P(t \leq n|\mathcal{F}_n) - P(\theta = n + 1|\mathcal{F}_n) \\ &= 1 - \pi_n - P(\theta = n + 1|\theta > n)P(\theta > n|\mathcal{F}_n) \\ &= (1 - \pi_n)(1 - \rho). \end{aligned}$$

We can write

$$\begin{aligned} E\{g(L_{n+1})|\mathcal{F}_n\} &= E_1\{g(L_{n+1})\}\, P(t \leq n + 1|\mathcal{F}_n) + E_0\{g(L_{n+1})\}\, P(t > n + 1|\mathcal{F}_n) \\ &= \int [g(x)[x(1 - (1 - \rho)(1 - \pi_n)) + 1 - \pi_n]]\, \mathrm{d}Q_1(x). \end{aligned}$$

Therefore we obtain,

$$E\{g(L_{n+1})(1 - \pi_{n+1})|\mathcal{F}_n\} = (1 - \rho)(1 - \pi_n) \int g(x)\mathrm{d}Q_1(x). \tag{5.102}$$

We now readily notice that (π_{n+1}, L_{n+1}) is a function of (π_n, L_n), and L_{n+1}, and that the conditional distribution of L_{n+1} given $\mathcal{F}_n$ depends only on π_n. Therefore the Markovian character of Y_n with respect to the filtration $\mathcal{F}_n$ is revealed.

From Chapter 3, we now know that the optimal stopping time is

$$T_0 = \inf\{n | h(\pi_n, L_n) = h^*(\pi_n, L_n)\}, \tag{5.103}$$

where $h^* = \lim_{k\to\infty} T^k h$, with the operator T defined as

$$Tf = \max\{f, Pf\},$$

and

$$Pf(x, y) = E\{f(\pi_{n+1}, L_{n+1}) | \pi_n = x, L_n = y\}.$$

Using (5.102), we have that

$$Ph(x, y) = \frac{\rho}{(1-\rho)}(1-x),$$

and

$$Th(x, y) = \frac{\rho}{(1-\rho)}(1-x)\max\{y, 1\}.$$

Hence by induction we have that

$$h^*(x, y) = \frac{\rho}{(1-\rho)}(1-x)\lim_{n\to\infty} r_n,$$

where r_n is defined in the statement of the theorem.

Therefore, using (5.103), the optimal stopping time is $T_0 = \inf\{n | L_n \geq r^*\}$, as long as it is finite. To examine this issue of finiteness, consider the quantity

$$a = \inf\{a' | P_1(L_1 \leq a') = 1\}.$$

If $a = \infty$ then $T^* < \infty$ trivially. Alternatively, if $a < \infty$, then P_1 is concentrated on $[0, a]$,

$$0 = P_1(L_1 > a) \geq a P_0(L_1 > a).$$

This means that the distribution of the L_n's is concentrated on $[0, a]$. By induction on the r_i's, it follows that $r^* \leq a$. However, there exist an n for which L_n will be arbitrarily close to a, and therefore T^* will be finite.

This concludes the proof of the theorem. ■

5.5 Other penalty functions

A number of penalty functions other than the ones considered in the preceding sections have also been treated in the literature. A list of many of these is found in [193]. Most of these, however, are similar to those we have considered in that they seek to penalize some combination of false alarms and detection delay. An example is a delay penalty of the polynomial type $(T - t)^p$ for fixed $p > 0$. This gives rise to the problem [167]:

$$\inf_{T\in\mathcal{T}} \left[P(T < t) + cE\left\{(T - t)^p\right\}\right].$$

Another is the exponential penalty, leading to the criterion [179]:

$$\inf_{T\in\mathcal{T}}\left[P(T<t)+cE\left\{e^{\alpha(T-t)^+}-1\right\}\right],$$

which has also been modified by replacing $P(T<t)$ with $P(T<t-\epsilon)$, for fixed $\epsilon>0$. (See [21].) And a further alternative delay penalty is

$$E\left\{|T-t|\right\},$$

which can be shown to be equivalent to the Shiryaev criterion [117].

Yet another penalty function of interest is

$$c_f P(T<t)+c_b E\left\{(t-T)^+\right\}+c_a E\left\{(T-t+1)^+\right\}, \tag{5.104}$$

which allows for a linear-time penalty on false detection in addition to (or in lieu of) the penalty on the probability of false detection. The result of Proposition 5.1 can easily be generalized to treat this case. In particular, using the methods of Proposition 5.1, it is easily seen that (5.104) is equal to

$$c_f+c_b E\{t\}+E\left\{(c_a-c_f)\pi_T+\sum_{m=0}^{T-1}\left[c_a\pi_m-c_b(1-\pi_m)\right]\right\}, \tag{5.105}$$

which is minimized by a stopping time of the form (5.25) with appropriately chosen threshold, π^*. The threshold π^* will be identically zero (i.e., $T_B\equiv 0$) if

$$c_a-c_f\geq c_b. \tag{5.106}$$

An interesting but completely different approach to the detection problem was first introduced and solved in [183] using the following penalty function

$$l(t,T)=c_1 1_{\{T<t\}}-c_2\min\{T,t-1\}+c_3(T-t+1)^+. \tag{5.107}$$

In the following section we examine this approach in detail.

5.6 A game theoretic formulation

An interesting alternative approach to the quickest detection problem arises from a game theoretic formulation as follows. Consider the observation model of Section 5.1, and a game consisting of two players in which one of the players ("the statistician") is attempting to quickly detect a random change point as in the preceding sections, while the other player ("nature") is attempting to choose the distribution of the change point (based on observations) in a way so as to foil the first player. Suppose, in particular, that the transition probability of the change point, i.e., $\rho_k=P(t=k|t\geq k)$, is allowed to be a (measurable) function of the past observations, $Z_1, Z_2, \ldots, Z_{k-1}$, and that this function can be chosen by the second player. This second player, knowing that the first player will attempt to minimize the cost function (5.107), would like to choose the change point t so as to achieve the supremum

$$\sup_t \inf_{T \in \mathcal{T}} l(t, T). \tag{5.108}$$

On the other hand, the first player, knowing of the second player's objective, would like to select a stopping time T so as to achieve the infimum

$$\inf_{T \in \mathcal{T}} \sup_t l(t, T). \tag{5.109}$$

As before, let

$$L_n = \frac{dQ_1}{dQ_0}(Z_n), \tag{5.110}$$

and define

$$S_n = L_n \max\{1, S_{n-1}\}, \quad S_0 = 0. \tag{5.111}$$

A strategy for nature is to specify $P(t = n | t \geq n, \mathcal{F}_{n-1})$. Suppose that nature chooses

$$P(t = n | t \geq n, \mathcal{F}_{n-1}) = \rho(1 - S_{n-1})^+, \tag{5.112}$$

for some $0 \leq \rho \leq 1$. Define

$$\frac{\rho_n}{1 - \rho_n} = \frac{\rho}{1 - \rho} S_n.$$

It is fairly easy to see that

$$P(t \leq n | \mathcal{F}_n) = \rho_n. \tag{5.113}$$

To see this notice first that the claim is true for $n = 1$; i.e.

$$\frac{P(t \leq 1 | \mathcal{F}_1)}{1 - P(t \leq 1 | \mathcal{F}_1)} = \frac{P(t = 1)}{1 - P(t = 1)} L_1 = \frac{\rho}{1 - \rho} S_1.$$

Now suppose that the claim is true for $n - 1$, that is, that $P(t \leq n - 1 | \mathcal{F}_{n-1}) = \rho_{n-1}$. Then,

$$\begin{aligned} P(t \leq n | \mathcal{F}_{n-1}) &= P(t \leq n - 1 | \mathcal{F}_{n-1}) \\ &= (1 - P(t \leq n - 1 | \mathcal{F}_{n-1})) P(t = n | \mathcal{F}_{n-1}, t \geq n) \\ &= \rho_{n-1} + (1 - \rho_{n-1}) P(t = n | \mathcal{F}_{n-1}, t \geq n). \end{aligned}$$

Hence

$$\begin{aligned} \frac{P(t \leq n | \mathcal{F}_{n-1})}{1 - P(t \leq n | \mathcal{F}_{n-1})} &= \frac{\rho_{n-1}}{1 - \rho_{n-1}} \frac{1}{1 - \rho(1 - S_{n-1})^+} + \frac{\rho(1 - S_{n-1})^+}{1 - \rho(1 - S_{n-1})^+} \\ &= \frac{\rho}{1 - \rho} \left[\frac{S_{n-1} + (1 - \rho)(1 - S_{n-1})^+}{1 - \rho(1 - S_{n-1})^+} \right] \\ &= \frac{\rho}{1 - \rho} \max\{1, S_{n-1}\} \end{aligned}$$

Therefore,

$$\frac{P(t \le n|\mathcal{F}_n)}{1 - P(t \le n|\mathcal{F}_n)} = \frac{P(t \le n|\mathcal{F}_{n-1})}{1 - P(t \le n|\mathcal{F}_{n-1})} L_n = \frac{\rho}{1-\rho} \max\{1, S_{n-1}\} L_n$$
$$= \frac{\rho}{1-\rho} S_n = \frac{\rho_n}{1-\rho_n},$$

and (5.113) is established by induction.

As it happens, both players can achieve their objective, as is seen from the following result, which is due to Ritov [183].

THEOREM 5.16. *Define*

$$T_\nu = \inf\{n | S_n \ge \nu\}. \tag{5.114}$$

Suppose that $c_1 - c_2 \ge c_3$, *and denote by* ν^* *the unique solution of*

$$c_1 - c_2 E_\infty \{T_{\nu^*}\} = c_3 E_0 \{T_{\nu^*}\}, \tag{5.115}$$

where the first expectation is taken with respect to the measure P_∞ *and the second is taken with respect to* P_0 *corresponding to the probability measures generated by the observations conditional on the events* $\{t = \infty\}$ *and* $\{t = 0\}$, *respectively. Then for some* $\rho \in [0, 1]$ $(t_\rho, \hat{T})$ *is a saddle-point solution to the problems* (5.108) *and* (5.109). *That is,*

$$\sup_t l(t, \hat{T}) = l(t_\rho, \hat{T}) = \inf_{T \in \mathcal{T}} l(t_\rho, T), \tag{5.116}$$

where $\hat{T} = T_{\nu^*}$.

Proof: Suppose that nature chooses ρ of (5.112). The best the statistician can now do is choose a strategy based on $P(t \le n|\mathcal{F}_n)$. From (5.113), it follows that $P(t \le n|\mathcal{F}_n) = \rho_n$. We will now argue that the optimal stopping time is a threshold stopping time of the type $\inf\{n|S_n \ge \nu\}$ or equivalently of the type $\inf\{n|\rho_n \ge \nu\}$. We call the interval $[k, l]$ a *cycle* if the sequence $\rho_n = \rho$, and $l = \inf\{n > k|\rho_n = \rho\}$. In other words, a cycle is completed the first time the sequence $\{\rho_n\}$ restarts. An arbitrary stopping time T is characterized by the following quantities:

(1) $t_0(T)$: the expected run length of the cycle under the assumption the change point occurs after the end of the cycle;
(2) $t_1(T)$: the expected run length of the cycle under the assumption the change point occurs before the beginning of the cycle;
(3) $q_0(T)$: the probability that the change will be declared during the cycle under the assumption the change point occurs after the end of the cycle; and
(4) $q_1(T)$: the probability that the change will be declared during the cycle under the assumption the change point occurs before the beginning of the cycle.

The cost incurred by a stopping time is the expected value of the loss during the cycle multiplied by the expected number of cycles. That is

$$\frac{(1-\rho)(c_1 q_0(T) - c_2 t_0(T)) + \rho c_3 t_1(T)}{(1-\rho) q_0(T) + \rho q_1(T)} \tag{5.117}$$

To solve the problem of (5.117) we need to minimize the numerator subject to a fixed value of the denominator. We can use the method of Lagrange multipliers, and consider

$$(1-\rho)(c_1 q_0(T) - c_2 t_0(T)) + \rho c_3 t_1(T) + \lambda\left[(1-\rho) q_0(T) + \rho q_1(T)\right].$$

This loss function is linear in ρ, and therefore

$$\inf_T (1-\rho)(c_1 q_0(T) - c_2 t_0(T)) + \rho c_3 t_1(T) + \lambda\left[(1-\rho) q_0(T) + \rho q_1(T)\right]$$

is concave in ρ. If $\rho = 1$ we should obviously stop at once, and the optimal procedure should have threshold one.

But how can we guarantee the existence of a suitable ν^* more generally? Notice that both $E_0\{T_\nu\}$ and $E_\infty\{T_\nu\}$ are increasing functions of ν, and

(1) $\lim_{\nu\to 0} E_\infty\{T_\nu\} = \lim_{\nu\to 0} E_0\{T_\nu\} = 0$, and
(2) $\lim_{\nu\to\infty} E_\infty\{T_\nu\} = \lim_{\nu\to\infty} E_0\{T_\nu\} = \infty$.

Moreover, as ρ varies from 1 to 0, $E_\infty\{T_\nu\}$ ranges continuously from 0 to ∞, and hence nature can choose a value of ν^* for which (5.115) holds.

Suppose that the statistician chooses ν^* as above. The best strategy for nature is then to choose t_ρ with $\nu(\rho) = \nu^*$, as follows.

Nature can choose either $t = \infty$ or $t < \infty$. If the former is chosen then the loss will be $c_1 - c_2 E_\infty\{T_{\nu^*}\}$. Otherwise, nature can choose any $t = n \le T_{\nu^*}$. If $S_{n-1} \le 1$, then the expected future loss is $c_3 E_0\{T_{\nu^*}\}$. This is larger than the expected loss when $S_{n-1} > 1$, since the expected time from n to T_{ν^*} is monotone in S_{n-1}. (This is easily seen since S_n is stochastically increasing in S_k for all $n \ge k$.) Nature should therefore choose $t = \infty$ in this case, and can randomize between any n and ∞ if $t \ge n$ and $S_{n-1} \le 1$. Hence $c_3 E_0\{T\}$ is the maximum that nature can achieve, and the suggested procedure guarantees it. ■

REMARK 5.17. The test $\hat{T} = T_{\nu^*}$ is known as the cumulative sum (CUSUM) test, or Page's test, and plays an important role in a non-Bayesian form of the quickest detection problem to be treated in the next chapter. The connection between this game-theoretic problem and the non-Bayesian quickest detection problem will be discussed further in Chapter 6.

Ritov's result has been extended to the case of Brownian observations in [25] and in [192]. For further details please refer to Chapter 6.

5.7 Discussion

In this chapter we have examined the problem of detecting a random change point in several fundamental models. In particular, we have shown the optimality of stopping times based on the first up-crossing of a threshold by the posterior probability of a change in the case of general independent discrete-time observations, in the case of continuous Brownian observations when the drift changes to a known value, and in the case of Poisson processes, when the intensity changes to a known value. These problems have been generalized to cases in which the post-change parameter value is not known exactly, but rather obeys a given distribution. Here, the optimal solution often becomes much more complex than the single threshold stopping times derived in this chapter. Such extensions are found in [23] and [68] and in further generality in [93]. The disorder problem has also recently been studied in the context of compound Poisson models and threshold stopping times of the type appearing in Section 5.3 are optimal in this case as well (see [71] and [69]). Finally, the probability maximizing approach of Section 5.4 has also been studied in a model of continuous Markov processes. For a full treatment of this case, as well as a description of the optimal stopping time, please refer to [231].

6 Non-Bayesian quickest detection

6.1 Introduction

In Chapter 5, we considered the quickest detection problem within the framework proposed by Kolmogorov and Shiryaev, in which the unknown change point is assumed to be a random variable with a given, geometric, prior distribution. This formulation led to a very natural detection procedure; namely, announce a change at the first upcrossing of a suitable threshold by the posterior probability of a change. Although the assumption of a prior on the change point is rather natural in applications such as condition monitoring, there are other applications in which this assumption is unrealistic. For example, in surveillance or inspection systems, there is often no pre-existing statistical model for the occurence of intruders or flaws.

In such situations, an alternative to the formulations of Chapter 5 must be found, since the absence of a prior precludes the specification of expected delays and similar quantities that involve averaging over the change-point distribution. There are several very useful such formulations, and these will be discussed in this chapter.

We will primarily consider a notable formulation due to Lorden, in which the average delay is replaced with a worst-case value of delay. However, other formulations will be considered as well.

As in the Bayesian formulation of this problem, optimal stopping theory plays a major role in specifying the optimal procedure, although (as we shall see) more work is required here to place the problems of interest within the standard optimal stopping formulation of Chapter 3.

6.2 Lorden's problem

We turn, then, to the situation in which the change point t is a fixed, non-random quantity that can be either ∞ or any value in the positive integers. To model this situation, we consider a measurable space $(\Omega, \mathcal{F})$, consisting of a sample space Ω and a σ-field $\mathcal{F}$ of events. We further consider a family $\{P_t | t \in [1, 2, \ldots, \infty]\}$ of probability measures on $(\Omega, \mathcal{F})$, such that, under P_t, $Z_1, Z_2, \ldots, Z_{t-1}$ are independent and identically distributed (i.i.d.) with a fixed marginal distribution Q_0, and $Z_t, Z_{t+1}, \ldots$ are i.i.d. with another marginal distribution Q_1 and are independent of $Z_1, Z_2, \ldots, Z_{t-1}$. For simplicity, we assume that Q_1 and Q_0 are mutually absolutely continuous, that the likelihood ratio $L(Z_k) = \mathrm{d}Q_1/\mathrm{d}Q_0(Z_k)$ has no atoms under Q_0, and that $2 < D(Q_1 \parallel Q_0) < \infty$, where $D(Q_1 \parallel Q_0)$ denotes the Kullback–Leibler divergence of Q_1 from Q_0:

$$D(Q_1 \parallel Q_0) = -\int \log L(x)\, \mathrm{d}Q_1(x). \tag{6.1}$$

For technical reasons, we also assume the existence of a random variable Z_0 that is uniformly distributed in [0, 1] and that is independent of $Z_1, Z_2, \ldots$ under each P_t.

As before, we would like to consider procedures that can detect the change point, if it occurs (i.e., if $t < \infty$), as quickly as possible after it occurs. As a set of detection strategies, it is natural to consider the set $\mathcal{T}$ of all (extended) stopping times with respect to the filtration $\{\mathcal{F}_k; k \geq 0\}$ where $\mathcal{F}_k$ denotes the smallest σ-field with respect to which $Z_0, Z_1, \ldots, Z_k$ are measurable. Thus, when the stopping time T takes on the value k, the interpretation is that T has detected the existence of a change point t at or prior to time k.

Following [139], it is of interest to penalize exponential detection delay via its worst-case value

$$d(T) = \sup_{t \geq 1} d_t(T) \tag{6.2}$$

with

$$d_t(T) = \operatorname{esssup} E_t\left\{(T - t + 1)^+ \middle| \mathcal{F}_{t-1}\right\}, \tag{6.3}$$

where $E_t\{\cdot\}$ denotes expectation under the distribution P_t.[1] Note that $d_t(T)$ is the worst-case average delay under P_t, where the worst case is taken over all realizations of $Z_0, Z_1, \ldots, Z_{t-1}$. The desire to make $d(T)$ small must be balanced with a constraint on the rate of false alarms. In Chapter 5, this constraint was enforced by penalizing the probability of false alarms. Although such a constraint can be imposed here as well, it turns out to be a rather severe constraint. (See Section 6.6 below.) In this non-Bayesian case, it is more practical to adopt the philosophy that false alarms will occur, but to impose some limit on the rate at which they occur. One can envision an inspection or surveillance scheme which is ongoing. On occasion, false alarms will be announced, which must be investigated and are thus undesirable. However, it is very difficult to avoid any such alarms in a sufficiently long observation interval without sacrificing a significant degree of sensitivity in detecting actual alarm conditions.

The rate of false alarms can be quantified by the mean time between false alarms:

$$f(T) = E_\infty\{T\}; \tag{6.4}$$

and a useful design criterion is then given by

$$\inf_{T \in \mathcal{T}} d(T) \text{ subject to } f(T) \geq \gamma, \tag{6.5}$$

[1] In defining the worst case delay in (6.3), we have used the essential supremum of the random variable $E_t\left\{(T - t + 1)^+ \middle| \mathcal{F}_{t-1}\right\}$. In Chapter 3, we defined the essential supremum of a family of random variables. The essential supremum of a single random variable is the least upper bound of the set of constants that bound the random variable with probability 1.

where γ is a positive, finite constant. That is, we seek a stopping time that minimizes the worst-case delay within a lower-bound constraint on the mean time between false alarms.

To examine the solution to (6.5), we consider Page's CUSUM test, introduced in Chapter 5 as the solution to Ritov's game-theoretic quickest detection problem. In particular, for $h \geq 0$ we define the CUSUM stopping time

$$T_h^c = \inf\{k \geq 0 \mid S_k \geq h\}, \tag{6.6}$$

where, as in (5.111),

$$S_k = \max_{1 \leq j \leq k} \left(\prod_{\ell=j}^{k} L(Z_\ell) \right) \equiv \max\{S_{k-1}, 1\} L(Z_k), \ k \geq 1, \tag{6.7}$$

and $S_0 = 0$. Although this test was proposed as a continuous inspection scheme by Page in the 1950s, it was not until 1971 that its optimality as such was established. In particular, Lorden [139] proved that this stopping time was optimal in the sense of (6.5), asymptotically as $\gamma \to \infty$. Some discussion of Lorden's result is found in Section 6.5.1 below. More than a decade later, Moustakides [150] proved that this test (or a slightly modified version of it) is optimal for all finite γ. More recently, Ritov [183] used his game-theoretic version of the Bayesian quickest detection problem (compare Section 5.6) to provide an alternate proof of this optimality.

Here, we will examine both Moustakides' and Ritov's methods of proof of the optimality of Page's test in the sense of (6.5).

To follow Moustakides' method of proof, it is convenient to rework the worst-case delay $d(T)$ into a form that can be optimized using the techniques of Markov optimal stopping theory. To do so, we state the following result.

PROPOSITION 6.1. *Suppose* $0 < E_\infty\{T\} < \infty$. *Then*

$$d(T) \geq \frac{E_\infty\left\{\sum_{m=0}^{T-1} \max\{S_m, 1\}\right\}}{E_\infty\left\{\sum_{m=0}^{T-1} (1 - S_m)^+\right\}} \stackrel{\Delta}{=} \overline{d}(T), \tag{6.8}$$

with equality if $T = T_h^c$ *for any* $h \geq 0$.

Proof: For integers $m \geq 1$, define the quantity

$$b_m(T) = E_m\left\{ (T - m + 1)^+ \middle| \mathcal{F}_{m-1} \right\}. \tag{6.9}$$

Since $Z_1, \ldots, Z_{m-1}$ have identical distributions under P_m and P_∞, it follows that $d(T) \geq d_m(T) \geq b_m(T)$ a.s. P_∞. Thus, we can write

$$d(T) \sum_{m=1}^{\infty} E_\infty\left\{1_{\{T \geq m\}}(1 - S_{m-1})^+\right\} \geq \sum_{m=1}^{\infty} E_\infty\left\{b_m(T) 1_{\{T \geq m\}}(1 - S_{m-1})^+\right\}. \tag{6.10}$$

The sum on the left-hand side of this inequality is given by

$$\sum_{m=1}^{\infty} E_\infty \left\{1_{\{T\geq m\}}(1 - S_{m-1})^+\right\} = E_\infty \left\{\sum_{m=1}^{\infty} 1_{\{T\geq m\}}(1 - S_{m-1})^+\right\}$$
$$= E_\infty \left\{\sum_{m=1}^{T} (1 - S_{m-1})^+\right\}$$
$$= E_\infty \left\{\sum_{m=0}^{T-1} (1 - S_m)^+\right\}, \tag{6.11}$$

where the interchange of sum and expectation is permitted by the monotone convergence theorem.

We have

$$b_m(T) = \sum_{k=m}^{\infty} (k - m + 1) P_m(T = k|\mathcal{F}_{m-1})$$
$$= \sum_{k=m}^{\infty} P_m(T \geq k|\mathcal{F}_{m-1})$$
$$= \sum_{k=m}^{\infty} E_m \left\{ 1_{\{T\geq k\}} \middle| \mathcal{F}_{m-1}\right\}$$
$$= \sum_{k=m}^{\infty} E_\infty \left\{ \prod_{\ell=m}^{k-1} L(Z_\ell) 1_{\{T\geq k\}} \middle| \mathcal{F}_{m-1}\right\}$$
$$= E_\infty \left\{ \sum_{k=m}^{T} \prod_{\ell=m}^{k-1} L(Z_\ell) \middle| \mathcal{F}_{m-1}\right\}. \tag{6.12}$$

Using (6.12) and the facts that $1_{\{T\geq m\}}$ and S_{m-1} are $\mathcal{F}_{m-1}$-measurable, the right-hand side of (6.10) becomes

$$\sum_{m=1}^{\infty} E_\infty \left\{b_m(T) 1_{\{T\geq m\}}(1 - S_{m-1})^+\right\} = \sum_{m=1}^{\infty} E_\infty \left\{ 1_{\{T\geq m\}}(1 - S_{m-1})^+ \sum_{k=m}^{T} \prod_{\ell=m}^{k-1} L(Z_\ell)\right\}$$
$$= E_\infty \left\{ \sum_{m=1}^{\infty} 1_{\{T\geq m\}}(1 - S_{m-1})^+ \sum_{k=m}^{T} \prod_{\ell=m}^{k-1} L(Z_\ell)\right\}$$
$$= E_\infty \left\{ \sum_{m=1}^{T} (1 - S_{m-1})^+ \sum_{k=m}^{T} \prod_{\ell=m}^{k-1} L(Z_\ell)\right\}$$
$$= E_\infty \left\{ \sum_{k=1}^{T} \sum_{m=1}^{k} (1 - S_{m-1})^+ \prod_{\ell=m}^{k-1} L(Z_\ell)\right\}$$
$$= E_\infty \left\{ \sum_{k=0}^{T-1} \max\{S_k, 1\}\right\}, \tag{6.13}$$

where we have used the representation

$$\max\{S_k, 1\} = \sum_{m=1}^{k+1} (1 - S_{m-1})^+ \prod_{j=m}^{k} L(Z_\ell), \tag{6.14}$$

which can easily be verified via induction. The inequality (6.8) thus follows from (6.10), (6.11) and (6.13). ■

To show that

$$d(T_h^c) = \bar{d}(T_h^c) \tag{6.15}$$

we consider more carefully the sequence

$$U_k = \max\{S_k, 1\}, \ k = 0, 1, \ldots \tag{6.16}$$

It is straightforward to prove that this sequence has the property that, for any $n > m \geq 1$ and for fixed $\{Z_{m+1}, \ldots, Z_n\}$, S_n is a non-decreasing function of U_m. (See for comparison Lemma 1 of [150].) It follows that, on the event $\{T_h^c \geq m\}$, T_h^c is non-increasing as a function of U_{m-1}. Thus, since $U_{m-1} \geq 1$, we have

$$\begin{aligned} d_m(T_h^c) &= \text{esssup } E_m \left\{ (T_h^c - m + 1)^+ \middle| \mathcal{F}_{m-1} \right\} \\ &= \text{esssup } E_m \left\{ (T_h^c - m + 1)^+ \middle| U_{m-1} = 1 \right\} \\ &= \text{esssup } E_m \left\{ (T_h^c - m + 1)^+ \middle| S_{m-1} \leq 1 \right\}. \end{aligned} \tag{6.17}$$

From the homogeneous Markovity of $\{U_k\}$, it follows in turn that

$$d_m(T_h^c) = d_1(T_h^c) \ \forall \, m \geq 1; \tag{6.18}$$

i.e., that the stopping time T_h^c is an *equalizer rule* (see for comparison [179]). Now, consider the summand in (6.10). Since $d_m(T_h^c) = d(T_h^c)$, and since esssup $b_m(T_h^c)$ is achieved on $\{T_h^c \geq m\} \cap \{S_{m-1} \leq 1\}$, we can conclude that

$$d(T_h^c) E_\infty \left\{ 1_{\{T_h^c \geq m\}} (1 - S_{m-1})^+ \right\} = E_\infty \left\{ b_m(T_h^c) 1_{\{T_h^c \geq m\}} (1 - S_{m-1})^+ \right\}, \tag{6.19}$$

for all $m \geq 1$. Thus, for $T = T_h^c$, the inequality (6.10) is an equality, and (6.15) follows.

Proposition 6.1 provides the means to convert the quickest detection problem (6.5) into a more traditional optimal stopping problem. This allows the proof of the following main result.

THEOREM 6.2. *Choose $h \geq 0$. Then, the stopping time T_h^c solves (6.5) with $\gamma = f(T_h^c)$. That is,*

$$f(T) \geq f(T_h^c) \Rightarrow d(T) \geq d(T_h^c). \tag{6.20}$$

Proof: In seeking optimal stopping times for the problem (6.5), it is sufficient to consider stopping times that satisfy the constraint $f(T) \geq \gamma$ with equality. To see this, we first note that we can ignore any T for which $f(T)$ is not finite. In particular, if $f(T) = \infty$, then we can choose a sufficiently large integer n such that $\gamma \leq f(\min\{T, n\}) < \infty$.

Since $d(\min\{T, n\}) \leq d(T)$, we do not need to consider such T. Now suppose $\gamma < f(T) < \infty$. Then, for $p = \gamma/f(T)$, the randomized stopping time

$$T' = \begin{cases} T & \text{with probability } p \\ 0 & \text{with probability } 1 - p, \end{cases} \tag{6.21}$$

satisfies $f(T') = \gamma$ and $d(T') \leq d(T)$. So, there is no need to consider T when T' also satisfies the constraint and has no larger worst-case delay. ■

It follows from the above comments and from Proposition 6.1 that, if we could show that T_h^c solves the problem

$$\min_{T \in \mathcal{T}} \overline{d}(T) \text{ subject to } f(T) = \gamma, \tag{6.22}$$

then the minimax optimality of T_h^c in the sense of Lorden would be established. That this is, in fact, the case is established by the following result.

PROPOSITION 6.3. *T_h^c solves the following maximization problem for all continuous non-increasing functions $g : [0, \infty) \to \mathbb{R}$:*

$$\sup_{T \in \mathcal{T}} E_\infty \left\{ \sum_{k=0}^{T-1} g(S_k) \right\} \text{ subject to } f(T) = \gamma. \tag{6.23}$$

Proof: We first note that T_h^c solves (6.23) if, and only if, it solves the same problem with g replaced by the function

$$\overline{g}(x) = \max\{g(x), g(h)\}. \tag{6.24}$$

This follows since $\overline{g} \geq g$ and

$$E_\infty \left\{ \sum_{k=0}^{T_h^c - 1} g(S_k) \right\} = E_\infty \left\{ \sum_{k=0}^{T_h^c - 1} \overline{g}(S_k) \right\}.$$

So, without loss of generality, we may replace g with $\overline{g}$ in the following proof.

To show that T_h^c solves (6.23) it is sufficient to show that there is a real λ such that T_h^c solves the unconstrained maximization problem

$$\sup_{T \in \mathcal{T}} E_\infty \left\{ \sum_{k=0}^{T-1} [\overline{g}(S_k) - \lambda] \right\}. \tag{6.25}$$

So, let us examine the solution to this problem as a function of λ.

In order to take advantage of the Markov property of $\{S_k\}$, we generalize the problem slightly. In particular, for fixed λ and for all $s \geq 0$, define the sequence

$$Y_n^s = \sum_{k=0}^{n-1} \overline{g}(S_k^s) - n\lambda, \; n = 0, 1, \ldots, \tag{6.26}$$

where

$$S_k^s = \max\{S_{k-1}^s, 1\} L(Z_\ell), \; k = 1, 2, \ldots, \; S_0^s = s. \tag{6.27}$$

Note that (6.23) can be rewritten as

$$\sup_{T \in \mathcal{T}} E_\infty \left\{ Y_T^s \right\}, \tag{6.28}$$

with $s = 0$.

We wish to consider the quantity (6.28) and the stopping time achieving it (if such exists), as functions of $s \geq 0$ and $\lambda \in \mathbb{R}$. To examine this problem, let us consider a sequence of stopping times, $\nu_0, \nu_1, \nu_2, \ldots$ defined recursively by

$$\nu_k = \inf \left\{ n > \nu_{k-1} \,|\, S_n^s \leq 1 \right\}, \; k = 1, 2, \ldots, \; \nu_0 = 0. \tag{6.29}$$

So, ν_k is the time of the kth entry of S_k^s into the interval [0, 1]. For future reference, we note that ν_1 is a random variable (i.e. it is almost surely finite) and that its distribution satisfies a bound of the form

$$P_\infty(\nu_1 = k) \leq K\rho^k, k = 1, 2, \ldots \tag{6.30}$$

These properties follow from the fact that ν_1 is the first return to zero of the simple random walk

$$\sum_{\ell=1}^{k} \log L(Z_\ell) \tag{6.31}$$

which has negative mean under P_∞. A consequence of (6.31) is that all moments of ν_1 are finite.

Define, further, the sequences

$$\xi_n = \sum_{k=\nu_{n-1}+1}^{\nu_n} \overline{g}(S_k^s), \; n = 1, 2, \ldots, \tag{6.32}$$

$$\eta_n = \nu_n - \nu_{n-1}, \; n = 1, 2, \ldots, \tag{6.33}$$

and

$$\omega_n = \xi_n - \lambda \nu_n. \tag{6.34}$$

It is easily seen from the definition of $\{S_k^s\}$ that $\{\xi_n\}$, $\{\eta_n\}$, and $\{\omega_n\}$ are i.i.d. sequences.

Define $\lambda_0 = E_\infty\{\xi_1\}/E_\infty\{\nu_1\}$, and consider the case $\lambda > \lambda_0$. (Note that the finiteness of λ_0 follows from the bound $|\xi_1| \leq D\nu_1$, where $D = |g(0)| < \infty$.) Here, since $E_\infty\{\omega_1\} < 0$, we can choose $\delta > 0$ such that $E_\infty\{\omega_k + \delta\} < 0$, $\forall k$. Choose such a δ and note that, for each n, we can choose a positive integer r such that

$$\begin{aligned}
(Y_n^s)^+ &\leq \left(\sum_{k=1}^{r} \omega_k + \eta_r D' \right)^+ \\
&\leq \left(\sum_{k-1}^{r} (\omega_k + \delta) \right)^+ + \left(\eta_r D' - \delta r \right)^+ \\
&\leq \left(\sum_{k=1}^{r} (\omega_k + \delta) \right)^+ + D' \eta_r 1_{\{\eta_r \geq \delta r / D'\}},
\end{aligned} \tag{6.35}$$

where $D' = D + \lambda$. From this inequality, we have that

$$E_\infty\left\{\sup_n (Y_n^s)^+\right\} \leq E_\infty\left\{\sup_r \left(\sum_{k=1}^r (\omega_k + \delta)\right)^+\right\} + D' \sum_{r=0}^\infty E_\infty\left\{\eta_r 1_{\{\eta_r \geq \delta r/D'\}}\right\}. \tag{6.36}$$

The first term on the right-hand side of (6.36) is the mean maximal positive excursion of a simple random walk with negative mean. A sufficient condition for the finiteness of this term is that the variance of ω_1 be finite (see for comparison page 92 of [57]). That this variance is, in fact, finite follows from the finiteness of the second moment of ν_1 and the bound $|\xi_1| \leq D\nu_1$. Since

$$1_{\{x \geq a\}} \leq x^2/a^2, \; a > 0, \tag{6.37}$$

and since $\{\eta_r\}$ is an i.i.d. sequence with $\eta_1 = \nu_1$, the second term on the right-hand side of (6.36) can be bounded as

$$D' \sum_{r=0}^\infty E_\infty\left\{\eta_r 1_{\{\eta_r \geq \delta r/D'\}}\right\} \leq \frac{D'^3 E_\infty\left\{(\nu_1)^3\right\}}{\delta^2} \sum_{r=1}^\infty \frac{1}{r^2} < \infty. \tag{6.38}$$

Thus, we conclude that, for all $\lambda > \lambda_0$ and all $s \geq 0$, we have

$$E_\infty\left\{\sup_n (Y_n^s)^+\right\} < \infty. \tag{6.39}$$

Continuing to restrict attention to the case $\lambda > \lambda_0$, it follows from (6.39) and Theorem 3.7 that (6.28) is solved by the stopping time

$$T_{\text{opt}}^s = \inf\{k \geq 0 | Y_k^s = \overline{\gamma}_k^s\} \tag{6.40}$$

where $\{\overline{\gamma}_k^s\}$ is the Snell envelope of $\{Y_k^s\}$:

$$\overline{\gamma}_k^s = \text{esssup}_{T \in \mathcal{T}_k} E_\infty\left\{Y_T^s \middle| \mathcal{F}_k\right\}, \; k = 0, 1, \ldots, \tag{6.41}$$

and where $\mathcal{T}_k$ denotes the subset of $\mathcal{T}$ satisfying $P_\infty(T \geq k) = 1$.

The homogeneous Markovity of $\{S_k^s\}$ allows us to represent $\overline{\gamma}_k^s$ as

$$\overline{\gamma}_k^s = Y_k + v(S_k^s), \tag{6.42}$$

where

$$v(s) = \sup_{T \in \mathcal{T}} E_\infty\left\{Y_T^s\right\}, \; s \geq 0. \tag{6.43}$$

Since S_k^s is non-decreasing in s, and g is non-increasing, it follows that v is a non-increasing function. Moreover, since the Snell envelope satisfies the equation (see for comparison [57])

$$\overline{\gamma}_k^s = \max\left\{Y_k^s, E_\infty\left\{\overline{\gamma}_{k+1}^s \middle| \mathcal{F}_k\right\}\right\} \text{ a.s. } P_\infty, \; \forall k \geq 0, \tag{6.44}$$

v satisfies the integral equation (see for comparison Theorem II.16 of [190])

$$v(s) = \max\{0, g(s) - \lambda + \mathcal{V}v(s)\}, \; s \geq 0, \tag{6.45}$$

with

$$\mathcal{V}v(s) = \int v\,(\max\{s, 1\}L(x))\, Q_1(\mathrm{d}x),\ s \geq 0. \tag{6.46}$$

Thus, v is continuous.

We can conclude from the above that the optimal stopping time for (6.23) is

$$T_{\text{opt}} = \inf\{k \geq 0 | v(S_k) \leq 0\} = \inf\{k \geq 0 | S_k \geq h_\lambda\}, \tag{6.47}$$

where

$$h_\lambda = \sup\{s \geq 0 | v(s) > 0\}. \tag{6.48}$$

With $\lambda > \lambda_0$, it follows from the fact that $E_\infty\{\omega_1\} < 0$, and the strong law of large numbers, that $Y_k \to -\infty$ almost surely under P_∞. This implies, via Theorem 3.7, that $P_\infty(T_{\text{opt}} < \infty) = 1$, and thus that $h_\lambda < \infty$.

We have established that, for every $\lambda > \lambda_0$ there is an h_λ such that T_{h_λ} solves (6.23). To finish the proof, it is sufficient to show that we can choose $\lambda > \lambda_0$ so that $h_\lambda = h$. To see that this is the case, let us denote the dependence of v on λ via the notation $v_\lambda(s)$. We know from the preceding analysis that $v_\lambda(s)$ is finite for all $\lambda > \lambda_0$. It is straightforward to show from its definition that $v_\lambda(s)$ is a continuous, convex function of (λ, s) in the range $s \geq 0$, $\lambda > \lambda_0$, and moreover that it is non-increasing in each of s and λ.

We now show that, for each $s \geq 0$, $v_\lambda(s) \uparrow \infty$ as $\lambda \downarrow \lambda_0$. We first note that, since $v_\lambda(s)$ is non-increasing in λ, its limit as $\lambda \downarrow \lambda_0$ is well defined. Let $\mathcal{R}$ denote the set of stopping times with respect to the filtration $\{\sigma\,(\xi_1, \xi_2, \ldots, \xi_r, \eta_1, \eta_2, \ldots, \eta_r)\,; r \geq 0\}$. Then, for any $R \in \mathcal{R}$, we can write

$$Y_{\nu_R}^s \geq \sum_{k=1}^{R} [\xi_k - \lambda\eta_k] - 2D', \tag{6.49}$$

from which it follows that

$$v_\lambda(s) \geq E_\infty\left\{\sum_{k=1}^{R} [\xi_k - \lambda\eta_k]\right\} - 2D',\ \forall\, R \in \mathcal{R}. \tag{6.50}$$

For an arbitrary $R \in \mathcal{R}$ with $E_\infty\{R\} < \infty$, we can write

$$\lim_{\lambda\downarrow\lambda_0} v_\lambda(s) \geq E_\infty\left\{\sum_{k=1}^{R} [\xi_k - \lambda_0\eta_k]\right\} - 2D'. \tag{6.51}$$

The right-hand side of (6.51) involves the mean stopped value of a simple random walk with zero mean. It follows from [57], p. 27, that this quantity can be made arbitrarily large by proper choice of R. Thus, $\lim_{\lambda\downarrow\lambda_0} v_\lambda(s)$ must be ∞.

Since it is obviously the case that $v_\lambda(s) \equiv 0$, for all $\lambda \geq g(0)$, we can conclude from the above properties that h_λ is a continuous, non-increasing function of λ with $h_{g(0)} = 0$ and with $h_{\lambda_0} \uparrow \infty$ as $\lambda \downarrow \lambda_0$. (See Figure 6.1.) Thus, we can find λ with $h_\lambda = h$, and the proposition follows.

Particularizing the result of Proposition 6.3 to the two cases $g(x) = (1 - x)^+$ and $g(x) = -\max\{x, 1\}$, we see that T_h^c simultaneously maximizes the denominator and

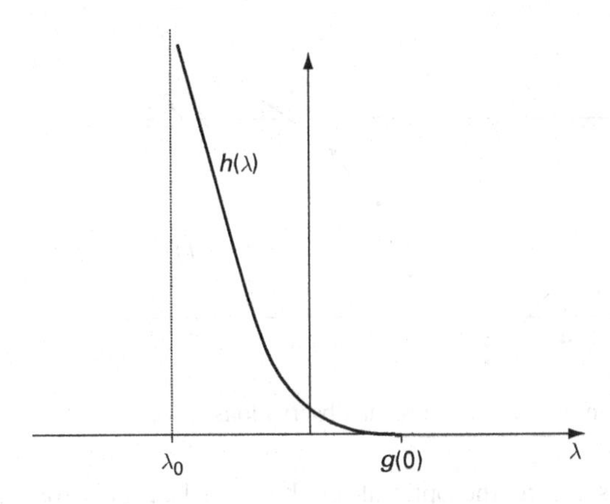

Fig. 6.1. An illustration of $h(\lambda)$.

minimizes the numerator of the lower bound $\overline{d}(T)$ within the constraint $f(T) = \gamma$. Thus, since $\overline{d}(T_h^c) = d(T_h^c)$, the theorem follows immediately. ■

Theorem 6.2 asserts the optimality of the stopping time based on the first exit of S_k from the interval $[0, h)$. This time is particularly interesting when $h \geq 1$, in which case T_h^c can be written equivalently as

$$T_h^c \quad = \quad \inf\{k \geq 0 | m_k \geq \log h\} \tag{6.52}$$

where

$$m_k = \log \max\{S_k, 1\}, \; k = 0, 1, \ldots \tag{6.53}$$

It is easily seen that the sequence $\{m_k\}$ can be computed recursively via

$$m_k = \max\{m_{k-1} + \log L(Z_k)\;,\; 0\}, \; k = 1, 2, \ldots \tag{6.54}$$

with $m_0 = 0$. That is, the test based on T_h^c accumulates the log-likelihoods, $\log L(Z_k)$, resetting the accumulation to zero whenever it goes negative. The alarm is sounded when this accumulation crosses the upper threshold $\log h$.

For illustrative purposes consider the following example.

Example 6.1: Consider the case in which $Q_0 = \mathcal{N}(0, 1)$ and $Q_1 = \mathcal{N}(\mu, 1)$. In this case, we have

$$\log L(Z_k) \quad = \quad \mu\left(Z_k - \frac{1}{2}\mu\right), \tag{6.55}$$

and (6.54) takes the form

$$m_k \quad = \quad \max\left\{m_{k-1} + \mu\left(Z_k - \frac{1}{2}\mu\right), 0\right\}. \tag{6.56}$$

An illustration of the CUSUM stopping time T_h^c of (6.52) in this case is given in Figure 6.2.

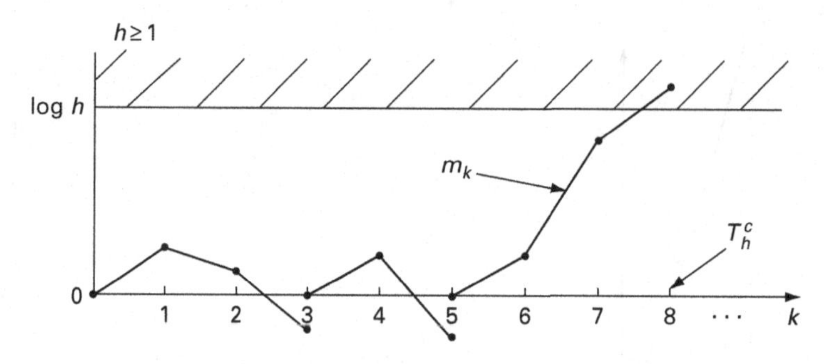

Fig. 6.2. An illustration of the CUSUM in the case of Gaussian observations

The appearance of Page's test as the optimal test both for Lorden's formulation of the quickest detection problem, and for Ritov's formulation of Section 5.6 is not coincidental. Analogously with the Wald–Wolfowitz proof of the non-Bayesian optimality of the sequential probability ratio test (Theorem 4.7), the decision theoretic optimality of Page's test can be used to prove its optimality with respect to Lorden's criterion. Such a proof appears in [183] and is presented as follows.

THEOREM 6.4. *A CUSUM stopping time (6.6) with $h > 1$ minimizes $d(T)$ among all $T \in \mathcal{T}$ for which $E_\infty\{T\} \geq E_\infty\{T_h^c\}$.*

Proof: Suppose that for some T, $E_\infty\{T\} \geq E_\infty\{T_h^c\}$ and $d(T) < E_1\{T_h^c\}$. Consider the following stopping time:

$$T' = \begin{cases} T+1 & \text{with probability } \epsilon \\ T & \text{with probability } 1-\epsilon \end{cases}$$

for some $0 < \epsilon < 1$. Then it follows that for T'

$$d(T') \geq d(T), \text{ and} \tag{6.57}$$
$$f(T') > f(T). \tag{6.58}$$

Therefore, there exists a stopping time T' such that

$$d(T') < E_1\{T_h^c\}, \text{ and} \tag{6.59}$$
$$f(T') > E_\infty\{T_h^c\}. \tag{6.60}$$

Let (t_ρ, T_h) be the saddle point (see (5.116) of Theorem 5.16) of the game with loss function

$$l^\rho(t, T) = 1_{\{T<t\}} - c_2^\rho \min\{T_h^c, t-1\} + c_3^\rho [T-(t-1)]^+ . \tag{6.61}$$

From Theorem 5.16 of Section 5.6, it follows that

$$c_2^\rho E_\infty\{T_h^c\} = 1 - c_3^\rho E_1\{T_h^c\}. \tag{6.62}$$

But

$$\lim_{\rho\to 0} E\left\{\min\{T_h^c, t_\rho - 1\}\right\} = E_\infty\left\{T_h^c\right\}, \text{ and} \tag{6.63}$$

$$\lim_{\rho\to 0} E\left\{\min\{T', t_\rho - 1\}\right\} = E_\infty\left\{T'\right\}. \tag{6.64}$$

Moreover,

$$\begin{aligned} E\left\{(T_h^c - t_\rho + 1)^+\right\} &= P(T_h^c \geq t_\rho)E\left\{(T_h^c - t_\rho + 1)|T_h^c \geq t_\rho\right\} \\ &= P(T_h^c \geq t_\rho)E_1\left\{T_h^c\right\}, \end{aligned} \tag{6.65}$$

and

$$\begin{aligned} E\left\{(T' - t_\rho + 1)^+\right\} &= P(T' \geq t_\rho)E\left\{(T' - t_\rho + 1)|T' \geq t_\rho\right\} \\ &\leq P(T' \geq t_\rho) && (6.66) \\ &< P(T' \geq t_\rho)d(T') && (6.67) \\ &< E_1\left\{T_h^c\right\}. && (6.68) \end{aligned}$$

Using (6.66) and (6.62), we have

$$\begin{aligned} 1 - E\left\{l^\rho(t_\rho, T')\right\} &> P(T' \geq t_\rho) + c_2^\rho E\left\{\min\{T', t_\rho - 1\}\right\} - c_3^\rho P(T' \geq t_\rho)E_1\left\{T_h^c\right\} \\ &= c_2^\rho\left[P(T' \geq t_\rho)E_\infty\left\{T_h^c\right\} + E\left\{\min\{T', t_\rho - 1\}\right\}\right]. \end{aligned} \tag{6.69}$$

Similarly, it follows that

$$1 - E\left\{l^\rho(t_\rho, T_h^c)\right\} = c_2^\rho\left[P(T' \geq t_\rho)E_\infty\left\{T_h^c\right\} + E\left\{\min\{T', t_\rho - 1\}\right\}\right]. \tag{6.70}$$

Using the fact that

$$\lim_{\rho\to 0} P(T' \geq t_\rho) = \lim_{\rho\to 0} P(T_h^c \geq t_\rho) = 0,$$

we obtain

$$\lim_{\rho\to 0} \frac{1 - E\left\{l^\rho(t_\rho, T_h^c)\right\}}{c_2^\rho} = E_\infty\left\{T_h^c\right\}, \tag{6.71}$$

and

$$\lim_{\rho\to 0} \frac{1 - E\left\{l^\rho(t_\rho, T')\right\}}{c_2^\rho} \geq E_\infty\left\{T'\right\}. \tag{6.72}$$

But (6.60) holds, and according to (6.71) and (6.72), there exists ρ small enough so that

$$E\left\{l^\rho(t_\rho, T')\right\} < E\left\{l^\rho(t_\rho, T_h^c)\right\}. \tag{6.73}$$

This is clearly a contradiction, since T_h^c achieves the smallest possible loss according to Theorem 5.16 of Section 5.6. This completes the proof. ∎

6.3 Performance of Page's test

In the preceding section, we showed that the stopping time T_h^c is optimal in the sense of Lorden. In this section, we consider the performance of this stopping time by determining the quantities $d(T_h^c)$ and $f(T_h^c)$.

These quantities can be computed via the following result.

THEOREM 6.5. *Suppose $h > 1$. Then*

$$f(T_h^c) = \frac{E_\infty\{N\}}{1 - P_\infty(F_0)} < \infty \tag{6.74}$$

and

$$d(T_h^c) = \frac{E_1\{N\}}{1 - P_1(F_0)} < \infty \tag{6.75}$$

where N is the stopping time

$$N = \min\left\{ n \geq 1 \,\middle|\, \sum_{\ell=1}^{n} \log L(Z_\ell) \notin (0, \log h) \right\}, \tag{6.76}$$

and where F_0 denotes the event

$$\left\{ \sum_{\ell=1}^{N} \log L(Z_\ell) \leq 0 \right\}. \tag{6.77}$$

Proof: We first examine the quantity $f(T_h^c)$. Under the distribution P_∞ the observations $Z_1, Z_2, \ldots$ are i.i.d. with marginal distribution Q_0. It is clear that T_h^c arises from a renewal process [187], with renewals occurring whenever the accumulated sum m_k of (6.54) is reset to zero, and with a termination when m_k exits from $[0, \log h)$. It follows that we can write

$$T_h^c = \sum_{j=1}^{J} N_j \quad \text{a.s. } P_\infty, \tag{6.78}$$

where $N_1, N_2, \ldots$ are i.i.d. repetitions (under P_∞) of the random variable N of (6.76) and where J denotes the number of repetitions of N that occur before the sum exits at the upper boundary $\log h$. Let M_j denote the indicator of the event that the jth repetition of N results in an exit at the upper boundary. Then J is a stopping time with respect to the sequence $(N_1, M_1), (N_2, M_2), \ldots,$ which is i.i.d. under P_∞. Since $E\{T_h^c\}$ clearly exists, the Wald identity (2.83) thus allows us to write

$$E_\infty\{T_h^c\} = E_\infty\{J\} E_\infty\{N\}. \tag{6.79}$$

It is easy to see that, under P_∞, J is a geometric random variable with

$$P_\infty(J = j) = [1 - P_\infty(F_0)]\,[P_\infty(F_0)]^{j-1}, \quad j = 1, 2, \ldots, \tag{6.80}$$

and the equality in (6.74) thus follows. To prove the inequality in (6.74), we note that the property $P_\infty(L(Z_1) > 1) > 0$, implies $P_\infty(F_0) < 1$, from which it follows that

$E_\infty\{J\} < \infty$. Furthermore, a lemma of Stein's (see for example [198], Proposition 2.19) implies that $E\{N\} < \infty$, and so the inequality in (6.74) follows as well.

To analyze $d(T_h^c)$ it is useful to recall two facts from the proof of Theorem 6.2. Namely,

$$\operatorname{esssup} E_t\left\{(T_h^c - t + 1)^+ \middle| \mathcal{F}_{t-1}\right\} = E_t\left\{(T_h^c - t + 1)^+ \middle| m_{t-1} = 0\right\}, \; t = 1, 2, \ldots \tag{6.81}$$

and

$$d_1(T_h^c) = d_2(T_h^c) = \cdots \tag{6.82}$$

That is, the worst-case pre-change sample paths are those that lead to a resetting of m_k just before the change point; and, as a consequence, the stopping time T_h^c is an equalizer rule.

From these two facts, it follows that the worst-case exponential delay to detection for T_h^c is given by

$$d(T_h^c) = d_1(T_h^c) = E_1\left\{T_h^c\right\}. \tag{6.83}$$

Since, under the measure P_1, $Z_1, Z_2, \ldots$ is an i.i.d. sequence drawn from Q_0, the analysis of this quantity proceeds in essentially the same fashion as that of $f(T_h^c)$, and (6.75) follows. ■

Since N of (6.76) is the first exit time of a random walk from an interval, its statistical behavior can be analyzed via the methods discussed in Section 4.3.

We can also consider the identity of Kennedy (see Section 4 of [121]) for certain cases. This will work later for Brownian motion, although in the case of Brownian motion we can use a differential equation approach to calculate the first moment of the CUSUM stopping time.

Another approach to the examination of $f(T_h^c)$ and $d(T_h^c)$ is to estimate them directly by approximating the behavior of T_h^c with that of the stopping time

$$\tilde{T}_h = \inf\left\{u \geq 0 \middle| Z_u - \min_{0 \leq s \leq u} Z_s \geq \log h\right\}, \tag{6.84}$$

where $\{Z_u;\, u \geq 0\}$ is a Brownian motion approximating the random walk

$$\sum_{\ell=1}^{n} \log L(Z_\ell), \; n = 1, 2, \ldots \tag{6.85}$$

To be more specific, under P_∞ we may use the model

$$Z_u = \sqrt{2D}\, W_u - Du, \; u \geq 0, \tag{6.86}$$

and under P_1 we may use the model

$$Z_u = \sqrt{2D}\, W_u + Du, \; u \geq 0, \tag{6.87}$$

where $\{W_u;\, u \geq 0\}$ is a standard Brownian motion, and where $D = D(Q_1 \parallel Q_0)$, with $D(Q_1 \parallel Q_0)$ given by (6.1). Such an approximation will be accurate under either P_∞ or P_1 if the two distributions Q_0 and Q_1 are sufficiently "close" to one another in the usual sense required by invariance theory (see for example, Section 4 of [121].) Approximation of this type for linear delay penalty and the classical Page test has been considered in [181]. Analogous results can be obtained for the exponential-penalty case.

In particular, the statistics of stopping times of the form (6.84) have been analyzed in several works, including [121], [123], [133] and [210].

6.4 The continuous-time case

As in Chapters 4 and 5, the results described in the preceding sections can be examined in the case of continuous-time observations as well. Some basic results along these lines are described in this section. As before, we treat the cases of Brownian and Poisson observations.

6.4.1 Brownian observations

In this section we will demonstrate the optimality of the CUSUM procedure in the sense of Lorden in the case of Brownian observations. This was done independently by Beibel [25] and Shiryaev [192]; however, we will follow a method similar to the one used in Section 6.2, whereby we establish a lower bound to the performance of a generic stopping time T that satisfies the false alarm constraint with equality. This lower bound is seen to be achieved by the CUSUM stopping time. This methodology was developed by Moustakides in [152].

To begin, we continuously observe the process $\{Z_s; s \geq 0\}$ with the following dynamics

$$dZ_s = \begin{cases} dW_s & s \leq t \\ \mu \, ds + dW_s & s > t, \end{cases} \tag{6.88}$$

where $\{W_s; s \geq 0\}$ is a standard Brownian motion.

We assume that the drift parameter μ is known and (without loss of generality) positive. No prior distribution on the change point t is assumed.

The probabilistic setting of the problem can be summarized as follows:

(1) $\Omega = C[0, \infty]$;
(2) $\mathcal{F} = \mathcal{F}_\infty = \cup_{u \geq 0} \mathcal{F}_u$, with $\mathcal{F}_u = \sigma\{Z_s; s \leq u\}$; and
(3) The family of probability measures P_t, $t \in [0, \infty)$, with P_t describing the model (6.88) and P_∞ denoting the Wiener measure.

As a set of detection strategies, it is natural to consider the set $\mathcal{T}$ of all stopping times with respect to the filtration $\{\mathcal{F}_u; u \geq 0\}$. As a delay penalty of a stopping time, we consider Lorden's criterion. That is, for every $T \in \mathcal{T}$ we consider the delay penalty

$$J(T) = \sup_t \operatorname{ess\,sup} E_t \left\{ (T - t)^+ | \mathcal{F}_t \right\}. \tag{6.89}$$

This gives rise to the following stochastic optimization problem:

$$\inf_{T \in \mathcal{T}} J(T), \quad \text{subject to} \quad E_\infty\{T\} \geq \gamma. \tag{6.90}$$

We first notice that in view of Theorem 6.2, in seeking optimal solutions to the above problem, we can confine ourselves to the stopping times that satisfy the false alarm constraint with equality. Otherwise, consider a stopping time T, with $E_\infty\{T\} > \gamma$ and let T' be a randomized stopping time that randomizes between T and the stopping time 0 (according to which we announce an alarm immediately), so that $E_\infty\left\{T'\right\} = \gamma$. Then

$$\sup_t \operatorname{esssup} E_t\left\{(T-t)^+|\mathcal{F}_t\right\} \geq \sup_t \operatorname{esssup} E_t\left\{(T'-t)^+|\mathcal{F}_t\right\},$$

and we have achieved a smaller detection delay, while still remaining within the class of permissible stopping times, as the constraint is satisfied.

In this continuous-time setting, the CUSUM process is defined as

$$\sup_{0\leq t\leq s} \left.\frac{\mathrm{d}P_t}{\mathrm{d}P_\infty}\right|_{\mathcal{F}_s}. \tag{6.91}$$

Notice that this reduces to the form that appears in (5.111) in the discrete-time case.

Therefore, the log of the CUSUM process can be written in this case as

$$\sup_{0\leq t\leq s} (U_s - U_t), \tag{6.92}$$

where

$$U_s = \mu Z_s - \frac{1}{2}\mu^2 s. \tag{6.93}$$

Therefore, by defining $M_s = \inf_{0\leq r\leq s} U_r$, we can rewrite (6.91) as

$$Y_s = U_s - M_s. \tag{6.94}$$

REMARK 6.6. From Girsanov's theorem (see Theorem 2.9)

$$\left.\frac{\mathrm{d}P_t}{\mathrm{d}P_\infty}\right|_{\mathcal{F}_s} = \mathrm{e}^{U_s - U_t} \tag{6.95}$$

is the likelihood ratio process, since

$$E_\infty\left\{\mathrm{e}^{\frac{1}{2}\mu^2 s}\right\} < \infty, \quad \forall\, s < \infty.$$

That is, the Novikov condition holds (see Theorem 2.9), and the process (6.95) is a martingale under P_∞. Notice that each member of the family of measures $\{P_t,\ t \in [0,\infty]\}$ is absolutely continuous with respect to P_∞.

In the rest of this section we will demonstrate that the CUSUM stopping time

$$T_h^c = \inf\{s \geq 0 | Y_s \geq h\}, \tag{6.96}$$

with h selected so that $E_\infty\left\{T_h^c\right\} = \gamma$, is optimal in the sense of (6.90).

To demonstrate this optimality we will begin by demonstrating a number of key intermediate results. The first one shows that the worst detection delay for a CUSUM stopping time occurs when the change occurs at time 0. The implication of this is that if T is a CUSUM stopping time then $\sup_t \operatorname{esssup} E_t\left\{(T-t)^+|\mathcal{F}_t\right\} = E_0\{T\}$. This result is demonstrated by showing that the CUSUM process that arises if the change is to occur at time 0 stochastically dominates the CUSUM process that arises if the change is at time $t > 0$ unless $Y_t = 0$. This result is summarized in the following proposition.

PROPOSITION 6.7. *Fix $t \in [0, \infty)$. Consider the process*

$$Y_{s,t} = U_s - U_t - \inf_{t \le r \le s} (U_r - U_t), \ s \ge t.$$

(Note that this is the CUSUM process when starting at time t.) Then $Y_s \ge Y_{s,t}$ with equality if $Y_t = 0$.

Proof: The proof is a matter of noticing that we can write

$$Y_s = Y_{s,t} + \left(\inf_{t \le r \le s} (U_r - U_t) + Y_t \right)^+ \ge Y_{s,t} \tag{6.97}$$

and that $\inf_{t \le r \le s}(U_r - U_t) \le 0$. ■

It is clear that $Y_{s,t}$ depends only on information received *after* time t. Relation (6.97), therefore, suggests that the worst detection delay *before t* occurs whenever $Y_t = 0$. In other words,

$$\text{esssup}\, E_t \left\{ (T_h^c - t)^+ | \mathcal{F}_t \right\} = E_t \left\{ (T_h^c - t)^+ | Y_t = 0 \right\} = E_0 \left\{ T_h^c \right\}. \tag{6.98}$$

Equation (6.98) states that the CUSUM stopping time is an equalizer rule over the change point t, in the sense that its performance does not depend on the value of this unknown constant.

We now proceed to two propositions, the first of which provides a closed-form formula for the expected delay of a CUSUM stopping time under each of the measures P_0 and P_∞ while asserting its a.s. finiteness under the whole family of measures $\{P_t\}$. The second proposition provides us with a representation of the expected value of a general almost surely finite stopping time of the form $T_h = T \wedge T_h^c$.

PROPOSITION 6.8. *We have*

$$P_t(T_h^c = \infty | \mathcal{F}_t) \quad = \quad 0 \ \ a.s. \ P_t, \ \textit{for all } t \ge 0, \tag{6.99}$$

and

$$P_\infty(T_h^c = \infty | \mathcal{F}_t) \quad = \quad 0 \ \ a.s. \ P_\infty. \tag{6.100}$$

Moreover,

$$E_0 \left\{ T_h^c \right\} \quad = \quad (2/\mu^2) g(h) \tag{6.101}$$

and

$$E_\infty \left\{ T_h^c \right\} \quad = \quad (2/\mu^2) g(-h), \tag{6.102}$$

where $g(h) = \mathrm{e}^{-h} + h - 1$ and $g(-h) = \mathrm{e}^{h} - h - 1$.

Proof: Consider the function g, which is twice continuously differentiable and satisfies

$$g'(y) + g''(y) = 1, \text{ with } g'(0) = g(0) = 0.$$

Using Itô's rule on the process $g(Y_s)$ (see Theorem 2.6), we obtain

$$g(Y_s) - g(0) = \int_0^s \mu g'(Y_u)\mathrm{d}Z_u - \int_0^s \frac{1}{2}\mu^2 g'(Y_u)\mathrm{d}u - \int_0^s g'(Y_u)\mathrm{d}M_u + \int_0^s \frac{1}{2}\mu^2 g''(Y_u)\mathrm{d}u. \tag{6.103}$$

Since $g'(0) = 0$, it follows that $\int_0^s g'(Y_u)\mathrm{d}M_u = 0$, by the definition of M_u. Now consider the sequence of stopping times $T_h^n = T_h^c \wedge n$, each of which is bounded. It follows that

$$E_0\left\{g(T_h^n)\right\} = \frac{1}{2}\mu^2 E_0\left\{T_h^n\right\} + \mu E_0\left\{\int_0^{T_h^n} \mathrm{d}W_u\right\}. \tag{6.104}$$

Now, since

$$\int_0^s \mathrm{d}u < \infty$$

and $E\left\{T_h^n\right\} < n < \infty$, using Theorem 2.5 and (2.98), it follows that the last term in (6.104) is 0.

Furthermore, since g is a strictly increasing function, it follows that

$$g(h) \geq E_0\left\{g(Y_{T_h^n})\right\} = E_0\left\{T_h^n\right\}.$$

By the bounded convergence theorem, taking the limit of both sides as $n \to \infty$ we obtain that $E_0\left\{T_h^c\right\} < \infty$. The same argument, but with conditional P_t expectations conditioned on $\mathcal{F}_t$, leads to the conclusion that (6.99) holds. Evaluating (6.103) at T_h^c, taking expectations, and using arguments similar to the above, we have that

$$g(h) = E_0\left\{g(T_h^c)\right\} = \frac{1}{2}\mu^2 E_0\left\{T_h^c\right\}.$$

Hence (6.101) follows. Similarly, we can show (6.102) and (6.100). ■

COROLLARY 6.9. *Let $T \in \mathcal{T}$ and consider $T_h = T \wedge T_h^c$. Then*

$$E_t\left\{(T_h - t)^+ | \mathcal{F}_t\right\} = E_t\left\{g(Y_{T_h}) - g(Y_t) | \mathcal{F}_t\right\} \mathbf{1}_{\{T_h > t\}} \tag{6.105}$$

and

$$E_\infty\left\{(T_h - t)^+ | \mathcal{F}_t\right\} = E_\infty\left\{g(-Y_{T_h}) - g(-Y_t) | \mathcal{F}_t\right\} \mathbf{1}_{\{T_h > t\}}. \tag{6.106}$$

Proof: The proof follows from an application of Itô's rule and noticing that $E\left\{T_h\right\} < E\left\{T_h^c\right\} < \infty$. Another application of Theorem 2.5 and (2.98) assures that the stochastic integral vanishes. ■

PROPOSITION 6.10. *Let $T \in \mathcal{T}$ and $T_h = T \wedge T_h^c$. Then for any function f continuous and bounded on $0 \leq y \leq h$, we have*

$$E_t \left\{ f(Y_{T_h}) | \mathcal{F}_t \right\} \mathbf{1}_{\{T_h > t\}} = E_\infty \left\{ e^{U_{T_h} - U_t} f(Y_{T_h}) | \mathcal{F}_t \right\} \mathbf{1}_{\{T_h > t\}} \ a.s. \ P_\infty. \tag{6.107}$$

Proof: We cannot readily apply Girsanov's theorem (see Theorem 2.9) because T_h is not bounded. But, for $M > 0$, we can write

$$\begin{aligned} E_t \left\{ f(Y_{T_h}) | \mathcal{F}_t \right\} &= E_t \left\{ \mathbf{1}_{\{T_h \leq M\}} f(Y_{T_h}) | \mathcal{F}_t \right\} + E_t \left\{ \mathbf{1}_{\{T_h > M\}} f(Y_{T_h}) | \mathcal{F}_t \right\} \\ &= E_\infty \left\{ e^{U_{T_h} - U_t} f(Y_{T_h}) | \mathcal{F}_t \right\} - E_\infty \left\{ \mathbf{1}_{\{T_h > M\}} e^{U_{T_h} - U_t} f(Y_{T_h}) | \mathcal{F}_t \right\} \\ &\quad + E_t \left\{ \mathbf{1}_{\{T_h > M\}} f(Y_{T_h}) | \mathcal{F}_t \right\}. \end{aligned}$$

We now notice that on $\{T_h > t\}$, we have $U_{T_h} - U_t \leq U_{T_h} - M_{T_h} = Y_{T_h} \leq h$, and thus

$$\begin{aligned} \left| E_t \left\{ \mathbf{1}_{\{T_h > M\}} f(Y_{T_h}) | \mathcal{F}_t \right\} \right| &\leq \max_{0 \leq y \leq h} |f(y)| P_t (T_h > M | \mathcal{F}_t) \\ &\leq \max_{0 \leq y \leq h} |f(y)| P_t \left(T_h^c > M | \mathcal{F}_t \right), \end{aligned}$$

and

$$\begin{aligned} E_\infty \left\{ \mathbf{1}_{\{T_h > M\}} e^{U_{T_h} - U_t} f(Y_{T_h}) | \mathcal{F}_t \right\} &\leq e^h \max_{0 \leq y \leq h} |f(y)| P_t (T_h > M | \mathcal{F}_t) \\ &\leq e^h \max_{0 \leq y \leq h} |f(y)| P_t \left(T_h^c > M | \mathcal{F}_t \right). \end{aligned}$$

Applying Proposition 6.5, we see that as $M \to \infty$ both of the above bounds tend to 0. ■

We are now in a position to establish a lower bound on the performance of any general stopping time T. The optimality of the CUSUM stopping time (6.96) will then be a matter of showing that this lower bound is achieved only by it.

PROPOSITION 6.11. *Choose $T \in \mathcal{T}$ and define $T_h = T \wedge T_h^c$. Then, we have*

$$J(T) \geq \frac{E_\infty \left\{ e^{Y_{T_h}} g(Y_{T_h}) \right\}}{E_\infty \left\{ e^{Y_{T_h}} \right\}}. \tag{6.108}$$

Proof: Since $T \geq T_h$, we can write

$$J(T) \geq J(T_h) \geq E_t \left\{ (T - t)^+ | \mathcal{F}_t \right\}, \tag{6.109}$$

and

$$J(T) \geq J(T_h) \geq E_0 \{T\}. \tag{6.110}$$

Applying Corollary 6.9 and Proposition 6.10 on both sides of (6.109) we obtain

$$J(T) E_\infty \left\{ e^{U_{T_h} - U_t} \middle| \mathcal{F}_t \right\} \geq E_\infty \left\{ e^{U_{T_h} - U_t} \left[g(Y_{T_h}) - g(Y_t) \right] \middle| \mathcal{F}_t \right\},$$

which is equivalent to

$$J(T)E_\infty\left\{\int_0^{T_h} e^{U_{T_h}-U_s}(-dM_s)\Big|\mathcal{F}_t\right\} \geq E_\infty\left\{\int_0^{T_h} e^{U_{T_h}-U_s}\left[g(Y_{T_h})-g(Y_s)\right](-dM_s)\Big|\mathcal{F}_t\right\}.$$

Taking expectations with respect to P_∞ we obtain

$$J(T)E_\infty\left\{\int_0^{T_h} e^{U_{T_h}-U_s}(-dM_s)\right\} \geq E_\infty\left\{\int_0^{T_h} e^{U_{T_h}-U_s}\left[g(Y_{T_h})-g(Y_s)\right](-dM_s)\right\}.$$

Since $g(0)=0$, and the process $\{M_s; s\geq 0\}$ is constant in time on the complement of the set $\{s\geq 0|Y_s=0\}=\{s\geq 0|U_s=M_s\}$, we can rewrite the above equation as

$$J(T)E_\infty\left\{\int_0^{T_h} e^{U_{T_h}-M_s}(-dM_s)\right\} \geq E_\infty\left\{\int_0^{T_h} e^{U_{T_h}-M_s}\left[g(Y_{T_h})-g(Y_s)\right](-dM_s)\right\}.$$

Carrying out the integration we obtain

$$J(T)E_\infty\left\{e^{Y_{T_h}-U_{T_h}}\right\} \geq E_\infty\left\{e^{Y_{T_h}-U_{T_h}}g(Y_{T_h})\right\}. \tag{6.111}$$

Moreover, using Proposition 6.8 and Corollary 6.9 we can also obtain

$$J(T)E_\infty\left\{e^{U_{T_h}}\right\} \geq E_\infty\left\{e^{U_{T_h}}g(Y_{T_h})\right\}. \tag{6.112}$$

Adding (6.111) and (6.112) we then have

$$J(T)E_\infty\left\{e^{Y_{T_h}}\right\} \geq E_\infty\left\{e^{Y_{T_h}}g(Y_{T_h})\right\},$$

and the result follows. ■

In order to finally establish the optimality of the CUSUM stopping time (6.96) we will need to compare its detection delay to the detection delay of a general T but for the same frequency of false alarms. To this effect, fix γ and choose h^* so that $E_\infty\left\{T_h^c\right\}=\gamma$. The objective now is to show that for any $T\in\mathcal{T}$ with $E_\infty\{T\}=E_\infty\left\{T_h^c\right\}=\gamma$, we have $J(T)\geq\frac{2}{\mu^2}g(h^*)$. Since for all T, we can define $T_h=T\wedge T_h^c$, and $J(T)\geq J(T_h)$, we need only to focus on stopping times of the form T_h. Note that the function $g(-h)=E_\infty\{T_h\}$ is a continuous function of the threshold h. This is a result of a technical lemma, which can be found in [152]. As a result of this, there exists an $\epsilon>0$ and a threshold h_ϵ such that for any $\gamma>0$ we have

$$\gamma\geq E_\infty\left\{T_{h_\epsilon}\right\}\geq\gamma-\epsilon, \tag{6.113}$$

where $T_{h_\epsilon}=T\wedge S_{h_\epsilon}$. In order to establish the optimality of (6.96), we will have to show that for every such ϵ, $J(T_{h_\epsilon})\geq\frac{2}{\mu^2}g(h^*)$. This is established in the following theorem.

THEOREM 6.12. *For any $T\in\mathcal{T}$ that satisfies $E_\infty\{T\}=\gamma$, we have $J(T)\geq\frac{2}{\mu^2}g(h^*)$.*

Proof: In view of Proposition 6.11, it suffices to show that for every $\epsilon>0$, we have

$$\frac{E_\infty\left\{e^{Y_{T_{h_\epsilon}}}g(Y_{T_{h_\epsilon}})\right\}}{E_\infty\left\{e^{Y_{T_{h_\epsilon}}}\right\}}\geq g(h^*)-\epsilon. \tag{6.114}$$

Consider the function $U(x) = e^x [g(x) - g(h^*)] - [g(-x) - g(-h^*)]$ for $x \geq 0$. Simple inspection shows that $U(x) \geq 0$ for all $x \geq 0$, with equality only at the point $x = h^*$. Therefore by letting $x = Y_{T_{h\epsilon}}$ we see that

$$e^{Y_{T_{h\epsilon}}} \left[g(Y_{T_{h\epsilon}}) - g(h^*)\right] - \left[g(-Y_{T_{h\epsilon}}) - g(-h^*)\right] \geq 0 \;\; P_\infty \; a.s. \tag{6.115}$$

Taking expectations with respect to P_∞ on both sides of (6.115) and rearranging terms, we obtain

$$\begin{aligned} E_\infty \left\{ e^{Y_{T_{h\epsilon}}} (g(Y_{T_{h\epsilon}})) \right\} &\geq g(h^*) E_\infty \left\{ e^{Y_{T_{h\epsilon}}} \right\} + E_\infty \left\{ g(-Y_{T_{h\epsilon}}) \right\} - g(-h^*) \\ &\geq g(h^*) E_\infty \left\{ e^{Y_{T_{h\epsilon}}} \right\} - \epsilon \\ &\geq \left[g(h^*) - \epsilon \right] E_\infty \left\{ e^{Y_{T_{h\epsilon}}} \right\}, \end{aligned}$$

where the last inequality follows from the fact that $e^x \geq 1$ for all $x \geq 0$. Hence inequality (6.114) holds and the theorem follows. ■

Theorem 6.12 establishes the optimality of the CUSUM stopping time (6.96) since it demonstrates that for any stopping time with the same frequency of false alarms as the CUSUM stopping time, the detection delay is at least as high as that achieved by the CUSUM stopping time. That is, the CUSUM stopping time is the stopping time that achieves the smallest possible detection delay for any given level of the mean time between false alarms.

6.4.2 Itô processes

In this section, we generalize the problem of the preceding section to the case of continuous-time Itô processes. In particular, suppose we observe the process $\{Z_s; s \geq 0\}$ with the following dynamics:

$$dZ_s = \alpha_s \mathbf{1}_{\{s \geq t\}} ds + dW_s, \;\; s \geq 0,$$

where, as usual, $\{W_s; s \geq 0\}$ denotes standard Brownian motion, and where $\{\alpha_s; s \geq 0\}$ is a known measurable function of the past observations. As before, the change-point t is assumed to be an unknown constant. The probabilistic setting of the problem is the same as the one described in the above section.

The objective here, as before, is to detect the unknown change point t as soon as possible, by means of a stopping time $T \in \mathcal{T}$ within a lower bound constraint on the mean time between false alarms.

It is possible to show that the CUSUM stopping time is optimal in this case as well, in the sense that it minimizes an alternative to Lorden's criterion subject to the usual bound in the mean time between false alarms. In particular, consider the following delay penalty

$$J(T) = \sup_t \text{esssup}\, E_t \left\{ \mathbf{1}_{\{T > t\}} \int_t^T \alpha_s^2 ds \;\middle|\; \mathcal{F}_t \right\},$$

which gives rise to the following optimization problem:

$$\begin{aligned} &\inf_{T\in\mathcal{T}} J(T), \text{ subject to} \\ &E_\infty\left\{\int_0^T \alpha_s^2 \mathrm{d}s\right\} \geq \gamma. \end{aligned} \tag{6.116}$$

The above criterion can be motivated by considering the Kullback–Leibler divergence

$$E_t\left\{\log \frac{\mathrm{d}P_t}{\mathrm{d}P_\infty}\bigg|_{\mathcal{F}_s}\bigg|\mathcal{F}_t\right\} = E_t\left\{\int_t^s \alpha_r \mathrm{d}W_r + \frac{1}{2}\int_t^s \alpha_r^2 \mathrm{d}r \bigg| \mathcal{F}_t\right\} = E_t\left\{\frac{1}{2}\int_t^s \alpha_r^2 \mathrm{d}r \bigg| \mathcal{F}_t\right\}, \tag{6.117}$$

for $s \geq t$, where the last equality follows as long as

$$E_t\left\{\int_t^s \alpha_r^2 \mathrm{d}r \bigg| \mathcal{F}_t\right\} < \infty \;\; a.s. \tag{6.118}$$

Thus (6.116) uses the Kullback–Leibler divergence as a weighting of time penalties. In what follows we will impose conditions on $\{\alpha_s\}$ that guarantee the existence of the Radon–Nikodym derivative of P_t with respect to P_∞. We will also impose conditions that will in turn guarantee the finiteness of the CUSUM stopping time and will present the form that the CUSUM stopping time takes in this case.

Let us first suppose that

$$E_\infty\left\{\mathrm{e}^{\frac{1}{2}\int_0^u \alpha_s^2 \mathrm{d}s}\right\} < \infty, \;\; u \in [0, \infty). \tag{6.119}$$

Of course this condition guarantees that

$$P_\infty\left(\frac{1}{2}\int_0^u \alpha_s^2 \mathrm{d}s < \infty\right) = 1,$$

and further that the Radon–Nikodym derivative exists:

$$\frac{\mathrm{d}P_t}{\mathrm{d}P_\infty}\bigg|_{\mathcal{F}_s} = \mathrm{e}^{U_s - U_t}, \;\; 0 \leq t \leq s < \infty, \tag{6.120}$$

where

$$U_s = \int_0^s \alpha_r \mathrm{d}Z_r - \frac{1}{2}\int_0^s \alpha_r^2 \mathrm{d}r. \tag{6.121}$$

Now since for all $s > 0$, (6.120) is positive, we have that P_t is absolutely continuous with respect to P_∞. Therefore,

$$P_t\left(\int_0^s \alpha_r^2 \mathrm{d}r < \infty\right) = 1, \tag{6.122}$$

for all $s < \infty$ and $t \in [0, \infty)$. (For the interested reader, we note that the event $\{\int_0^s \alpha_r^2 \mathrm{d}r < \infty\} \in \mathcal{F}_s$, where $\mathcal{F}_s = \sigma\{Z_u; 0 \leq u \leq s\}$.) The probability measure P_∞ is unique and condition (6.119), which guarantees that the process (6.120) is a martingale, also implies that P_t and P_∞ are equivalent measures for all $0 \leq t < \infty$. Notice that (6.122) also implies (6.118).

Using (6.91), (6.120) and (6.121), it follows that the log of the CUSUM process in this case can be written as

$$Y_s = U_s - M_s, \tag{6.123}$$

with U_s defined as in (6.121) and $M_s = \inf_{0 \le r \le s} U_r$. The CUSUM stopping time is then

$$T_h^c = \inf\{s \ge 0 | Y_s \ge h\}. \tag{6.124}$$

To ensure the a.s. finiteness of the CUSUM stopping time we need to impose an additional condition (see [135]), whose physical interpretation is that the signal received after the change point has sufficient energy (see Section 4.5):

$$P_t\left(\int_0^\infty \alpha_s \mathrm{d}s = \infty\right) = P_\infty\left(\int_0^\infty \alpha_s \mathrm{d}s = \infty\right) = 1. \tag{6.125}$$

Using the exact same steps as in the proof of the optimality of the CUSUM in the Brownian model with constant drift, it is possible to show that in this case too, the CUSUM stopping time (6.96), with h uniquely determined by the equality

$$E_\infty\left\{T_h^c\right\} = g(-h) = \gamma,$$

is optimal.

For further details on this case, please refer to [152].

REMARK 6.13. Notice that the criterion (6.117) reduces to the Lorden criterion in the case $\alpha_s = \mu$, and the problem then becomes exactly the one treated in the previous section.

6.4.3 Brownian motion with an unknown drift parameter

In this section, we consider the situation in which a continuous-time observation process $\{Z_s; s \ge 0\}$ obeys the following model:

$$\mathrm{d}Z_s = \begin{cases} \mathrm{d}W_s & s \le t \\ \mu\, \mathrm{d}s + \mathrm{d}W_s & s > t, \end{cases}$$

where $\{W_s; s \ge 0\}$ is a standard Brownian motion, t, as before, is assumed to be an unknown constant, and μ is known only to lie in the interval $[m, \infty)$ for some $m > 0$. (By symmetry this problem can be treated in the same way if μ lies in $(-\infty, m]$ with $m < 0$.)

The space Ω, the σ algebra $\mathcal{F}$ and the filtration $\{\mathcal{F}_s\}$ are defined in the same way as in Section 6.4.1. The family of probability measures is now indexed not only by the change point but also by the drift parameter μ. In other words we have the family of measures $\{P_t^\mu | t \in [0, \infty),\ \mu \in [m, \infty)\}$, with P_∞ the Wiener measure. We again consider the family of stopping times $\mathcal{T}$, as before, and the objective is again to detect the change point as soon as possible while controlling the mean time between false alarms. In order to incorporate the different possible changes in the drift, we consider the following extension of Lorden's criterion as a delay penalty for a stopping time $T \in \mathcal{T}$:

$$J_L(T) = \sup_\mu \sup_t \operatorname{esssup} E_t^\mu\left\{(T - t)^+ | \mathcal{F}_t\right\}, \tag{6.126}$$

which results in the corresponding optimization problem of the form:

$$\inf_{T \in \mathcal{T}} J_L(T) \text{ subject to} \qquad E_\infty\{T\} \geq \gamma. \tag{6.127}$$

We will show that the one-sided CUSUM stopping time that detects the smallest drift in absolute value is the solution of the problem in (6.127). We follow [101]. To this effect define the following quantities:

(1) $U_s(\rho) = \rho Z_s - \frac{1}{2}\rho^2 ds$,
(2) $M_s(\rho) = \inf_{0 \leq r \leq s} U_r(\rho)$, and
(3) $Y_s(\rho) = U_s(\rho) - M_s(\rho)$.

The process $\{Y_s(\rho); s \geq 0\}$ is the log of the CUSUM statistic process for the case in which the drift parameter assumed after the change is ρ.

PROPOSITION 6.14. *For every path of the Brownian motion $\{W_s; s \geq 0\}$, the process $\{Y_s(\rho); s \geq 0\}$ is an increasing (decreasing) function of the drift of the observation process $\{Z_s; s \geq 0\}$ when $\rho > 0$ ($\rho < 0$).*

Proof: Consider two possible drift values $\mu_1, \mu_2 \in [m, \infty)$ with $\mu_1 < \mu_2$. We define two observation processes, $Z_s(\mu_i) = \mu_i(s-t)^+ + W_s$, $i = 1, 2$, and consider the following processes

$$U_s(\rho, \mu_i) = \rho Z_s(\mu_i) - \frac{1}{2}\rho^2 s = \rho\{W_s + \mu_i(s-t)^+\} - \frac{1}{2}\rho^2 s,$$

$$M_s(\rho, \mu_i) = \inf_{0 \leq r \leq s} U_r(\rho, \mu_i),$$

and

$$Y_s(\rho, \mu_i) = U_s(\rho, \mu_i) - M_s(\rho, \mu_i).$$

Consider the difference $Y_s(\rho, \mu_2) - Y_s(\rho, \mu_1) = \delta(s-t)^+ - M_s(\rho, \mu_2) + M_s(\rho, \mu_1)$ where $\delta = \rho(\mu_2 - \mu_1)$. Notice now that $\rho > 0$ implies $\delta > 0$ and so we can write

$$U_s(\rho, \mu_2) = U_s(\rho, \mu_1) + \delta(s-t)^+ \leq U_s(\rho, \mu_1) + \delta(s-t)^+.$$

Taking the infimum over $0 \leq r \leq s$, we obtain $M_s(\rho, \mu_2) \leq M_s(\rho, \mu_1) + \delta(s-t)^+$ from which, by rearranging terms, it follows that $Y_s(\rho, \mu_2) \geq Y_s(\rho, \mu_1)$. The case $\rho < 0$ can be shown similarly. ∎

Now, suppose that $\rho = \mu_1$. The CUSUM stopping time for the case in which μ_1 is the change is $T_h^c(\mu_1) = \inf\{s \geq 0 | Y_s(\mu_1) \geq h\}$. From Proposition 6.14 it follows that $0 \leq \mu_1 \leq \mu_2$ implies $E_0^{\mu_1}\{T_h^c(\mu_1)\} \geq E_0^{\mu_2}\{T_h^c(\mu_1)\}$. As a direct consequence of this fact comes our first optimality result concerning drifts with the same sign.

THEOREM 6.15. *Suppose $0 < \mu_1 \leq \mu_2$ or $\mu_2 \leq \mu_1 < 0$. Then the one-sided CUSUM stopping time $T_h^c(\mu_1)$ with h_1 satisfying $E_\infty\{T_h^c(\mu_1)\} = \gamma$ solves the optimization problem defined in (6.127).*

Proof: For all T that satisfy the false alarm constraint with equality, we have

$$\begin{aligned} J_L(T) &= \sup_{\mu} \sup_{t} \operatorname{essup} E_t^{\mu} \left\{ (T-t)^+ | \mathcal{F}_t \right\} \\ &\geq \sup_{t} \operatorname{essup} E_t^{\mu_1} \left\{ (T-t)^+ | \mathcal{F}_t \right\} \\ &\geq E_0^{\mu_1} \{ T_h^c(\mu_1) \} = \sup_{\mu} E_0^{\mu} \{ T_h^c(\mu_1) \}. \end{aligned}$$

The last inequality follows from the optimality of the one-sided CUSUM stopping time, and the last three equalities are due to Proposition 6.14 and the definition of the delay penalty $J_L(T)$ in (6.126). ■

Thus the problem of a (one-sided) unknown post-change drift has a rather straightforward solution.

For another approach to this problem see [27,28].

6.4.4 Poisson observations

In this section we consider the problem of change-point detection in the case of Poisson observations. That is, we will now consider the situation in which the observation process $\{Z_s; s \geq 0\}$ is a Poisson process with rate λ_0 before the unknown change point t and it is a homogeneous Poisson process with rate λ_1 after the change. More specifically, $\Omega = D[0, \infty)$ is the space of càdlàg functions and $\mathcal{F} = \mathcal{B}(D[0, \infty))$. The filtration of the observations is $\{\mathcal{F}_s; s \geq 0\}$ with $\mathcal{F}_s = \sigma\{Z_u, u \leq s\}$. Our model is described by the probability measures $\{P_t; t \in [0, \infty]\}$ with P_0 and P_∞ corresponding to the measure generated on Ω by a homogeneous Poisson process with parameters λ_1 and λ_0, respectively. The measure P_t with $t \in (0, \infty)$ describes the model when the change takes place at time t.

Using (6.91) we see that the log of the CUSUM process reduces in this case to the form

$$Y_s = U_s - M_s,$$

where

$$U_s = as + bZ_s,$$

and

$$M_s = \inf_{0 \leq r \leq s} U_r,$$

where $a = -(\lambda_1 - \lambda_0)$ and $b = \log(\lambda_1/\lambda_0)$, and the CUSUM stopping time is

$$T^c = \inf\{t | y_t \geq h\}. \tag{6.128}$$

We distinguish two cases:

(1) $a > 0, b < 0$ corresponding to the case in which $\lambda_1 < \lambda_0$; and
(2) $a < 0, b > 0$ corresponding to the opposite case.

In what follows we will go through the basic results that assert the optimality of the CUSUM stopping time of (6.128) in the sense of Lorden. That is, in the sense of (6.90). We follow [154].

In doing so we will first begin by evaluating $E\{T^c\}$ under each of the measures P_∞ and P_0. Note that both in the discrete-time case (6.18) and in the continuous-time case (6.98), we have that the CUSUM stopping time is an equalizer rule with respect to the unknown change point t and therefore it is only necessary to calculate $E\{T^c\}$ under the two measures mentioned above.

After we arrive at a closed form expression for $E\{T^c\}$ we will provide a global lower bound on $J(T)$. We will then outline the steps that lead to the fact that this lower bound is achievable exactly by the CUSUM stopping time of (6.128) for any given mean time between false alarms γ, thus asserting its optimality.

The method for evaluating $E\{T^c\}$ is common to both of the above cases. It involves expressing $E\{T^c|Y_0 = y\}$ as a function $g(y)$ that satisfies a given DDE from which $E\{T^c\}$ can be easily extracted by evaluating $g(y)$ at $y = 0$.

In the case $\lambda_1 < \lambda_0$, the above function g satisfies the forward DDE

$$\begin{aligned} ag'(y) + \lambda\left[g(y+b) - g(y)\right] &= -1, \ y \in [0, h) \\ g(y) = g(0), \ y \leq 0, \ &\text{and } g(h) = 0. \end{aligned} \tag{6.129}$$

It is possible to arrive to (6.129) by applying the generalized Itô rule (see Theorem 2.7) to $\{g(Y_s); s \geq 0\}$, evaluating the resulting stochastic differential equation at T^c, taking expectations and using the martingale property of the Poisson process (see the next-to-last bullet point of Section 2.6). In doing so, we also need to pay close attention to two properties of the paths of $\{Y_s; s \geq 0\}$ in the case in which $\lambda_1 < \lambda_0$ (see Figure 6.3). They are:

(1) the paths of $\{M_s; s \geq 0\}$ are constant except on a subset of the jump times $\{V_n\}$;
(2) $Y_{V_n} = (Y_{V_{n-}} + b)^+$ which leads to the boundary condition $g(y) = g(0), \ y \leq 0$; and
(3) the CUSUM process $\{Y_s; s \geq 0\}$ hits the threshold h exactly. That is, there is no overshoot of the boundary.

The solution to (6.129) evaluated at 0 is given by

$$E\{T^c\} = \frac{1}{\lambda}\sum_{n=0}^{\left[\frac{h}{|b|}\right]}\left\{e^{\frac{\lambda(h-n|b|)}{a}}\left[\sum_{k=0}^{n}\frac{\left(\frac{-\lambda(h-n|b|)}{a}\right)^k}{k!}\right] - 1\right\}, \tag{6.130}$$

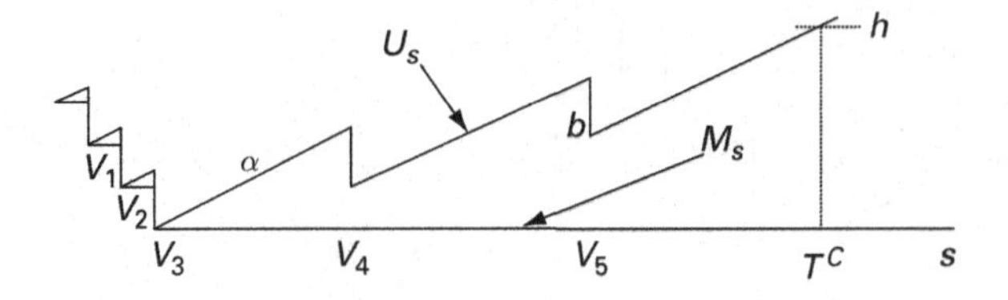

Fig. 6.3. The case $a > 0$ and $b < 0$.

where λ equals λ_1 and λ_0 under the measures P_0 and P_∞, respectively.

In the case $\lambda_1 > \lambda_0$, the function g above satisfies the backward DDE

$$\begin{aligned} a g'(y) + \lambda\left[g(y+b) - g(y)\right] &= -1, \;\; y \in [0, h) \\ g'(0) = 0, \;\text{and}\; g(h) &= 0, \; for \;\; y \geq h. \end{aligned} \tag{6.131}$$

The key properties of the paths of $\{Y_s; s \geq 0\}$ can be summarized in the following (see Figure 6.4):

(1) the paths of $\{M_s; s \geq 0\}$ are no longer constant even off the set of jump times $\{U_n\}$ which leads to the boundary condition $g'(0) = 0$;

(2) $Y_{V_n} = (Y_{V_{n-}} + b)$; and

(3) the CUSUM process $\{Y_s; s \geq 0\}$ does not necessarily hit the threshold h exactly. That is, there may be an overshoot of the boundary. This leads to the boundary condition $g(y) = 0, \;\; y \geq h$.

The solution of (6.131) evaluated at 0 is given by

$$E\left\{T^c\right\} = \frac{1}{\lambda} \sum_{n=0}^{\left[\frac{h}{|b|}\right]} \left\{ 1 - e^{\frac{\lambda(h-nb)}{|a|}} \left[\sum_{k=0}^{n} \frac{\left(\frac{-\lambda(h-nb)}{|a|}\right)^k}{k!} \right] \right\} + \frac{p}{\lambda} \sum_{n=0}^{\left[\frac{h}{|b|}\right]} e^{\frac{\lambda(h-nb)}{|a|}} \frac{\left(\frac{-\lambda(h-nb)}{|a|}\right)^n}{n!} \tag{6.132}$$

where $p = A/(A - B)$ with

$$A = \sum_{n=0}^{\left[\frac{h}{|b|}\right]} e^{\frac{\lambda(h-nb)}{|a|}} \frac{\left(\frac{-\lambda(h-nb)}{|a|}\right)^n}{n!},$$

and

$$B = \sum_{n=0}^{\left[\frac{h}{|b|}\right]} e^{\frac{\lambda(h-nb)}{|a|}} \frac{\left(\frac{-\lambda(h-nb)}{|a|}\right)^{n-1}}{(n-1)!}.$$

Using arguments similar to those used in Proposition 6.11 it is possible to show that for all $T \in \mathcal{T}$

$$J(T) \geq \frac{E_\infty\left\{\int_0^T e^{Y_s} ds\right\}}{E_\infty\{e^{Y_T}\}}. \tag{6.133}$$

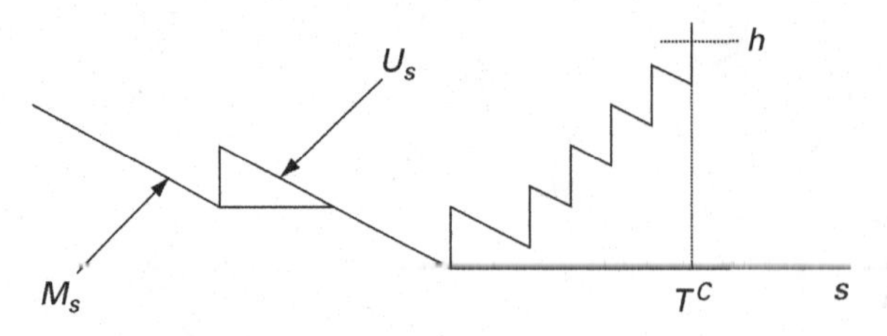

Fig. 6.4. The case $a < 0$ and $b > 0$.

The proof of the optimality of (6.128) is then a matter of showing that for all $T \in \mathcal{T}$ such that $E_\infty\{T\} \geq E_\infty\{T^c\}$ we have $J(T) \geq J(T^c) = E_0\{T^c\}$.

Denote $E_0\{T^c\}$ and $E_\infty\{T^c\}$ by $g_{\lambda_1}(0)$ and $g_{\lambda_0}(0)$, respectively. (Note that the functions $g_{\lambda_1}(0)$ and $g_{\lambda_0}(0)$ are different under cases 1 and 2 above.)

The objective is to demonstrate that

$$\frac{E_\infty\left\{\int_0^T \mathrm{e}^{Y_s}\mathrm{d}s\right\}}{E_\infty\{\mathrm{e}^{Y_T}\}} \geq g_{\lambda_1}(0), \tag{6.134}$$

for all T for which $E_\infty\{T\} = g_{\lambda_0}(0)$. Or equivalently, that

$$E_\infty\left\{\int_0^T (\mathrm{e}^{Y_s} - 1)\mathrm{d}s - g_{\lambda_1}(0)\mathrm{e}^{Y_T} + g_{\lambda_0}(0)\right\} \geq 0. \tag{6.135}$$

The above inequality can be shown by considering a function similar to the one used in the proof of Theorem 6.12, namely $U(y) = \mathrm{e}^y\left[g_{\lambda_1}(0) - g_{\lambda_1}(y)\right] - \left[g_{\lambda_0}(0) - g_{\lambda_0}(y)\right]$. It is then possible to obtain the representation

$$E_\infty\{\int_0^T (\mathrm{e}^{Y_s} - 1)\mathbf{1}_{\{Y_s<h\}}\mathrm{d}s\} = E_\infty\{U(Y_T)\}. \tag{6.136}$$

Using (6.136), we can rewrite (6.135) as

$$E_\infty\left\{\int_0^T (\mathrm{e}^{Y_s} - 1)\mathbf{1}_{\{Y_s\geq h\}}\mathrm{d}s\right\} + E_\infty\{g_{\lambda_0}(Y_T) - \mathrm{e}^{Y_T}g_{\lambda_1}(Y_T)\} \geq 0. \tag{6.137}$$

Inequality (6.137) clearly holds since the function $g_{\lambda_0}(y) - \mathrm{e}^y g_{\lambda_1}(y) > 0$ for all $0 \leq y < h$, and is equal to 0 for all $y \geq h$, while $(\mathrm{e}^{Y_s} - 1)\mathbf{1}_{\{Y_s\geq h\}} \geq 0$ pathwise. Therefore (6.137) holds generally and it holds with equality when $T = T^c$, as can be easily verified.

For more details on the above derivation please refer to [154].

6.5 Asymptotic results

In this section, we consider some asymptotic results concerning the CUSUM and related tests. The asymptotic regime of interest here is that in which the mean time between false alarms increases without bound.

We begin with a result on the asymptotic optimality of the one-sided CUSUM stopping time. This result was shown by Lorden in his pioneering paper [139], in which he proposed Lorden's criterion as a delay penalty on stopping times T. Although, in Section 6.1 and 6.2 the strictly optimal character of the CUSUM stopping time is demonstrated with respect to this (non-asymptotic) criterion, the reader may also find Lorden's asymptotic approach interesting, as it explores the connections between the SPRT and the CUSUM.

We also refer to the special case of a parametric exponential family and discuss the procedure suggested by Lorden for dealing with problems of many alternatives on opposite sides of the null hypothesis. Finally we turn our attention to the case of two-sided alternatives in the Brownian motion model and mention some asymptotic results for this situation.

6.5.1 Lorden's approach

Let us return to the discrete-time case; i.e. consider the probabilistic setting of Section 6.2. We begin with a preliminary result that will help illuminate the relationship between the CUSUM and the SPRT.

THEOREM 6.16. *Suppose T is a stopping time with respect to $\sigma\{Z_1, Z_2, \ldots, Z_k\}$, $k = 1, 2, \ldots$, such that*

$$P_\infty(T < \infty) \leq a, \quad 0 < a < 1. \tag{6.138}$$

For each $k = 1, 2, \ldots$, let T_k denote the stopping time obtained by applying T to $Z_k, Z_{k+1}, \ldots$, and define

$$T^* = \inf\{T_k + k - 1 | k = 1, 2, \ldots\}. \tag{6.139}$$

Then T^* is also a stopping time, and it satisfies

$$E_\infty\left\{T^*\right\} \geq \frac{1}{a}, \tag{6.140}$$

and for any alternative distribution Q_1,

$$d(T^*) \leq E_1\{T\}, \tag{6.141}$$

where $d(T)$ is as defined in (6.2).

Proof: To show that T^* is a stopping time it suffices to notice that

$$\{T^* \leq n\} = \{T_1 \leq n\} \cup \{T_2 \leq n-1\} \cup \ldots \cup \{T_n \leq 1\} \in \sigma\{Z_1, \ldots, Z_n\}. \tag{6.142}$$

For $m = 1, 2, \ldots$ we have

$$\begin{aligned} E_m\left\{(T^* - (m-1))^+ | Z_1, \ldots, Z_{m-1}\right\} &\leq E_m\{T_m | Z_1, \ldots, Z_{m-1}\} \\ &= E_m\{T_m\} = E_1\{T\}, \end{aligned}$$

where the next-to-last equality follows from the fact that T_m is independent of $Z_1, \ldots, Z_{m-1}$ and the last equality from the fact that T_m has the same distribution under P_m as T_1 does under P_1. Therefore, this inequality holds for every path $Z_1, \ldots, Z_{m-1}$ and every m. Hence

$$\sup_m \operatorname{esssup} E_m\left\{(T^* - (m-1))^+ | Z_1, \ldots, Z_{m-1}\right\} \leq E_1\{T\}. \tag{6.143}$$

Thus the validity of (6.141) is established. To prove (6.140), consider

$$\xi_k = \begin{cases} 1 & \text{if } T_k < \infty \\ 0 & \text{if } T_k = \infty. \end{cases}$$

The strong law of large numbers implies

$$\lim_{n\to\infty} \frac{\sum_{k=1}^{n} \xi_k}{n} = E_\infty\{\xi_1\} \ a.s.P_\infty.$$

But $E_\infty\{\xi_1\} = P_\infty(T_1 < \infty) \le a$. Therefore,

$$\lim_{n\to\infty} \frac{\sum_{k=1}^{n} \xi_k}{n} \le a \ a.s.P_\infty. \tag{6.144}$$

If $E_\infty\{T^*\} = \infty$ then (6.140) holds trivially. We need only consider the case in which $E_\infty\{T^*\} < \infty$. Define the increasing sequence of stopping times $\{T^n\}$ as follows:

If $T^{m-1} = n$, then for each $r = 1, 2, \ldots$, apply T to $Z_{n+r}, Z_{n+r+1}, \ldots$, and let T^m be the first time stopping occurs for some r.

Then $T^1 = T^*$ and $T^1, T^2 - T^1, T^3 - T^2$, are i.i.d. due to the fact that they each depend on a different sequence of the Z_i's which under P_∞ are i.i.d.

From the definition of the T^i's, it follows that

$$\xi_{T^m+1} + \cdots + \xi_{T^{m+1}} \ge 1, \tag{6.145}$$

since $\xi_{T^m+r} = 1$ for some r causing the stop at T^{m+1}. Hence,

$$\xi_1 + \cdots + \xi_{T^m} \ge m, \ \ m = 0, 1, \ldots, \tag{6.146}$$

and therefore,

$$\frac{\xi_1 + \cdots + \xi_{T^m}}{T^m} \ge \frac{m}{T^m}. \tag{6.147}$$

Taking the limit as $m \to \infty$ of both sides of this equation, it follows from the strong law of large numbers and (6.144) that

$$a \ge \frac{1}{E_\infty\{T^*\}}, \tag{6.148}$$

which establishes (6.141). ∎

Suppose the stopping time T is a one-sided SPRT, that is an SPRT with $A = 0$. Then using Proposition 4.10 from Chapter 4, it follows that $\gamma = 0$ and

$$\alpha = P_\infty(L(Z_T) \ge B) = P_\infty(T < \infty) \le \frac{1}{B}.$$

Therefore if $B = 1/a$, we have that

$$P_\infty(T < \infty) \le a. \tag{6.149}$$

Moreover, using Wald's identity (2.83), we have

$$E_1\{T\} E_1\{\log L(Z_1)\} = E_1\left\{\sum_{n=1}^{T} \log L(Z_n)\right\}. \tag{6.150}$$

Since the left log boundary is 0, and Proposition 4.10 implies that $\gamma = 0$, it follows that, as $a \to 0$,

$$E_1\{T\} \sim \frac{|\log a|}{I_1}, \tag{6.151}$$

where

$$I_1 = E_1 \{\log L(Z_1)\}.$$

Using the above theorem it follows further that the CUSUM stopping time T^* can be considered to be the result of applying repeated SPRT's with left boundary 0, and hence satisfies

$$E_\infty \{T^*\} \geq \frac{1}{a} \quad \text{and} \tag{6.152}$$

$$d(T^*) \leq \frac{|\log a|}{I_1}, \quad \text{as } a \to 0. \tag{6.153}$$

The next theorem establishes that (6.152) and (6.153) are asymptotically the best one can do.

THEOREM 6.17. *Suppose $n(\gamma) = \inf_T d(T)$, where the infimum is taken over all stopping times T satisfying $E_\infty \{T\} \geq \gamma$. If $I_1 < \infty$, then $n(\gamma) \sim \log \gamma / I_1$ as $\gamma \to \infty$.*

Proof: From the previous theorem it suffices to show that $n(\gamma)$ is asymptotically no smaller than $\log \gamma / I_1$. That is, it suffices to show that for every $\epsilon \in (0, 1)$, there is a $C(\epsilon) < \infty$, independent of T, such that for all stopping times T,

$$I_1 d(T) \geq (1 - \epsilon) \log E_\infty \{T\} - C(\epsilon). \tag{6.154}$$

Fix ϵ and define an increasing sequence $\{T^n\}$ of stopping times, with T^{i+1} equal to the smallest n such that $n > T^i$ and

$$\frac{\prod_{k=T^i+1}^{n} q_1(Z_k)}{\prod_{k=T^i+1}^{n} q_0(Z_k)} \leq \epsilon.$$

Using arguments similar to the ones used in the proof of Proposition 4.10, we obtain $P_1(T^1 < \infty) \leq \epsilon$. Hence, provided that $P_1(T^{r-1} = k < T) > 0$, we have that $P_{k+1}(T^r < \infty | T^{r-1} = k) \leq \epsilon$. Notice that $\{T^{r-1} = k\} \in \sigma\{Z_1, \ldots, Z_k\}$.

Consider all subsets $\{T^{r-1} = k\}$ that have positive P_∞ probability. Since, under P_{k+1} and P_∞ the marginal distribution of $Z_1, \ldots, Z_k$ is Q_0, the above subsets will also have positive P_{k+1} probability.

Given $\{T^{r-1} = k\}$, T and T^r determine the outcome of the following sequential test based on $Z_{k+1}, \ldots$: Stop at $\min\{T, T^r\}$, and

(1) decide in favor of P_{k+1} if $T \leq T^r$; or
(2) decide in favor of P_∞ if $T > T^r$.

The number of observations taken is $\min\{T, T^r\} - k$, and

$$E_{k+1} \left\{ \min\{T, T^r\} - k \middle| T^{r-1} = k \right\} \leq d(T).$$

We can now apply Proposition 4.11 with

(1) $\alpha = P_\infty(T \leq T^r | T^{r-1} = k)$; and
(2) $\gamma = P_{k+1}(T \geq T^r | T^{r-1} = k)$,

to obtain

$$\begin{aligned} I_1 d(T) &\geq P_{k+1}(T \leq T^r | T^{r-1} = k) |\log P_\infty(T \leq T^r | T^{r-1} = k)| - \log 2 \\ &\geq (1-\epsilon) |\log P_\infty(T \leq T^r | T^{r-1} = k)| - \log 2, \end{aligned} \tag{6.155}$$

where we have used the fact that

$$P_{k+1}(T \leq T^r | T^{r-1} = k) \geq P_{k+1}(T^r = \infty | T^{r-1} = k).$$

Let

$$R = \inf\{r | T^r > T\}.$$

We have that

$$P_\infty\left(T \leq T^r | T^{r-1} < T\right) = P_\infty(R < r+1 | R \geq r). \tag{6.156}$$

Therefore, $P_\infty(R \geq r) > 0$ and (6.155) imply

$$I_1 d(T) \geq (1-\epsilon) |\log P_\infty(R \leq r | R \geq r)| - \log 2.$$

Hence

$$\begin{aligned} P_\infty(R < r+1 | R \geq r) \geq H &\Leftrightarrow P_\infty(R \geq r+1) \leq (1-H)^r \\ &\Leftrightarrow E_\infty\{R\} \leq \frac{1}{H}, \end{aligned}$$

from which (6.155) becomes

$$I_1 d(T) \geq (1-\epsilon) \log E_\infty\{R\} - \log 2. \tag{6.157}$$

Applying Wald's identity (2.83), we obtain

$$E_\infty\left\{T^R\right\} = E_\infty\{R\} E_\infty\left\{T^1\right\}. \tag{6.158}$$

Therefore

$$\log E_\infty\{T\} \leq \log E_\infty\left\{T^R\right\} = \log E_\infty\{R\} + \log E_\infty\left\{T^1\right\}. \tag{6.159}$$

Combining (6.158) with (6.159), and letting $C(\epsilon) = \log E_\infty\left\{T^1\right\}$, the theorem follows. ■

The above proof of asymptotic optimality of the CUSUM stopping time, which is due to Lorden [139], provides the basis of many asymptotic results that followed it including ones that treated the case of an unknown post-change parameter. The most commonly used of these is the generalized CUSUM stopping time that will be discussed in Section 7.3. Its asymptotic optimality is also shown in [139]. Another asymptotically optimal approach that is also based on the CUSUM and is less computationally intensive was further introduced and analyzed in [204]. This approach is based on the assumption that there are countably many alternatives that describe the post-change distribution of the observations. The sequential decision rule suggested in [204] consists of a stopping time of the CUSUM type and a random variable that takes one of a countable number of values, once the stopping time declares an alarm, each corresponding to one of the possible post-change alternatives.

It is also worth mentioning a connection between the Shiryaev–Bayes optimal solution of Chapter 5 and a min–max asymptotic result of Pollak [173], in which a min–max solution is constructed as a limit of Bayesian stopping times solving the Shiryaev problem.

More specifically, denote by $B(\pi, \rho, c)$ the (Bayesian) solution to the Shiryaev problem of Section 5.2. Define

(1) $R_{n+1} = L(Z_{n+1})(1 + R_n), \quad R_0 = 0;$ (6.160)

(2) $R_n^{\rho} = \frac{1}{1-\rho} \sum_{k=1}^{n} \prod_{i=k}^{n} L(Z_i);$ and (6.161)

(3) $\Delta = \inf\{x | P_\infty(L(Z_i) \leq x) > 0\} < 1.$

REMARK 6.18. The threshold stopping time defined in terms of (6.160) is known as the *Shiryaev–Roberts* stopping time. That is, the Shiryaev–Roberts stopping time is defined as

$$T_h^{\text{SR}} = \inf\{n | R_n \geq h\}. \tag{6.162}$$

A very interesting stopping time known as the *Shiryaev–Roberts–Pollak* stopping time is an extension of (6.160) in that it is the same as the stopping time (6.160) but with randomization involved in the choice of R_0.

We have the following preliminary result due to Pollak [173].

THEOREM 6.19. *Suppose that the P_∞ distribution of $L(Z_1)$ is atomless. Then all of the following statements are true.*

(1) *For any $\Delta < h < \infty$ there exists a constant $0 < c^* < \infty$, and a sequence of constants $\{\rho_i, c_i\}$ with $\lim_{i\to\infty} \rho_i = 0$ and $\lim_{i\to\infty} c_i = c^*$, such that the stopping time*

$$T_h = \inf\{n \geq 1 | R_n \geq h\}$$

is a limit as $i \to \infty$ of Bayesian stopping times $B(0, \rho = \rho_i, c = c_i)$.

(2) *For any set of Bayesian solutions $B(\pi, \rho, c)$, with $\pi = 0$, $\rho \to 0$, $c \to c^*$,*

$$\limsup_{\rho\to 0,\ c\to c^*} \frac{1 - \{\text{Cost using the Bayes stopping time for problem } B(0,\rho,c)\}}{1 - \{\text{Cost using } T_h \text{ for problem } B(0,\rho,c)\}} = 1$$

(3) *For any $1 \leq \gamma < \infty$, there exists a unique $\Delta \leq A < \infty$ such that $\gamma = E_\infty\{T_h\}$.*

Consider the following alternative min–max criterion as a delay penalty for the stopping time $T \in \mathcal{T}$, suggested by Shiryaev–Roberts–Pollak:

$$J_{\text{SRP}}(T) = \sup_{1\leq t<\infty} E_t\{T - t | T \geq t\}, \tag{6.163}$$

which gives rise to the min-max optimization problem

$$\inf_{\mathcal{T}} J_{\mathrm{SRP}}(T) \text{ subject to} \quad E_\infty\{T\} \geq \gamma . \tag{6.164}$$

It turns out that the stopping time T_h with R_0 having a specific distribution ψ_h (rather than being equal to 0 as previously defined), is asymptotically optimal as $\gamma \to \infty$ in the sense described by the following theorem from [173].

THEOREM 6.20. *Denote by T_{h,ψ_h} the above-described stopping time, and assume that $\frac{\Delta}{1-\Delta} < h < \infty$ as well as the existence of a distribution ψ_h such that $E_\infty\{T_{h,\psi_h}\} = \gamma > 1$. Then for all $T \in \mathcal{T}$ satisfying $E_\infty\{T\} \geq \gamma$, we have*

$$\sup_{1 \leq t < \infty} E_t\left\{T_{h,\psi_h} - t | T_{h,\psi_h} \geq t\right\} - \sup_{1 \leq t < \infty} E_t\{T - t | T \geq t\} \leq o(1) \text{ as } \gamma \to \infty, \tag{6.165}$$

where ψ_h satisfies a functional equation given on page 216 of [173].

For a proof of this result please refer to [173].

An interesting connection between the Shiryaev–Roberts–Pollak stopping time, the CUSUM stopping time and the Shiryaev posterior density stopping time is well-explained in [16].

To elaborate on this connection, let s and u be two times and suppose P_∞ and P_u denote the restrictions of the probability measure P of Section 5.2 to the σ-algebra $\mathcal{F}_s$ in the cases in which $\{t = \infty\}$ and $\{t = u\}$, respectively, and let $\left.\frac{dP_t}{dP_\infty}\right|_{\mathcal{F}_s}$ denote the Radon–Nikodym derivative between the two measures. We follow [16].

We begin with Shiryaev's stopping time, which is equivalent to a stopping time defined in terms of the odds ratio process (5.82), as seen in Section 5.3.2.

The odds ratio process can be written as

$$\begin{aligned} \Phi_s &= \frac{1}{1 - P(t = s)} \sum_{u=1}^{s} \left.\frac{dP_t}{dP_\infty}\right|_{\mathcal{F}_s} P(t = u) \\ &= \frac{1}{1 - P(t = s)} \sum_{k=1}^{s} \prod_{i=1}^{k} L(Z_i) P(t = i), \ s = 1, 2, \ldots \end{aligned} \tag{6.166}$$

The Shiryaev–Roberts process of Remark 6.18 (6.160) is then seen to be equal to

$$R_s = \sum_{t=1}^{s} \left.\frac{dP_t}{dP_\infty}\right|_{\mathcal{F}_s} = \lim_{\rho \to 0} \frac{\Phi_s}{\rho}, \tag{6.167}$$

where ρ is the parameter in the distribution of the prior in (5.23) and (5.24) in the special case in which the initial probability π of a change at time 0 equals 0. We can readily see that $\{R_s; s \geq 0\}$ is an $\mathcal{F}_s$ submartingale with

$$E_\infty \left\{ \sum_{t=1}^{s} \frac{\mathrm{d}P_t}{\mathrm{d}P_\infty} \bigg|_{\mathcal{F}_s} \right\} = s. \tag{6.168}$$

Thus, we can apply inequality (2.58) of Chapter 2 to obtain

$$P_\infty(T_h^{\mathrm{SR}} < s) = P_\infty\left(\max_{0 \le u \le s} R_u \ge h \right) \le \frac{E_\infty\{T_h^{\mathrm{SR}}\}}{h}.$$

The CUSUM process is

$$C_s = \max_{0 \le t \le s} \frac{\mathrm{d}P_t}{\mathrm{d}P_\infty} \bigg|_{\mathcal{F}_s}. \tag{6.169}$$

REMARK 6.21. The above analysis can also be carried out in the continuous-time case in which all sums become integrals and the Shiryaev–Roberts process of (6.167) is seen to be equal to

$$\lim_{\alpha \to 0} \frac{\Phi_s}{\alpha}, \tag{6.170}$$

where α is the exponential parameter in the distribution of the prior that appears in equations (5.57) and (5.79), in the special case in which $\pi = 0$.

REMARK 6.22. The CUSUM stopping time of Section 6.4.1 is the first passage time of the logarithm of (6.169). Since the logarithm is a strictly increasing function the first passage time of (6.169) is exactly the same as that for its logarithm, with appropriately modified thresholds.

Notice that the odds ratio process can be thought of as an average of the likelihood ratio process over all possible values of the unknown change point t while the CUSUM process is a maximum of the likelihood ratio process over all possible change points t. On the other hand, the odds ratio process is a weighted average of the likelihood ratio process weighted by the probability density of the change point.

The threshold stopping time based on the odds ratio process (or, equivalently, the posterior probability of a change) is the optimal solution to the Shiryaev problem presented in Section 5.2. The Shiryaev–Roberts stopping time and the Shiryaev–Roberts–Pollak stopping time are asymptotically optimal in the sense of Pollak (6.163), as summarized in Theorems 6.19 and 6.20, as the mean time between false alarms increases without bound. Finally, the CUSUM process is optimal in the sense of Lorden (6.89) subject to a constraint on the mean time between false alarms.

Recently, an interesting strict optimality result of the Shiryaev–Roberts stopping time was proven in [177]. In particular, the Shiryaev–Roberts stopping time of (6.167) minimizes an average delay to detection criterion, namely

$$\sum_{k=1}^{\infty} E_k\{(T-k)^+\},$$

under the usual lower bound constraint on the mean time between false alarm, namely $E_\infty\{T\} \ge \gamma$. The ratio of the average delay to detection to the mean time between false alarms γ can be interpreted as the average delay to detection of a change that occurs in the far horizon after repeated false alarms and applications of the Shiryaev–Roberts stopping rule. Thus the optimality of the Shiryaev–Roberts rule in this sense is interesting in the context of detecting a real change as quickly as possible after its occurrence, even at the price of raising many false alarms. It is interesting that the average delay to detection was first introduced in continuous time in [192] where it was referred to as the stationary average delay time. The equivalent of this optimality result in continuous time was proven in [82]. In what follows we will draw a connection between the traditional criteria of Lorden, Pollak and Shiryaev. We follow [155].

An interesting connection between the Pollak criterion (6.164), Lorden's criterion (7.75) and Shiryaev's Bayesian criterion of Chapter 5 (i.e. (5.3)) is developed in [155], where the appropriateness of each of these criteria is discussed in detail.

Here we outline the connections among these criteria in the discussion that follows. Let us begin with the Shiryaev criterion, which minimizes the sum of the probability of false alarm and the expected delay. That is, it penalizes the performance of stopping times T by $P(T < t) + cE\{(T - t)^+\}$. The optimal stopping time thus minimizes this quantity. This is equivalent to the minimization of the quantity $E_t\{T - t|T > t\}$, where $E_t\{\cdot\}$ denotes expectation under P_t, the restriction of the probability measure P of Section 5.2 to the filtration $\{\mathcal{F}_k; k = 1, 2, \ldots\}$ when the change point is t, subject to a constraint on the probability of false alarms:

$$P(T < t) = \sum_{k=0}^{\infty} P(t = k)P(T < k) \leq \alpha,\ \alpha \in (0, 1). \tag{6.171}$$

Note that we have

$$E_t\{T - t|T > t\} = \frac{\sum_{k=0}^{\infty} P(t = k)P(T \geq k)E_k\{T - k|T > k\}}{P(T > t)}. \tag{6.172}$$

In Shiryaev's problem it is assumed that the change point t is a random variable. Pollak's criterion (6.164), on the other hand, assumes no random mechanism behind the change point and can be seen as an equivalent to Shiryaev's criterion for the case in which t is an unknown constant. This naturally suggests the quantity $\sup_t E_t\{T - t|T > t\}$, where the worst conditional expected delay is considered over all possible values of t. Lorden's criterion penalizes stopping times T according to $J(T) = \sup_t \operatorname{esssup} E_t\{(T - t)^+|\mathcal{F}_t\}$. The above discussion already suggests that $J(T) > J_{\mathrm{SRP}}(T) > J_{\mathrm{S}}(T)$, where $J(T)$ is Lorden's delay penalty as it appears in (6.89), $J_{\mathrm{SRP}}(T)$ is the Pollak delay penalty, which appears in (6.163), and $J_{\mathrm{S}}(T)$ is the Shiryaev delay penalty, which appears in (5.42) (see for comparison [155]). Although Lorden's criterion seems overly pessimistic at first glance, it is the natural choice in the general case in which the change point mechanism depends on the history of the observations.

To give a motivation of such a change point mechanism consider a simple example of a linear state-space model with independent noise.

Example 6.2: Let the observations process $\{Z_k\}$ be generated as follows

$$Z_k = X_k + W_k,$$

and

$$X_k = \alpha X_{k-1} + V_k,$$

where $\{W_k\}$, $\{V_k\}$ are independent i.i.d. sequences of mean 0 random variables and $|\alpha| < 1$. The process $\{X_k\}$ is a state process underlying the observations. A natural disorder in such systems arises the first time that the state variable enters an undesirable set A (see [154]). That is, $t = \inf\{k | X_k \in A\}$.

This naive example illustrates that in most practical applications (see for example [208]) in which the observation processes involve general dependencies, it may be useful to think of the change point as a dynamic entity. This suggests the idea that nature at each instant of time k consults the information $\mathcal{F}_k$ and decides whether to introduce a disorder or not based on some type of randomization. That is, this suggests the notion of a randomization probability for the disorder time which can be summarized by $s_k = P(t = k | \mathcal{F}_k)$.

REMARK 6.23. Although in the usual setting (e.g. Section 5.2), the change point t is considered to be the first point in time at which an observation under the alternative regime appears, in the current setting t is the last point in time at which an observation under the initial regime is produced. This interpretation is appropriate here as there is a decision by nature on whether to change the distribution of the observations, and should that decision be taken at a time instant k, it will affect the distribution of the subsequent observation.

In [155] it is seen that the process $\{s_k\}$ defined above can be decomposed into the product

$$s_u = \omega_u p_u, \quad u = 0, 1, \ldots, \tag{6.173}$$

where the process $\{\omega_k\}$ is an $\mathcal{F}_0$-measurable process representing the aggregate probability of the event $\{t = k\}$; i.e. $\{\omega_k\}$ is independent of the observation process $\{Z_k; k = 1, 2, \ldots\}$. Moreover, p_k is defined on the event $\{\omega_k > 0\}$ and describes how the probability of the event $\{t = k\}$ is distributed amongst the events of $\mathcal{F}_k$.

In [155] it is shown that

$$\omega_k = E_\infty\{s_k\}, \tag{6.174}$$

from which it follows that

$$\sum_{k=0}^{\infty} \omega_k = P_k(k < \infty) = 1. \tag{6.175}$$

Notice that (6.174) implies that

$$E_\infty\{p_k\} = 1. \tag{6.176}$$

In many models that arise in practice (see Example 6.2) the notion of a change-point mechanism devised by nature and depending on the filtration $\{\mathcal{F}_k\}$ is natural. In turn, if the mechanism of the change point summarized in $\{\omega_k\}$ and $\{p_k\}$ is not known exactly, the quantity by which the statistician should penalize the performance of a stopping time T should include $\sup_{\{\omega_k\},\{p_k\}} E_t\{T-t|T>t\}$. That is, from the point of view of the statistician, if there is uncertainty as to nature's choice of $\{\omega_k\}$ and $\{p_k\}$, then the statistician should assume the worst choice of a change point on the part of nature, which suggests the use of $\sup_{\{\omega_k\},\{p_k\}} E_t\{T-t|T>t\}$ as a delay penalty for T.

Shiryaev's delay penalty for a stopping time corresponds to the case in which ω_k is given by (5.24) and $p_k = 1$ for all $k = 0, 1 \ldots$ Pollak's delay penalty on the other hand corresponds to the case $p_k = 1$ for all $k = 0, 1, \ldots$ and $\{\omega_k\}$ is arbitrary. That is,

$$\sup_{\{\omega_k\}} E_t\{T-t|T>t\} = \sup_{0\le t<\infty} E_t\{T-t|T>t\}.$$

(The details of this analysis can be found in [154] and [155].) Finally, Lorden's delay penalty correspond to the case in which $\{\omega_k\}$ and $\{p_k\}$ are both unknown to the statistician, in which context we also have ([154,155])

$$\sup_{\{\omega_k\},\{p_k\}} E_t\{T-t|T>t\} = \sup_t \operatorname{esssup} E_t\{(T-t)^+|\mathcal{F}_t\},$$

the right-hand side of which is immediately recognized as Lorden's delay penalty.

6.5.2 Brownian motion with two-sided alternatives

We now turn to the problem of detecting a change in the drift of Brownian motion but for the two-sided case in which the drift can either increase or decrease after the change.

This problem is considerably more difficult than the problem of one-sided alternatives. This is reminiscent of the fact that even in the case of fixed-sample testing between two hypotheses, a uniformly most powerful test usually exists only in the case of one-sided alternatives [178]. This, of course, does not exclude the possibility of the existence of an optimal solution in this case, but does indicate that this problem, while easy to state, is nevertheless considerably more difficult to solve. In what follows we will provide only an asymptotically optimal solution to this problem in the limit as the mean time between false alarms increases without a bound.

We begin by considering the observation process $\{Z_s; s \ge 0\}$ with the following dynamics:

$$dZ_s = \begin{cases} dW_s & , s \le t \\ \\ \mu_1 ds + dW_s & \\ \text{or} & , s \ge t \\ -\mu_2 ds + dW_s & \end{cases},$$

where $\{W_s; s \ge 0\}$ is a standard Brownian motion, and the μ_i are assumed to be known and positive. The probabilistic setting of this problem is identical to the one in Section 6.4.3, with P_t^i corresponding to the probability measure generated on the space of continuous functions Ω, when the change occurs at time t and the drift changes to μ_1 ($i = 1$) or $-\mu_2$ ($i = 2$). In order to incorporate the different possibilities for μ_i we extend Lorden's delay penalty to the following:

$$J_{pm}(T) = \max_i \sup_t \operatorname{esssup} E_t^i \left\{(T - t)^+ | \mathcal{F}_t\right\}, \tag{6.177}$$

which results in a corresponding optimization problem of the form:

$$\begin{aligned} &\inf_T J_{pm}(T) \quad \text{subject to} \\ &E_\infty\{T\} \geq \gamma. \end{aligned} \tag{6.178}$$

Notice that $J_{pm}(T)$ can be re-expressed as

$$J_{pm}(T) = \max\{J_1(T), J_2(T)\}, \tag{6.179}$$

where $J_i(T) = \sup_t \operatorname{esssup} E_t^i \left\{(T - t)^+ | \mathcal{F}_t\right\}$.

An interesting special case, namely that in which we know the amplitude of the post-change drift but not its sign, falls within this setting. What has traditionally been done in the literature, dating as far back as 1959 [15], is to stop at the minimum of the CUSUM stopping times S_{h_1} and S_{h_2} each tuned to detect the respective changes μ_1 and $-\mu_2$.

We begin our analysis of this problem with a simple argument that demonstrates that the optimal stopping time for problem (6.127) must satisfy $J_1(T) = J_2(T)$ and thus must be an equalizer rule. This argument is summarized in the following two remarks.

REMARK 6.24. Consider a stopping time T_{U} such that $J_1(T_{\mathrm{U}}) > J_2(T_{\mathrm{U}})$ and let T_{V} be a new stopping time that declares an alarm at exactly the same instant as T_{U} if and only if at each instant u it receives the observations $\{-Z_s,\ s \leq u\}$. Then $J_2(T_{\mathrm{V}}) > J_1(T_{\mathrm{V}})$ with $J_1(T_{\mathrm{U}}) = J_2(T_{\mathrm{V}})$ and $J_2(T_{\mathrm{U}}) = J_1(T_{\mathrm{V}})$. Now construct $T = \frac{1}{2}T_{\mathrm{U}} + \frac{1}{2}T_{\mathrm{V}}$. It is clear that for this stopping time $J_{\mathrm{L}}(T) < J_{\mathrm{L}}(T_{\mathrm{U}}) = J_{\mathrm{L}}(T_{\mathrm{V}})$ and $E_\infty\{T\} = E_\infty\{T_{\mathrm{U}}\} = E_\infty\{T_{\mathrm{V}}\}$, while $J_1(T) = J_2(T)$. That is, we have found another stopping time T that, for the same mean time between false alarms, has achieved a lesser detection delay.

REMARK 6.25. It is important to point out that T of Remark 6.24 is not technically a stopping time with respect to $\{\mathcal{F}_u\}$, since randomization is involved in its construction. In other words, we flip a fair coin and if it comes up $Heads$, then we stop according to the stopping time T_{U}, while if it comes up $Tails$, we stop according to the stopping time T_{V}. Hence T is a stopping time with respect to the enlarged filtration $\{\overline{\mathcal{F}}_u\}$ that consists of everything in $\{\mathcal{F}_u\}$ plus the two possible outcomes of the random experiment of tossing the fair coin, i.e. $\overline{\mathcal{F}_u} = \mathcal{F}_u \cup \{\{Heads\}, \{Tails\}\}$.

Although the optimal solution to problem (6.178) is unknown, it has been shown in the literature (see [100,200]) that the so-called classical 2-CUSUM stopping time and its modified drift counterpart are asymptotically optimal as $\gamma \to \infty$. In what follows, we will introduce the classical 2-CUSUM stopping time as well as the modified drift 2-CUSUM stopping time and discuss the problem of finding the best in the class of such stopping times. We will also discuss their asymptotic optimality properties.

We begin by introducing the *normalized* CUSUM stopping times. Define

(1) $U_s^+ = \log \frac{\mathrm{d}P_0^1}{\mathrm{d}P_\infty}\Big|_{\mathcal{F}_s} = \mu_1 Z_s - \frac{1}{2}\mu_1^2 s$;
(2) $U_s^- = \log \frac{\mathrm{d}P_0^2}{\mathrm{d}P_\infty}\Big|_{\mathcal{F}_s} = -\mu_2 Z_s - \frac{1}{2}\mu_2^2 s$;
(3) $M_s^+ = \inf_{u\le s} \frac{U_s^+}{\mu_1}$; and
(4) $M_s^- = \inf_{u\le s} \frac{U_u^-}{\mu_2}$.

The *normalized* CUSUM processes are defined respectively as follows:

(1) $Y_s^+ = \frac{U_s^+}{\mu_1} - M_s^+$; and
(2) $Y_s^- = \frac{U_s^-}{\mu_2} - M_s^-$.

We now define the *normalized* classical 2-CUSUM stopping times. The 2-CUSUM stopping time with threshold parameters $h_1 > 0$ and $h_2 > 0$ is defined as follows:

$$T^c(h_1, h_2) = T_1^c(h_1) \wedge T_2^c(h_2),$$

where

(1) $T_1^c(h_1) = \inf\{s > 0 | Y_s^+ > h_1\}$; and
(2) $T_2^c(h_2) = \inf\{s > 0 | Y_s^- > h_2\}$.

The above definition suggests the following classification of 2-CUSUM stopping times according to their threshold parameters.

(1) $\mathcal{G} = \{T^c(h_1, h_2); h_1 = h_2\}$, the harmonic mean 2-CUSUM stopping times; and
(2) $\mathcal{C}_i = \{T^c(h_i, h_j) \mid h_i > h_j > 0,\ i \ne j\}$, the non-harmonic mean 2-CUSUM stopping times.

Similarly, the modified drift 2-CUSUM stopping times are defined in term of the stopping times $T_1^c(\rho_1, h_1)$ and $T_2^c(\rho_2, h_2)$, for $\rho_1 > 0$ and $\rho_2 > 0$, which are given below:

(1) $U_s^+(\rho_1) = Z_s - \frac{1}{2}\rho_1 s$;
(2) $U_s^-(\rho_2) = -\xi_s - \frac{1}{2}\rho_2 s$;
(3) $M_s^+(\rho_1) = \inf_{u\le s} U_u^+(\rho_1)$;
(4) $M_s^-(\rho_2) = \inf_{u\le s} U_u^-(\rho_2)$;
(5) $Y_s^+(\rho_1) = U_s^+(\rho_1) - M_s^+(\rho_1)$;
(6) $Y_s^-(\rho_2) = U_s^-(\rho_2) - M_s^-(\rho_2)$;
(7) $T_1^c(\rho_1, h_1) = \inf\{s > 0; Y_s^+(\rho_1) \ge h_1\}$; and
(8) $T_2^c(\rho_2, h_2) = \inf\{s > 0; Y_s^-(\rho_2) \ge h_2\}$.

The modified drift 2-CUSUM stopping times are then defined as follows for drift parameters ρ_1 and ρ_2 and threshold parameters h_1 and h_2:

$$T^c(\rho_1, \rho_2, h_1, h_2) = T_1^c(\rho_1, h_1) \wedge T_2^c(\rho_2, h_2).$$

Similarly we distinguish the classes

(1) $\mathcal{G}_M = \{T^c(\rho_1, \rho_2, h_1, h_2); h_1 = h_2\}$, the modified drift harmonic mean 2-CUSUM stopping times; and
(2) $\mathcal{C}_i^M = \{T^c(\rho_1, \rho_2, h_i, h_j) \mid h_i > h_j > 0,\ i \neq j\}$, the modified drift non-harmonic mean 2-CUSUM stopping times.

It is easy to see that for any 2-CUSUM stopping time T^c (modified or not)

$$\begin{aligned} J_{pm}(T^c) &= \max_i \sup_t \operatorname{esssup} E_t^i \left\{(T^c - t)^+ | \mathcal{F}_t\right\} \\ &= \max\left\{E_0^1\left\{T^c\right\}, E_0^2\left\{T^c\right\}\right\}. \end{aligned} \tag{6.180}$$

Therefore, the problem of finding the best 2-CUSUM stopping time reduces to that of finding a 2-CUSUM stopping time that satisfies

$$E_0^1\{T^c\} = E_0^2\{T^c\}. \tag{6.181}$$

Equation (6.181) suggests that in order to find the best 2-CUSUM stopping time we need an expression for its first moment. This problem is treated in [102], where an expression is derived for the mean of a general 2-CUSUM stopping time. This expression however, is quite complex in the case $h_1 \neq h_2$ and will not be repeated here. However, the first moment of a 2-CUSUM harmonic mean stopping time takes the following simple form:

$$E\left\{T^c\right\} = \frac{E\left\{T_1^c\right\} E\left\{T_2^c\right\}}{E\left\{T_1^c\right\} + E\left\{T_2^c\right\}}, \tag{6.182}$$

which holds both for modified drift 2-CUSUM stopping times and classical 2-CUSUM stopping times.

Using Itô's rule much along the lines of Proposition 6.8, it is easy to derive the following formulae for the expected delay of T_1^c and T_2^c (see [101,198]):

$$E_\infty\{T_i^c(h_i)\} = 2 f_{h_i}(\mu_i), \quad i = 1, 2, \tag{6.183}$$

$$E_0^i\{T_i^c(h_i)\} = 2 f_{h_i}(-\mu_i), \quad i = 1, 2, \tag{6.184}$$

and

$$E_0^i\{T_j^c(h_j)\} = 2 f_{h_j}(\mu_j + 2\mu_i), \quad i \neq j,\ i, j \in \{1, 2\}. \tag{6.185}$$

Moreover, according to [100], we also have

$$E_\infty\{T^c(\rho_i, h)\} = 2 f_h(\rho_i), \tag{6.186}$$

$$E_0^i\{T^c(\rho_i, h)\} = 2 f_h(\rho_i - 2\mu_i), \tag{6.187}$$

and

$$E_0^i\{T^c(\rho_j, h)\} = 2 f_h(\rho_j + 2\mu_i), \tag{6.188}$$

where $f_h(y) = (e^{hy} - hy - 1)/y^2$. It is thus easily seen that in the case of a symmetric change $\mu_1 = \mu_2$, the classical harmonic mean 2-CUSUM stopping time is the best amongst all classical 2-CUSUM stopping times. Its unique optimality in the class of classical 2-CUSUM stopping times is asserted in [102]. Of course, it is also easily seen that any modified drift harmonic mean 2-CUSUM stopping time with $\rho_1 = \rho_2$ is the best in the class of modified drift 2-CUSUM stopping times, which is a more general result.

In the case of classical 2-CUSUM stopping times, we also have a strong asymptotic result, the details of which can be found in [101].

THEOREM 6.26. *The difference in the detection delay between the optimal classical 2-CUSUM stopping time $T^c \in \mathcal{G}$ and the globally optimal stopping time is bounded above by a quantity that tends to the constant* $(2 \log 2)/\mu^2$, *as the false alarm constraint γ increases without bound.*

In the case of a non-symmetric change it readily follows by the symmetry in distribution of Brownian motion that we need only consider the case $\mu_1 < \mu_2$. In [102] it is seen that the best in the class of classical 2-CUSUM stopping times, that is the classical 2-CUSUM stopping time for which (6.182) holds, is unique and satisfies $\nu_1 < \nu_2$. Moreover, in [100], it is seen that the best 2-CUSUM stopping time in the class of modified drift 2-CUSUM harmonic mean rules is given for the choice of parameters $\rho_2 - \rho_1 = 2(\mu_2 - \mu_1)$ as this is the choice for which (6.182) holds. In [101] it is also shown that the best among all modified drift 2-CUSUM stopping times satisfies $\rho_1 = \mu_1$, which implies $\rho_2 = 2\mu_2 - \mu_1$. It is for this choice of a 2-CUSUM stopping time that a very strong asymptotic result holds. This is summarized in the next theorem, a proof of which appears in [101].

THEOREM 6.27. *The difference in the detection delay between the optimal modified 2-CUSUM equalizer stopping time with $\rho_1 = \mu_1$ and $\rho_2 = 2\mu_2 - \mu_1$, and the globally optimal stopping time is bounded above by a quantity that tends to* 0, *as the false alarm constraint γ increases without bound.*

6.6 Comments on the false-alarm constraint

In this section we comment on the false-alarm penalty used in the min–max formulations discussed and solved in this chapter. In particular, these formulations involve a minimization of a penalty for detection delay subject to a lower bound constraint on the mean time between false alarms, as opposed to a constraint on the probability of false alarms. In particular, the CUSUM stopping time is optimal in the sense of Lorden as discussed in Sections 6.2 and 6.4, and the Shiryaev–Roberts–Pollak stopping time enjoys asymptotic optimality as the mean time between false alarms tends to ∞. It is only the Shiryaev (Bayesian) solution that minimizes the detection delay (6.172) with respect to a constraint in the probability of false alarms (6.171).

The equivalent of the constraint (6.171) in the min–max setting would be to require that $P_k(T < k) \leq \alpha \in (0, 1)$, for all $k = 1, \ldots$. This is further equivalent to requiring that $\sup_k P_k(T < k) \leq \alpha$, which can be seen as $P_\infty(T < \infty) \leq \alpha$ (see for example

[207]). Unfortunately, this constraint is very restrictive for both the CUSUM and the Shiryaev–Roberts stopping times. To see this, let us denote by T_h^c the CUSUM stopping time and by T_h^{SR} the Shiryaev–Roberts stopping time. Since $E_\infty\{T_h^c\} < \infty$ and $E_\infty\{T_h^{\text{SR}}\} < \infty$, we have that $P_\infty(T_h^c < \infty) = P_\infty(T_h^{\text{SR}} < \infty) = 1$ (compare Sections 6.3 and 6.4 and equation (6.168)). So, a direct false-alarm constraint is meaningless in these formulations.

However, there are ways in which we can treat the problem

$$\begin{aligned} &\inf_T J_{\text{SRP}}(T) \quad \text{subject to} \\ &\quad P_\infty(T < \infty) \le \alpha, \end{aligned} \tag{6.189}$$

where J_{SRP} is the Shiryaev–Roberts–Pollak delay penalty.

For example, to guarantee the condition $P_\infty(T < \infty) \le \alpha$, one can consider a modification of the usual CUSUM and Shiryaev–Roberts stopping times with strictly increasing curved boundaries as opposed to the usual fixed threshold. In [43] it is seen that, although with such modifications the desired probability of false alarm constraint is satisfied, the conditional average delay time $E_k(T - k|T > k)$ increases as $O(\log k)$ as $k \to \infty$. Therefore, under this constraint neither minimax nor uniform solutions are feasible asymptotically as $\alpha \to 0$ [207]. An alternative solution to this problem is to consider a constraint on the local conditional probability of false alarms. That is, we require $\sup_k P_\infty(k \le T \le k + d + 1|T \ge k) \le \alpha$ for some window d, which may go to ∞ at a certain rate. Then the CUSUM and Shiryaev–Roberts stopping times have uniformly asymptotic optimality properties. That is, they minimize J_{SRP} (see for example [128,129,206]). A further approach, based on dynamic sampling, has been proposed by Assaf and others ([9–11,223]).

6.7 Discussion

In this chapter we have examined the problem of quickest detection in the case in which the change point is modeled as an unknown constant, rather than as a random variable. We have introduced various performance criteria for stopping times meant to detect such a change in distribution. Particular attention has been focused on the performance criterion of Lorden with respect to which we have proved optimality of the CUSUM stopping time both in the case of a discrete-time model and in the case of a Brownian motion model. We also have proved optimality of the CUSUM stopping time tuned to detect the smallest (in absolute value) drift in the case of a one-sided change of unknown magnitude in the drift with respect to an extended Lorden criterion. It is with respect to the same criterion that we also presented two asymptotic optimality results for two carefully chosen modified drift 2-CUSUM stopping times in the case of two-sided alternatives. The CUSUM has also been seen to be optimal with respect to a generalized Lorden criterion motivated by the Kullback–Leibler divergence in the case of more general continuous processes, namely Itô processes.

In this chapter we have also described the celebrated result of Lorden on the asymptotic optimality of the one-sided CUSUM stopping time. Moreover, we have examined the connection between the Bayesian solution of Chapter 5 and a min–max solution to the problem of quickest detection with respect to a delay penalty known as the Shiryaev–Roberts–Pollak criterion. Finally we concluded by mentioning briefly several non-Bayesian approaches based on a penalty on false-alarm probability, rather than mean time between false alarms.

7 Additional topics

7.1 Introduction

In Chapter 4, we considered the problem of optimally deciding between two hypotheses on the state of the environment given a constant cost of sampling for each additional observation. Within each of these two models the data are homogeneous; that is, the data obey only one of the two alternative statistical models during the entire period of observation. In Chapters 5 and 6 on the other hand, we considered the problem of optimally detecting an abrupt change in the mechanism generating the data from one regime to another. In this chapter, we examine several generalizations and modifications of these basic sequential decision-making problems, in which various of the assumptions are relaxed so as to provide more practical solutions.

We will first address the problem of decentralized sequential detection. In this setting information becomes available sequentially at distinct sensors, rather than at a common location, as in the models considered in the preceding chapters. In general, these sensors communicate a summary message to a central fusion center (which may or may not also be receiving information on its own), which must ultimately decide about the state of the environment. Various sensor configurations are possible for decentralized detection. See for example [215]. One of the main advantages of the decentralized setting over its centralized counterpart is the reduced communication requirements of such a configuration. This comes as a result of the fact that each sensor transmits only a quantized version of its message to the fusion center, which requires transmission of only a few bits of information, whereas transmission of the original observations in total require considerably greater communications resources. Of course, the main disadvantage is that in the decentralized setting the fusion center must base its decision on less information, which may result in reduced performance.

We will then revisit the problem of quickest detection in the situation in which there is uncertainty in the post-change distribution. However, unlike the cases treated in Chapters 5 and 6, here we consider more general types of uncertainty, leading to procedures for robust and adaptive quickest detection.

We finally examine the problems of sequential detection and change-point detection in the case of discrete-time processes with dependent observations. We first identify the special dependence structures for which we can extend the optimality for the CUSUM stopping time in the change-point detection problem. We then introduce a special class of discrete-time processes that exhibit local asymptotic normality and present techniques for the solution of the problems of sequential and quickest detection for observation processes in this class.

Note that all of the results in this chapter are presented in a discrete-time setting.

7.2 Decentralized sequential and quickest detection

We begin by considering two possible configurations of a decentralized system.

(1) Each sensor sends a sequence of summary messages to a fusion center, where a sequential test is carried out to determine the true hypothesis.
(2) Each sensor first performs a sequential test based on its observations and arrives at a final local decision; these local decisions are then used for a common purpose at a site remote to all sensors.

For simplicity of exposition, we consider the case of two sensors, which we will denote by S_1 and S_2. At time k sensor S_i makes an observation $Z_k^{(i)}$, and forms a summary message $U_k^{(i)}$ of the information available for decision at time k at the specific sensor. The message of each of the mappings $U_k^{(i)}$ takes on one of a finite set of values and, to simplify matters further, we consider the set $\{0, 1\}$. In general, two-way communication is allowed between the fusion center and each of the sensors (see Figure 7.1).

In the first of the above two cases, there are many possible structures for information flow, but tractability becomes an issue with some of them. (See [218].) For this reason we consider a particular structure for the flow of information that admits a dynamic programming solution to the optimal sequential decision problem. In particular, let us suppose that at each point in time k, the fusion center has access to quantized information of the sensors. The sensors however, base their decisions about how to quantize the new observation $Z_k^{(i)}$ not only on this observation itself but also on information that the

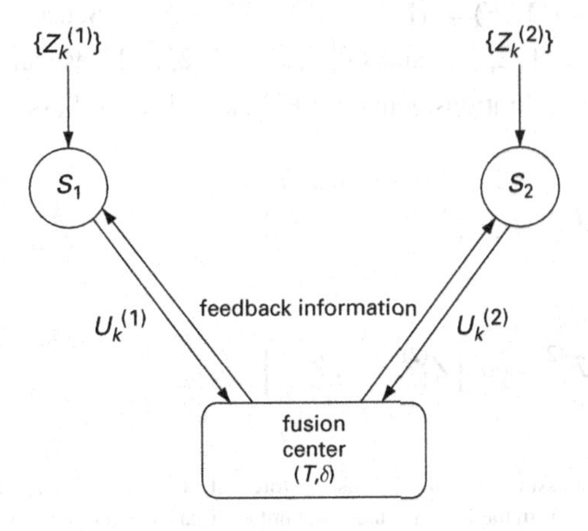

Fig. 7.1. Communication model.

fusion center has sent to them at the previous time step $k-1$. This information, in particular, is the quantized decision of the other sensors. To formalize this model, consider the σ-algebra generated by the quantized information, i.e. let

$$\mathcal{I}_k = \sigma\left\{U_i^{(1)}, U_i^{(2)}, i \leq k\right\},$$

with its corresponding filtration $\{\mathcal{I}_k; k = 1, \ldots\}$. Then the local decision function is a mapping[1]

$$\phi_k^{(i)} : \mathbb{R} \times \{\{0,1\} \times \{0,1\}\}^{k-1} \to \{0,1\}, \tag{7.1}$$

and the corresponding local decision is then of the form

$$U_k^{(i)} = \phi_k^{(i)}(Z_k^{(i)}; \mathcal{I}_{k-1}).$$

The fusion center makes its decision based on the filtration $\{\mathcal{I}_k; k = 1, \ldots\}$.

In the second case, each sensor first performs a sequential test on its observations and arrives at a final local decision. Subsequently the local decisions are used for a common purpose at a site possibly remote to all the sensors. In this setting, since the decisions of the individual sensors are used for a common goal, it is natural to assume that decision errors are penalized through a common decision cost function. The choice of the time penalty is, however, not as unambiguous. If we are concerned with processing cost at the sensors, then we associate a positive cost with each observation taken by each sensor. On the other hand, there may be situations in which we may wish to limit the time it takes for both decisions to be available at the remote site. In this case it may be more reasonable to associate a positive cost with each time step taken by the sensors as a *team*.

We now turn to the development of optimal solutions in these two cases.

7.2.1 Decentralized sequential detection with a fusion center

Consider the measurable space $(\Omega, \mathcal{F}) = \left(\mathbb{R}^\infty \times \mathbb{R}^\infty, \mathcal{B}^\infty \times \mathcal{B}^\infty\right)$. Sensors S_1 and S_2 receive observations $\{Z_k^{(1)}; k = 1, 2, \ldots\}$ and $\{Z_k^{(2)}; k = 1, 2, \ldots\}$, respectively. This suggests the introduction of two filtrations, namely $\{\mathcal{F}_k^{(1)}; k = 1, 2, \ldots\}$ and $\{\mathcal{F}_k^{(2)}; k = 1, 2, \ldots\}$, where

$$\mathcal{F}_k^{(1)} = \sigma\left\{Z_1^{(1)}, \ldots, Z_k^{(1)}\right\}, \tag{7.2}$$

and

$$\mathcal{F}_k^{(2)} = \sigma\left\{Z_1^{(2)}, \ldots, Z_k^{(2)}\right\}. \tag{7.3}$$

[1] Please note a subtlety here: we are assuming that the sensors store only their past local decisions (and those of the other sensors as received from the fusion center), but not their past observations. This models a situation in which local memory at the sensors is limited.

There are two hypotheses about the state of the environment, H_0 and H_1. The prior probability that the latter is true is π while the prior probability that the former is true is $1-\pi$. Conditioned on the true hypothesis being H_i, the $Z_k^{(j)}$'s are assumed to be i.i.d. with marginal probability density[2] q_i. Moreover, under either hypothesis, the $Z_k^{(1)}$'s and $Z_k^{(2)}$'s are independent. As noted above, each of the sensors S_j at each time step k sends a quantized version $U_k^{(j)} \in \{0, 1\}$ of its observation $Z_k^{(j)}$.

The fusion center devises a sequential decision rule (T, δ) (as in Chapter 4), comprising a stopping time T adapted to the filtration $\{\mathcal{I}_k\}$ and an $\mathcal{I}_T$ measurable δ_T. We denote by $\mathcal{T}$ the class of all T described above. At each time step k the flow of information is described in the following list.

(1) At time $k-1$, the fusion center sends updates to the sensors of all the quantized information that it has gathered from the sensors.
(2) Each of the sensors S_i receives a new observation $X_k^{(i)}$ and uses this observation along with the information received from the fusion center ($\mathcal{I}_{k-1}$) to make a decision on whether to send a 0 or a 1 to the fusion center, through the function $\phi_k^{(i)}$ of (7.1).
(3) The fusion center then, based on the new information summarized in $\mathcal{I}_k$ makes the decision on whether to stop sampling and declare that either H_1 or H_0 is true, or to continue sampling.

This flow of information is seen in Figure 7.1 and justifies the introduction of the above-described augmentation $\{\mathcal{I}_k^j\}$ of the filtrations, where $\mathcal{I}_k^j = \sigma\{U_1, \ldots, U_k, Z_k^{(j)}\}$ contains the internal information available at sensor S_j at time step k exclusive of feedback from the fusion center. We also introduce the following probability measures.

(1) The measure $P_\pi = \pi P_1 + (1-\pi)P_0$, where P_j denotes the probability measure on $(\Omega, \mathcal{F})$ describing the observations under H_j, for $j = 0$ and 1.
(2) The conditional measures $P_0^{\mathcal{I}_k}$ and $P_1^{\mathcal{I}_k}$ restricted to the filtration $\{\mathcal{I}_k\}$, under which the conditional joint probability mass functions of the observations $\left(U_{k+1}^{(1)}, U_{k+1}^{(2)}\right)$, $k = 1, 2, \ldots$ given $\mathcal{I}_k$ are denoted by $q_0^{\mathcal{I}_k}$ and $q_1^{\mathcal{I}_k}$.
(3) The conditional measures $\{P_\pi^{j,k}\}$ on the augmented filtration $\{\mathcal{I}_k^j; k = 1, 2, \ldots\}$ for $j = 1, 2$.

Similarly to the formalism of Chapter 4, decision errors are penalized through a cost function $C(\delta, H)$, for which $C(0, H_0) = C(1, H_1) = 0$, while $C(0, H_1) = c_1$ and $C(1, H_0) = c_0$. Also each time-step taken for detection is penalized by a constant amount c.

The system described above features full feedback but local memory is restricted to past decisions. This structure facilitates the use of dynamic programming to find optimal sequential solutions. More specifically, the fact that the information available

[2] For simplicity of exposition, we assume that the two sensors are statistically identical; i.e., that the observations have the same distribution at both sensors. The more general case in which the sensors are not statistically identical, is treated in [218].

at all the sensors and the fusion center (one-step delayed) is the same, and the fact that the final decision depends only on the sensor messages makes this possible. For a discussion of other decentralized systems and the difficulties arising by considering different information flows, please refer to [218].

The objective now becomes to minimize the total Bayes cost

$$E_\pi\{cT + C(\delta, H)\} \tag{7.4}$$

over all (T, δ) and $\{\phi_k^{(j)}\}$. Note that the dependence on $\phi_k^{(j)}$ is implicit through the information structure $\mathcal{I}_k$.

Similarly to Proposition 4.1 we have the following result.

PROPOSITION 7.1. *For any T in $\mathcal{T}$ and any set of local decision rules $\{\phi_k^{(j)}\}$, we have*

$$\inf_\delta C(\delta, H) = E_\pi\{\min\{c_1\pi_T, c_0(1-\pi_T)\}\},$$

where the sequence $\pi_k = P_\pi(H = H_1|\mathcal{I}_k)$ is defined through the recursion

$$\pi_{k+1} = \frac{\pi_k q_1^{\mathcal{I}_k}(U_{k+1}^{(1)}, U_{k+1}^{(2)})}{\pi_k q_1^{\mathcal{I}_k}(U_{k+1}^{(1)}, U_{k+1}^{(2)}) + (1-\pi_k) q_0^{\mathcal{I}_k}(U_{k+1}^{(1)}, U_{k+1}^{(2)})}, \tag{7.5}$$

and the terminal decision rule

$$\delta_T = \begin{cases} 1 & \text{if } \pi_T \geq c_0/(c_0+c_1) \\ 0 & \text{if } \pi_T < c_0/(c_0+c_1). \end{cases} \tag{7.6}$$

Proof: Equation (7.5) follows by a simple application of Bayes' formula. Note that in this case,

$$E_\pi\left\{\frac{dP_1}{dP_0}\middle|\mathcal{I}_k\right\} = \frac{\prod_{j=1}^k q_1^{\mathcal{I}_{j-1}}(U_j^{(1)}, U_j^{(2)})}{\pi\prod_{j=1}^k q_1^{\mathcal{I}_{j-1}}(U_j^{(1)}, U_j^{(2)}) + (1-\pi)\prod_{j=1}^k q_0^{\mathcal{I}_{j-1}}(U_j^{(1)}, U_j^{(2)})}.$$

Using this fact and steps similar to those used in the proof of Proposition 4.1, the result follows. ■

From the above proposition one immediately realizes that the π_k's do not necessarily form a Markov process with respect to the filtration $\{\mathcal{I}_k\}$, unless of course we can show that the π_{k+1}'s depend on $\mathcal{I}_k$ only through π_k. The value of π_{k+1} depends on the choice of the decision functions $\phi_i^{(1)}$ for $i = 1, \ldots, k+1$, and $\phi_i^{(2)}$ for $i = 1, \ldots, k+1$. This fact raises the question of what is the optimal choice of the local sensor decision functions.

To find the structure of the optimal local sensor decision rules, we begin by fixing all decision rules (including (T, δ)) except $\phi_k^{(1)}$ for some $k < \infty$.

We can rewrite the expectation in (7.4) in the following way:

$$\begin{aligned} E_\pi\{cT + C(\delta; H)\} &= E_\pi^{1,k}\left\{E_\pi\left\{cT + C(\delta; H)|H, \mathcal{F}_{k-1}^{(1)}, \mathcal{F}_k^{(2)}\right\}\right\} \\ &= E_\pi^{1,k}\{E_\pi\{cT + C(\delta; H)|H\}\}, \end{aligned}$$

where the second equality follows from the conditional independence of the observations at each sensor and across sensors given the true hypothesis. The outer expectation is taken with respect to $\mathcal{I}_{k-1}$, because the respective local decision functions are fixed. The result of the inner expectation is a function that will depend on $\phi_k^{(1)}(Z_k^{(1)}; \mathcal{I}_{k-1})$, $\mathcal{I}_{k-1}$, and the true hypothesis H, say $K(\phi_k^{(1)}(Z_k^{(1)}; \mathcal{I}_{k-1}), \mathcal{I}_{k-1}, H)$.

On substituting the above into (7.4), we have the criterion

$$\inf_{\phi_k^{(1)}} E_\pi^{1,k}\left\{K(\phi_k^{(1)}(Z_k^{(1)}; \mathcal{I}_{k-1}), \mathcal{I}_{k-1}, H)\right\}, \tag{7.7}$$

or equivalently,

$$\inf_{\phi_k^{(1)}} E_\pi^{1,k}\{K(\phi_k^{(1)}(Z_k^{(1)}; \mathcal{I}_{k-1}), \mathcal{I}_{k-1}, H_0)P_\pi^{1,k}(H = H_0)$$
$$+K(\phi_k^{(1)}(Z_k^{(1)}; \mathcal{I}_{k-1}), \mathcal{I}_{k-1}, H_1)P_\pi^{1,k}(H = H_1)\}. \tag{7.8}$$

Minimizing the above expectation is equivalent to minimizing the quantity inside the expectation a.s.

Now, $\phi_k^{(1)}$ can take only the value 0 or 1. Therefore, finding the minimum in (7.8) is equivalent to comparing the following two quantities:

(1) $K(1, \mathcal{I}_{k-1}, H_0)P_\pi^{1,k}(H = H_0) + K(1, \mathcal{I}_{k-1}, H_1)P_\pi^{1,k}(H = H_1)$; and
(2) $K(0, \mathcal{I}_{k-1}, H_0)P_\pi^{1,k}(H = H_0) + K(0, \mathcal{I}_{k-1}, H_1)P_\pi^{1,k}(H = H_1)$,

and choosing $\phi_k^{(1)} = 1$ if the former is smaller than the latter while choosing $\phi_k^{(1)} = 0$ if the latter is smaller than the former.

This gives rise to the following person-by-person (see [215]) optimal local decision rule:

$$\hat{\phi}_k^{(1)} = \begin{cases} 1 & \text{if } \frac{P_\pi^{1,k}(H=H_1)}{P_\pi^{1,k}(H=H_0)} > \frac{K(1,\mathcal{I}_{k-1},H_0)-K(0,\mathcal{I}_{k-1},H_0)}{K(0,\mathcal{I}_{k-1},H_1)-K(1,\mathcal{I}_{k-1},H_1)} \\ 0 & \text{otherwise} \end{cases}. \tag{7.9}$$

From Bayes' formula and the conditional independence of $Z_k^{(i)}$ from $\mathcal{I}_{k-1}$ given the true hypothesis, it follows that[3]

$$P_\pi^{(1,k)}(H = H_1) = \frac{q_1(Z_k^{(1)})\pi_{k-1}}{q_1(Z_k^{(1)})\pi_{k-1} + q_0(Z_k^{(1)})(1-\pi_{k-1})}, \tag{7.10}$$

and

$$P_\pi^{(1,k)}(H = H_0) = \frac{q_0(Z_k^{(1)})(1-\pi_{k-1})}{q_1(Z_k^{(1)})\pi_{k-1} + q_0(Z_k^{(1)})(1-\pi_{k-1})}. \tag{7.11}$$

Substituting (7.10) and (7.11) into (7.9), we obtain

$$\hat{\phi}_k^{(1)} = \left\{ \begin{array}{ll} 1 & \text{if } L(Z_k^{(1)}) > \frac{1-\pi_k}{\pi_k} \frac{K(1,\mathcal{I}_{k-1},H_0)-K(0,\mathcal{I}_{k-1},H_0)}{K(0,\mathcal{I}_{k-1},H_1)-K(1,\mathcal{I}_{k-1},H_1)} \\ 0 & \text{otherwise} \end{array} \right\} \tag{7.12}$$

where $L = q_1/q_0$ is the local likelihood ratio.

[3] Note that each sensor knows π_{k-1} at time k, through the feedback from the fusion center.

In other words, the structure of the optimal local sensor decision functions is

$$\hat{\phi}_k^{(1)} = \left\{ \begin{array}{cc} 1 & \text{if } L(Z_k^{(1)}) > \lambda_k^{(1)}(\mathcal{I}_{k-1}) \\ 0 & \text{otherwise} \end{array} \right\} \tag{7.13}$$

for a threshold that, for each k, depend on $\mathcal{I}_{k-1}$, through the function K.

Using (7.13), and the conditional independence of the observations across sensors, we can rewrite

$$\begin{aligned} &P_j(U_{k+1}^{(1)}, U_{k+1}^{(2)}|\mathcal{I}_k)P_j(U_{k+1}^{(1)}|\mathcal{I}_k)P_j(U_{k+1}^{(1)}|\mathcal{I}_k) \\ &= \prod_{i=1}^{2}\left[P_j(L(Z_{k+1}^{(i)}) > \lambda_{k+1}^{(i)}(\mathcal{I}_k))\right]^{U_{k+1}^{(i)}}\left[1 - P_j(L(Z_{k+1}^{(i)}) > \lambda_{k+1}^{(i)}(\mathcal{I}_k))\right]^{1-U_{k+1}^{(i)}}, \end{aligned} \tag{7.14}$$

for each $j = 0, 1$. Using (7.14), (7.5) becomes

$$\pi_{k+1} = \frac{\pi_k g_1(U_{k+1}^{(1)}, U_{k+1}^{(2)}, \lambda_{k+1}^{(1)}, \lambda_{k+1}^{(2)})}{\pi_k g_1(U_{k+1}^{(1)}, U_{k+1}^{(2)}, \lambda_{k+1}^{(1)}, \lambda_{k+1}^{(2)}) + (1-\pi_k)g_0(U_{k+1}^{(1)}, U_{k+1}^{(2)}, \lambda_{k+1}^{(1)}, \lambda_{k+1}^{(2)})}, \tag{7.15}$$

where

$$\begin{aligned} g_1(U_{k+1}^{(1)}, U_{k+1}^{(2)}, \lambda_{k+1}^{(1)}, \lambda_{k+1}^{(2)}) &= \prod_{i=1}^{2}\left[P_1(L(Z_{k+1}^{(i)}) > \lambda_{k+1}^{(i)}(\mathcal{I}_k))\right]^{U_{k+1}^{(i)}} \\ &\times \left[P_1(L(Z_{k+1}^{(i)}) \le \lambda_{k+1}^{(i)}(\mathcal{I}_k))\right]^{1-U_{k+1}^{(i)}}, \end{aligned}$$

and

$$\begin{aligned} g_0(U_{k+1}^{(1)}, U_{k+1}^{(2)}, \lambda_{k+1}^{(1)}, \lambda_{k+1}^{(2)}) &= \prod_{i=1}^{2}\left[P_0(L(Z_{k+1}^{(i)}) > \lambda_{k+1}^{(i)}(\mathcal{I}_k))\right]^{U_{k+1}^{(i)}} \\ &\times \left[P_0(L(Z_{k+1}^{(i)}) \le \lambda_{k+1}^{(i)}(\mathcal{I}_k))\right]^{1-U_{k+1}^{(i)}}. \end{aligned}$$

We now wish to show that the optimal thresholds $\lambda_{k+1}(\mathcal{I}_k)$ depend only on π_k. To this effect we consider the problem of minimizing the expected Bayes cost $E_\pi\{cT + C(\delta, H)\}$ over the finite horizon $[0, \tau]$. In other words, in this finite horizon problem, the fusion center is obliged to stop and make a decision by time τ, which implies that T can take values only in the set $\{0, 1, \ldots, \tau\}$. Let us denote the minimal expected cost to go at time k by $J_k^\tau(\mathcal{I}_k)$. It is obvious that

(1) For $\{T = \tau\}$, we have $J_T^\tau(\mathcal{I}_T) = \min\{c_1 P_\pi(H = H_1|\mathcal{I}_T), c_0 P_\pi(H = H_0|\mathcal{I}_T)\}$; and

(2) For $\{T = k\}, k = 0, 1, \ldots, \tau - 1$,

$$J_k^\tau(\mathcal{I}_k) = \min \left\{ \begin{array}{c} c_1 P_\pi(H = H_1|\mathcal{I}_T), \\ c_0 P_\pi(H = H_0|\mathcal{I}_T), \\ c \inf_{\phi_{k+1}^{(1)}, \phi_{k+1}^{(2)}} E_\pi \left\{ J_{k+1}^\tau(\mathcal{I}_{k+1})|\mathcal{I}_k \right\} \end{array} \right\}. \tag{7.16}$$

Notice that the interpretation of (7.16) is fairly straightforward. The first term is the cost of stopping at time k and deciding in favor of hypothesis H_1. The second term is the cost of stopping at time k and deciding in favor of hypothesis H_0. Finally, the last term is the cost of continuing to sample. As usual, the constant c reflects a cost of c units for each additional sample. In connection with this cost function we have the following result.

PROPOSITION 7.2. *The function $J_k^\tau(\mathcal{I}_k)$ depends on $\mathcal{I}_k$ only through π_k for all $0 \leq k \leq \tau$, and the optimal sensor thresholds at time $k + 1$ depend on $\mathcal{I}_k$ only through π_k.*

Proof: It is obvious that the claim holds for $\{T = \tau\}$. We will prove the result by induction. Suppose that $J_{k+1}^\tau(\mathcal{I}_{k+1}) = J_{k+1}^\tau(\pi_{k+1})$. Then from (7.16), it suffices to show that for a given choice of $\lambda_{k+1}^{(1)}$ and $\lambda_{k+1}^{(2)}$ $E_\pi \left\{ J_{k+1}^\tau(\pi_{k+1})|\mathcal{I}_k \right\}$ is a function only of π_k. We have

$$E_\pi \left\{ J_{k+1}^\tau(\pi_{k+1})|\mathcal{I}_k \right\} = W_{J_{k+1}^\tau}(\lambda_{k+1}^{(1)}, \lambda_{k+1}^{(2)}; \pi_k),$$

with

$$W_{J_{k+1}^\tau}(\lambda_{k+1}^{(1)}, \lambda_{k+1}^{(2)}; \pi_k) = \sum_{i=0}^{1} \sum_{j=0}^{1} J_{k+1}^\tau \left(\frac{\pi_k g_1(i, j, \lambda_{k+1}^{(1)}, \lambda_{k+1}^{(2)})}{A(\pi_k, g_1, g_0)} \right) \times A(\pi_k, g_1, g_0), \tag{7.17}$$

where

$$A(\pi_k, g_1, g_0) = \pi_k g_1(i, j, \lambda_{k+1}^{(1)}, \lambda_{k+1}^{(2)}) + (1 - \pi_k) g_0(i, j, \lambda_{k+1}^{(1)}, \lambda_{k+1}^{(2)}).$$

Therefore, whichever the pair of optimal $(\lambda_{k+1}^{(1)}, \lambda_{k+1}^{(2)})$ is, it will depend on $\mathcal{I}_k$ only through π_k. That is, if the inductive hypothesis holds for $j = k + 1$, it also holds for $j = k$, and thus by induction for all $j = 0, \ldots, \tau$. This completes the proof of the proposition. ■

REMARK 7.3. As a result of the fact that the optimal pair $\{\lambda_{k+1}^{(1)}, \lambda_{k+1}^{(2)}\}$ depend only on π_k, we have that the sequence $\{\pi_k\}$ is a Markov process with respect to the filtration $\mathcal{I}_k$.

Our problem however, is one in which the restriction of a finite horizon is removed. That is we need to make sure that $\lim_{\tau \to \infty} J_k^\tau$ exists. If it does, then this limit will be the value function of our original problem. We first notice that, for all k, $J_k^{\tau+1} < J_k^\tau$. Moreover, due to the i.i.d. nature of the observations, the value J_k^τ is time-shift invariant. That is to say, $J_{k+1}^{\tau+1} = J_k^\tau$. Therefore, $\lim_{\tau \to \infty} J_k^\tau$ is independent of k. Hence the limit exists and we have that $\lim_{\tau\to\infty} J_k^\tau = \inf_{\tau>n} J_k^\tau = J$. Taking the limit in (7.16) we have that J satisfies the Bellman equation:

$$J(\pi) = \min\{c_1\pi, c_0(1-\pi), c + W_J(\lambda^{(1)}, \lambda^{(2)}, \pi)\},\ 0 \le \pi \le 1. \tag{7.18}$$

In what follows we will establish the uniqueness of this limit. This will in turn establish the uniqueness of the optimal pair $(\lambda^{(1)}, \lambda^{(2)})$.

PROPOSITION 7.4. *Suppose* $f(\pi) = \inf\{c_1\pi, c_0(1-\pi)\}$. *Define the operator* Q *as follows:*

$$Qh(\pi) = \min\{f(\pi), \inf_{\{\lambda^{(1)},\lambda^{(2)}\}} W_h(\lambda^{(1)}, \lambda^{(2)}, \pi)\}.$$

Then, J is the unique fixed point of the mapping Q.

Proof: Let g be a fixed point of the operator Q, and choose the pair $\left(\lambda^{(1)}, \lambda^{(2)}\right)$ so that the minimum in $W_G(\lambda^{(1)}, \lambda^{(2)}, \pi)$ is achieved. Define π_k through the recursion (7.15), with $\pi_0^{(1)} = \pi$. Now define a stopping time S and a fusion center decision rule δ_S, in the following way:

$$S = \inf\{k | f(\pi_k) \le c + W_g(\lambda^{(1)}, \lambda^{(2)}, \pi_k)\}, \tag{7.19}$$

and

$$\delta_S = \begin{cases} 1 & \text{if } c_1\pi_k \ge c_0(1-\pi_k) \\ 0 & \text{if } c_1\pi_k < c_0(1-\pi_k). \end{cases} \tag{7.20}$$

From (7.19) and the fact that g is a fixed point of Q, we obtain

$$\begin{aligned} g(\pi) &= c + E_\pi\{g(\pi_1)|\mathcal{I}_0\} \\ g(\pi_1) &= c + E_\pi\{g(\pi_2)|\mathcal{I}_1\} \\ &\vdots \\ g(\pi_{n-1}) &= c + E_\pi\{g(\pi_n)|\mathcal{I}_{n-1}\} \end{aligned}$$

and

$$g(\pi_S) = f(\pi_S).$$

Substituting backwards and taking expectations, we obtain, for $0 \le \pi \le 1$,

$$g(\pi) = E_\pi\{cS + C(\delta_S, H)\} \ge \inf_{T\in\mathcal{T}} E\{cT + C(\delta_T, H)\} = J(\pi).$$

To show the reverse equality, we note that for any $\pi \in [0, 1]$ we have that

$$g(\pi) \le f(\pi) = J_\tau^\tau(\pi),\ \ \forall\, \tau.$$

Now we fix τ for some integer $m < T - 1$, such that $J^{\tau}_{m+1} \geq g(\pi)$, for all $0 \leq \pi \leq 1$. Then, for $0 \leq \pi \leq 1$,

$$\begin{aligned} J^{\tau}_m(\pi) &= \min\{f(\pi), c + \inf_{\lambda^{(1)},\lambda^{(2)}} W^{\tau}_{J_{m+1}}(\lambda^{(1)}, \lambda^{(2)}, \pi)\} \\ &\geq \min\{f(\pi), c + \inf_{\lambda^{(1)},\lambda^{(2)}} W^{\tau}_{g}(\lambda^{(1)}, \lambda^{(2)}, \pi)\} \\ &= g(\pi). \end{aligned}$$

That is, the inequality holds for m as well, and by induction it holds for every $m \in [0, \tau]$. That is

$$J^{\tau}_k(\pi) \geq g(\pi) \text{ for all } \tau \geq 1, \text{ and } k \leq \tau.$$

Fixing k and taking the limit as $\tau \to \infty$,we obtain

$$J(\pi) \geq g(\pi), \ 0 \leq \pi \leq 1.$$

This establishes the uniqueness of the fixed point and the proposition. ∎

As a consequence of Proposition 7.4, the uniqueness of the optimal pair of thresholds $\{\lambda^{(1)}, \lambda^{(2)}\}$ is established.

Fix $\{\lambda^{(1)}, \lambda^{(2)}\}$ at their optimal values. Then, for $0 \leq \pi \leq 1$,

$$\begin{aligned} J(\pi) &= \inf_T E_\pi \{\min\{c_1\pi_T, c_0(1-\pi_T)\} + cT\} \\ &= \inf_{(T,\delta)} E_\pi \{cT + C(\delta, H)\}. \end{aligned} \tag{7.21}$$

From Proposition 7.1, it is evident that the function $C(\delta, H)$ is linear in π. Of course, we can also write $E_\pi\{cT\} = \pi E_1\{cT\} + (1-\pi)E_0\{cT\}$. That is, J is the minimum of linear functions of π and therefore is concave. Moreover from (7.21), it follows that

$$0 \leq J(\pi) \leq \min\{c_1\pi, c_0(1-\pi)\}, \ \ 0 \leq \pi \leq 1,$$

where the inequalities become equalities at $\pi = 1$ and $\pi = 0$.

Using the Markov optimal stopping theory of Chapter 3, in view of Remark 7.3, or arguments similar to those used in Chapter 4, it follows that the optimal stopping time for the fusion center is of the form

$$T_{\text{opt}} = \inf\{k | \min\{c_1\pi_k, c_0(1-\pi_k)\} = J(\pi_k)\}. \tag{7.22}$$

The concavity of J therefore implies that the optimal stopping time is

$$T_{\text{opt}} = \inf\{k | \pi_k \notin (\pi_{\text{L}}, \pi_{\text{U}})\}, \tag{7.23}$$

where

$$\pi_{\text{L}} = \sup\left\{0 \leq \pi \leq \frac{1}{2} \middle| J(\pi) = c_1\pi\right\}, \tag{7.24}$$

and

$$\pi_{\text{U}} = \inf\left\{\frac{1}{2} \leq \pi \leq 1 \middle| J(\pi) = c_0(1-\pi)\right\}. \tag{7.25}$$

Finally, the optimal decision rule at the fusion center is

$$\delta_k = \begin{cases} 1 & \text{if } \pi_k \geq \pi_{\text{U}} \\ 0 & \text{if } \pi_k \leq \pi_{\text{L}}. \end{cases} \tag{7.26}$$

Thus, (7.23) and (7.26) define the fusion center's optimal sequential decision rules, while the optimal local decision rules are given by (7.13) with thresholds depending only on π_k.

7.2.2 Decentralized quickest detection with a fusion center

In the previous section, we considered the decentralized version of the sequential detection problem of Chapter 4. In this section we consider a decentralized version of the Bayesian quickest problem of Chapter 5 for the situation in which there is feedback from the fusion center. So, as before, at each time step the sensors receive new observations sequentially and they also receive information accumulated by the fusion center as feedback from the fusion center. At some unknown change point t the statistical behavior generating the sensor observations changes abruptly. The fusion center, based on information that it receives from the sensors, wishes to detect this change as quickly as possible while maintaining a low probability of a false alarm.

To formalize this problem we consider the measurable space

$$(\Omega, \mathcal{F}) = \left(\mathbb{R}^\infty \times \mathbb{R}^\infty \times \mathcal{N}, \mathcal{B}^\infty \times \mathcal{B}^\infty \times \mathcal{B}_{\mathcal{N}}\right),$$

where $\mathcal{N}$ denotes the non-negative integers, and $\mathcal{B}_{\mathcal{N}}$ the σ-field consisting of all subsets of $\mathcal{N}$. As before, sensors S_1 and S_2 make observations $\{Z_k^{(1)}; k = 1, 2, \ldots\}$ and $\{Z_k^{(2)}; k = 1, 2, \ldots\}$, respectively. The change point t is a random variable, and conditional on $\{t = k\}$, the random variables $Z_1^{(j)}, \ldots, Z_{k-1}^{(j)}$ are i.i.d. with marginal distribution[4] Q_0 and density q_0 and $Z_k^{(j)}, \ldots,$ are i.i.d. with marginal distribution Q_1 and density q_1. As before, we assume that the distribution of the change point t is as follows: $P(t = 0) = \pi$ and $P(t = k|t > 0) = \rho$, $k = 1, 2, \ldots$, where π and ρ lie in $[0, 1]$. Moreover, it is assumed that the observations at each sensor and across sensors are independent conditioned on the value of the change point.

The flow of information is summarized once again, in Figure 7.1. In other words, at each time step k the sensors send quantized information to the fusion center through the decision functions ϕ_k^j as in the previous section, which take values in a finite alphabet. For simplicity of exposition we again assume in this case that this is the set $\{0, 1\}$. The fusion center then makes a decision through a stopping time T as to whether to stop sampling and declare that an abrupt change has occurred, or to continue sampling. In the latter case, it sends the information summarized in $\mathcal{I}_k$ back to the sensors which after receiving their new observations $Z_{k+1}^{(1)}$ and $Z_{k+1}^{(2)}$, respectively, send new binary messages to the fusion center, through decision functions ϕ_{k+1}^j.

[4] We again assume, for convenience, that the sensors are statistically identical. The more general case of non-identical sensors is treated in [218].

It is up to the fusion center to stop and declare that a change has occurred based on the quantized observations it receives from the sensors, that is, through the filtration $\{\mathcal{I}_k\}$. More specifically, let $\mathcal{T}$ again denote the set of all stopping times with respect to the filtration $\{\mathcal{I}_k\}$, $\mathcal{D}$ denote the set of all decision functions $\{\phi_k^{(1)}, \phi_k^{(2)}; k = 1, \ldots\}$ and $\underline{\phi}^{(j)}$ denote the vector of decision functions used by sensor S_j, $j = 1, 2$. Then, as in Chapter 5, we are interested in finding a stopping time that minimizes a weighted sum of an expected delay and the probability of false alarms. That is, our objective is to solve the following stochastic optimization problem:

$$\inf_{T \in \mathcal{T}, \{\underline{\phi}^{(1)}, \underline{\phi}^{(2)} \in \mathcal{D}\}} \left[P(T < t) + cE\left\{(T - t + 1)^+\right\}\right], \tag{7.27}$$

where the choice of a stopping time of the fusion center depends on the choice of the sensor decision functions as defined in (7.1) through the filtration $\{\mathcal{I}_k\}$.

To solve (7.27), we begin with the following preliminary result.

PROPOSITION 7.5. *Suppose that $E\{T\} < \infty$ and define the sequence $\{\pi_k\}$ by $\pi_k = P(t \leq k|\mathcal{I}_k)$ for all $k = 0, 1, \ldots$. Then*

$$\pi_k = \frac{[\pi_k + (1 - \pi_k)\rho]\, B}{[\pi_k + (1 - \pi_k)\rho]\, B + [(1 - \rho)(1 - \pi_k)]\, C}, \tag{7.28}$$

where

$$B = P(U_{k+1}^{(1)}, U_{k+1}^{(2)}|t \leq k + 1, \mathcal{I}_{k-1})$$

and

$$C = P(U_{k+1}^{(1)}, U_{k+1}^{(2)}|t > k + 1, \mathcal{I}_{k-1}).$$

Moreover, for each $T \in \mathcal{T}$ we can write

$$\left[P(T < t) + cE\left\{(T - t + 1)^+\right\}\right] = E\left\{1 - \pi_T + \sum_{m=0}^{T} \pi_m\right\}. \tag{7.29}$$

The proof of Proposition 7.5 mimics the proof of Proposition 5.1. The validity of (7.29) similarly does not rely on the specific choice of an exponential prior for t.

Proof: For every stopping time T, we have that

$$P(T < t) = E\left\{1_{\{T<t\}}\right\} = E\left\{1 - \pi_T\right\}. \tag{7.30}$$

Since T can take only countably many values, the above equality holds for every T. Now, it remains to show that

$$E\left\{(T - t)^+\right\} = E\left\{\sum_{m=0}^{T-1} \pi_m\right\}. \tag{7.31}$$

We can write

$$E\left\{(T - t)^+\right\} = E\{D_T\}, \tag{7.32}$$

where

$$D_k = E\left\{(k-t)^+ \middle| \mathcal{I}_k\right\}. \tag{7.33}$$

We have

$$\begin{aligned} D_k &= \sum_{m=0}^{k-1}(k-m)P\left(t = m|\mathcal{I}_k\right) \\ &= \sum_{m=0}^{k-1} P\left(t \le m|\mathcal{I}_k\right) \\ &= \sum_{m=0}^{k-1} \pi_m + M_k, \end{aligned} \tag{7.34}$$

where

$$M_k = \sum_{m=0}^{k-1}\left[P\left(t \le m|\mathcal{I}_k\right) - \pi_m\right],\ k = 1, \ldots \tag{7.35}$$

Consider the sequence $\{M_k\}$. Since $\mathcal{I}_{k-1} \subset \mathcal{I}_k$, the iterative property of conditional expectation (2.31) implies that

$$\begin{aligned} E\left\{P\left(t \le m|\mathcal{I}_k\right)|\mathcal{I}_{k-1}\right\} &= E\left\{E\left\{1_{\{t\le m\}}\middle|\mathcal{I}_k\right\}\middle|\mathcal{I}_{k-1}\right\} \\ &= E\left\{1_{\{t\le m\}}\middle|\mathcal{I}_{k-1}\right\} \\ &= P\left(t \le m|\mathcal{I}_{k-1}\right). \end{aligned} \tag{7.36}$$

Similarly, since $\mathcal{I}_m \subset \mathcal{I}_{k-1},\ m = 1, \ldots, k-1$, we have

$$E\left\{P(t \le m|\mathcal{I}_k)|\mathcal{I}_m\right\} = \pi_m, \forall\, m < k. \tag{7.37}$$

Thus,

$$E\left\{M_k|\,\mathcal{I}_{k-1}\right\} = M_{k-1}, \tag{7.38}$$

and so $\{M_k\}$ is an $\mathcal{I}_k$-martingale.

Now consider the quantity

$$M = \sum_{m=0}^{\infty}(1-\pi_m) - t. \tag{7.39}$$

Since,

$$E\left\{\sum_{m=0}^{k-1}(1-\pi_m)\right\} = \sum_{m=0}^{k-1} P(t > m) \uparrow E\{t\} < \infty, \tag{7.40}$$

it follows from Fatou's lemma that $E\{|M|\} < 2E\{t\} < \infty$, and thus that M is an integrable random variable. On writing,

$$t = \sum_{m=0}^{\infty} 1_{\{t>m\}}, \tag{7.41}$$

we have

$$M = \sum_{m=0}^{\infty} \left[1 - \pi_m - 1_{\{t>m\}}\right], \tag{7.42}$$

which by dominated convergence has conditional expectation

$$\begin{aligned} E\{M|\mathcal{I}_k\} &= \sum_{m=0}^{\infty} [E\{P(t > m|\mathcal{I}_m)|\mathcal{I}_k\} - P(t > m|\mathcal{I}_k)] \\ &= \sum_{m=0}^{k-1} [P(t > m|\mathcal{I}_m) - P(t > m|\mathcal{I}_k)] = M_k, \end{aligned} \tag{7.43}$$

where the second equality follows from the fact that $\mathcal{I}_k \subset \mathcal{I}_m,\ m > k$, and the iteration property of expectation. Thus, $\{M_k\}$ is also regular, and so optional sampling (compare Section 2.3.2) implies

$$E\{M_T\} = E\{M_0\} = 0. \tag{7.44}$$

Since $D_k = \sum_{m=0}^{k-1} \pi_m + M_k$, (7.29) follows. Again we note that, to this point in the proof, we have used only the finiteness of $E\{t\}$ and not the detailed structure of the exponential prior. Thus, (7.29) is a more general result.

To prove (7.28), we notice that from a straightforward application of Bayes' formula it follows that

$$\pi_{k+1} = \frac{P(t \le k+1|\mathcal{I}_k)B}{P(t \le k+1|\mathcal{I}_k)B + P(t > k+1|\mathcal{I}_k)C}.$$

Using the above equation and the equality we have

$$\begin{aligned} P(t \le k+1|\mathcal{I}_k) &= P(t \le k|\mathcal{I}_k) + P(t = k+1|\mathcal{I}_k) \\ &= P(t \le k|\mathcal{I}_k) + P(t = k+1|\mathcal{I}_k, t \ge k+1)P(t \ge k+1|\mathcal{I}_k), \end{aligned}$$

the result follows. ∎

Using arguments similar to those in the previous subsection, we can show that the person-by-person optimal sensor decision functions take the form

$$\hat{\phi}_k^{(i)} = \left\{ \begin{array}{ll} 1 & \text{if } L(Z_k^{(i)}) > \lambda_k^{(i)}(\mathcal{I}_{k-1}) \\ 0 & \text{otherwise} \end{array} \right\}, \tag{7.45}$$

where $L = q_1/q_0$ and the $\lambda_k^{(i)}(\mathcal{I}_{k-1})$'s are decision thresholds to be chosen optimally. We wish to show that the dependence of the optimal thresholds on $\mathcal{I}_{k-1}$ is only through π_{k-1}. To this end, we proceed along the lines of the previous subsection and consider the finite horizon problem of minimizing the expected Bayes risk given that we cannot stop beyond the interval $[0, \tau]$. Let us again denote by $J_k^\tau(\mathcal{I}_k)$ the minimal expected cost to go. Then we have the following conditions.

(1) When $\{T = \tau\}$, we have $J_T^\tau(\mathcal{I}_T) = 1 - \pi_T$.
(2) When $\{T = k\}$, $k = 1, \ldots, \tau - 1$, we have

$$J_k^\tau(\mathcal{I}_k) = \min\left\{1 - \pi_k,\, c\pi_k + \inf_{\phi_{k+1}^{(1)},\phi_{k+1}^{(2)}} E\left\{J_{k+1}^\tau(\mathcal{I}_{k+1})|\mathcal{I}_k\right\}\right\}. \tag{7.46}$$

Notice that the interpretation of (7.46) is fairly straightforward. The first term is the cost of stopping falsely (that is, when $t > k$) at time k. The second term is the cost of not stopping at time k when in fact $t \leq k$, and sampling continues. The argument on the dependence of the optimal pair of thresholds on π_k follows by induction and by noticing that for any given pair of thresholds $\{\lambda_{k+1}^{(1)}, \lambda_{k+1}^{(2)}\}$,

$$E\left\{J_{k+1}^\tau(\pi_{k+1})|\mathcal{I}_k\right\} = W_J(\lambda_{k+1}^{(1)}, \lambda_{k+1}^{(2)}, \pi_k),$$

where

$$W_J(\lambda_{k+1}^{(1)}, \lambda_{k+1}^{(2)}, \pi_k) = \sum_{i=0}^{1}\sum_{j=0}^{1} J\left(\frac{[\pi_k + (1-\pi_k)\rho]\, g_1\left(i, j, \lambda_{k+1}^{(1)}, \lambda_{k+1}^{(2)}\right)}{C(\pi_k, \rho, g_1, g_0)}\right) \times C(\pi_k, \rho, g_1, g_0),$$

with

$$g_j(U_1, U_2, \lambda_1, \lambda_2) = \prod_{i=1}^{2}\left[P_j(L(Z^{(i)})) > \lambda^{(i)}\right]^{U_i}\left[P_j(L(Z^{(i)})) \leq \lambda^{(i)}\right]^{1-U_i},$$

for $j = 0, 1$ and

$$C(\pi_k, \rho, g_1, g_0) = [\pi_k + (1-\pi_k)\rho]\, g_1\left(i, j, \lambda_{k+1}^{(1)}, \lambda_{k+1}^{(2)}\right) + (1-\rho)(1-\pi_k) g_0\left(i, j, \lambda_{k+1}^{(1)}, \lambda_{k+1}^{(2)}\right).$$

Therefore, the minimum in (7.46) will occur for a pair $\{\lambda_{k+1}^{(1)}, \lambda_{k+1}^{(2)}\}$ that depends on $\mathcal{I}_k$ only through π_k.

Letting τ tend to infinity, and using arguments similar to those used in the previous subsection, it follows that the minimal cost does not depend on k from which it further follows that the optimal pair of thresholds corresponding to the local decision functions do not depend on k either. In fact the optimal local decision function thresholds consist of the unique pair $\{\lambda^{(1)}, \lambda^{(2)}\}$ for which the minimum of $W_J(\lambda_1, \lambda_2, \pi)$ is achieved, where J is the unique solution to the equation

$$J(\pi) = \min\left\{1 - \pi,\, c\pi + \inf_{(\lambda^{(1)},\lambda^{(2)})} W_J(\lambda^{(1)}, \lambda^{(2)}, \pi)\right\}, \quad 0 \leq \pi \leq 1.$$

REMARK 7.6. As a result of the fact that the optimal pair $\{\lambda_k^{(1)}, \lambda_k^{(2)}\}$ depends on π_{k-1}, we have that the sequence $\{\pi_k\}$ is a Markov process with respect to the filtration $\{\mathcal{I}_k\}$.

Once the selection of the optimal pair $\{\lambda^{(1)}, \lambda^{(2)}\}$ is done as described above, we note that for fixed T we have

$$P(T < t) + c\,E\left\{(T-t)^+\right\} = (1-\pi)\left[P_0(T < t) + cE_0\{(T-t)^+\}\right] + c\pi E_1\{T\},$$

where subscripts refer to values of π under which the corresponding quantities are computed and from which it follows that J is a concave function satisfying

$$0 \leq J(\pi) \leq 1 - \pi, \; 0 \leq \pi \leq 1.$$

In view of Remark 7.6, it follows that the optimal stopping time for the fusion center is

$$T_{\text{opt}} = \inf\{k | J(\pi_k) = 1 - \pi_k\}. \tag{7.47}$$

From the concavity of J, it now readily follows that (7.47) can be rewritten as

$$T_{\text{opt}} = \inf\{k | \pi_k \geq \pi^*\}, \tag{7.48}$$

where $\pi^* = \inf\{\pi | J(\pi) = 1 - \pi\}$.

It is worth recalling at this point that in this analysis it has been assumed that the change point t is a random variable. Other approaches have been developed under which the assumption that t is an unknown constant is used. Please refer to [64], [145], and [153] for discussion of these cases. As an example, in [145] it is shown that a monotone likelihood ratio quantizer (MLRQ) is asymptotically optimal as the frequency of false alarms tends to ∞.

7.2.3 Decentralized sequential detection without fusion

In this subsection, we turn to the problem of decentralized sequential detection in the absence of a fusion center. More specifically, in this setting the final decision about the state of the environment is made locally by the sensors and not at a fusion center. To formalize this problem, consider a random variable that represents the true hypothesis H about the state of the environment. This random variable can take the value H_1 with probability π and H_0 with probability $1 - \pi$.

Each of the sensors S_1 and S_2 receives a sequence of observations $\{Z_k^{(1)}\}$ and $\{Z_k^{(2)}\}$, respectively. Under each hypothesis, the sequences $\{Z_k^{(i)}\}$ are i.i.d. and are independent of one another. The observations at sensor i are assumed to have marginal probability density functions[5] $q_j^{(i)}$, given hypothesis H_j.

The probability space of interest here is $(\Omega, \mathcal{F}) = (\mathbb{R}^\infty \times \mathbb{R}^\infty, \mathcal{B}^\infty \times \mathcal{B}^\infty)$ equipped with the probability measure $P = \pi P_1 + (1-\pi)P_0$, where $P_1 = P_1^{(1)} \times P_1^{(2)}$ and $P_0 = P_0^{(1)} \times P_0^{(2)}$ and where $P_j^{(1)}$ and $P_j^{(2)}$ denote the restrictions of P_j to the filtrations

(1) $\{\mathcal{F}_k^{(1)}\}$ with $\mathcal{F}_k^{(1)} = \sigma\{Z_1^{(1)}, \ldots, Z_k^{(1)}\}$; and
(2) $\{\mathcal{F}_k^{(2)}\}$ with $\mathcal{F}_k^{(2)} = \sigma\{Z_1^{(2)}, \ldots, Z_k^{(2)}\}$,

[5] Note that here we do not make the assumption of identicality of the sensors, as they largely operate autonomously in this formalism.

respectively. Each of the sensors S_i devises a sequential decision rule $(T^{(i)}, \delta^{(i)})$ as described in Chapter 4. Furthermore, we introduce a cost function $C(\delta^{(1)}, \delta^{(2)}; H)$ that represents the cost of making a mistake in none, either or both of the decisions made by the sensors. We assume that

$$\begin{aligned} C(0, \delta^{(2)}; H_1) &\geq C(1, \delta^{(2)}; H_1), \\ C(1, \delta^{(2)}; H_0) &\geq C(1, \delta^{(2)}; H_1), \\ C(1, \delta^{(2)}; H_0) &\geq C(0, \delta^{(2)}; H_0), \quad \text{and} \\ C(0, \delta^{(2)}; H_1) &\geq C(0, \delta^{(2)}; H_1). \end{aligned}$$

It is further assumed that similar inequalities hold for $\delta^{(1)}$. The above set of inequalities mean that at least one error is more or equally costly than at most one. Finally it is assumed that for each additional observation a cost of c is incurred by the corresponding individual sensor. This gives rise to the following Bayesian decision problem (see [211]):

$$\inf_{\{(T^{(i)}, \delta^{(i)})\}} E\left\{cT^{(1)} + cT^{(2)} + C(\delta^{(1)}, \delta^{(2)}; H)\right\}. \tag{7.49}$$

REMARK 7.7. An alternative to the above problem, in which a positive cost c is associated with each additional time step (i.e. each observation) taken by the sensors as a team is also of interest. The solution of this problem, which is treated in [221], can be found using the same method as the one we use here to solve (7.49).

To address (7.49), let us consider the common structure of all person-by-person optimal solutions. That is, let us fix $(T^{(2)}, \delta^{(2)})$, possibly at the optimum. Then sensor S_1 is faced with the following stochastic optimization problem:

$$J(\pi) = \inf_{\{(T^{(1)}, \delta^{(1)})\}} E\left\{cT^{(1)} + C(\delta^{(1)}, \delta^{(2)}; H)\right\}. \tag{7.50}$$

We note that in the special case in which $C(\delta^{(1)}, \delta^{(2)}, H) = C(\delta^{(1)}, H) + C(\delta^{(2)}, H)$, this problem degenerates to the classical sequential detection problem treated in Chapter 4. It is precisely the fact that the cost function is possibly coupled between the two sensors that makes this problem interesting.

Before we proceed to the solution of (7.50), we present a sufficient statistic for its solution, namely

$$\pi_k^{(1)} = P(H = H_1 | \mathcal{F}_k^{(1)}), \tag{7.51}$$

and note the following recursion resulting from Bayes' formula:

$$\pi_{k+1}^{(1)} = \frac{\pi_k^{(1)} q_1^{(1)}(x)}{\pi_k^{(1)} q_1^{(1)}(x) + (1 - \pi_k^{(1)}) q_0^{(1)}(x)}, \tag{7.52}$$

with $\pi_0^{(1)} = \pi$. From (7.52), it becomes obvious that $\{\pi_k^{(1)}\}$ forms a Markov process with respect to the filtration $\{\mathcal{F}_k^{(1)}\}$.

To solve (7.50), we consider the finite horizon problem, in which sensor S_1 stops sampling and makes a decision not later than time τ. Let us denote by J_k^τ the minimal expected cost to go at time k for this problem. Then we have the following dynamic programming equation.

(1) When $\{T^{(1)} = \tau\}$, we have

$$J^\tau_{T^{(1)}}(\pi^{(1)}_{T^{(1)}}) = \inf\left\{E\left\{C(0,\delta^{(2)};H)|\mathcal{F}^{(1)}_{T^{(1)}}\right\}, E\left\{C(1,\delta^{(2)};H)|\mathcal{F}^{(1)}_{T^{(1)}}\right\}\right\}. \tag{7.53}$$

(2) When $\{T^{(1)} = k\}, k = 1, \ldots, \tau-1$, we have

$$J^\tau_{T^{(1)}}(\pi^{(1)}_{T^{(1)}}) = \inf\left\{E\left\{C(0,\delta^{(2)};H)|\mathcal{F}^{(1)}_{T^{(1)}}\right\}, E\left\{C(1,\delta^{(2)};H)|\mathcal{F}^{(1)}_{T^{(1)}}\right\},\right.$$
$$\left. c + C^\tau_k(\pi^{(1)}_k)\right\}, \tag{7.54}$$

where

$$C^\tau_k(\pi^{(1)}_k) = E\left\{J^\tau_{k+1}(\pi^{(1)}_{k+1})|\mathcal{F}^{(1)}_k\right\}.$$

We note that J^τ_0 is the minimal expected cost for the finite horizon problem. Equations (7.53) and (7.54) explicitly show the dependence of the minimal expected cost to go on the sufficient statistic $\pi^{(1)}_k$. This becomes evident by expanding the right-hand side of (7.53), according to

(1) $E\left\{C(0,\delta^{(2)};H)|\mathcal{F}^{(1)}_{T^{(1)}}\right\} = \sum_{d=0}^1\sum_{j=0}^1 P_j(\delta^{(2)} = d)C(0,d;H_j)$
$\times P\left(H = H_j|\mathcal{F}^{(1)}_{T^{(1)}}\right),$

(2) $E\left\{C(1,\delta^{(2)};H)|\mathcal{F}^{(1)}_{T^{(1)}}\right\} = \sum_{d=0}^1\sum_{j=0}^1 P_j(\delta^{(2)} = d)C(1,d;H_j)$
$\times P\left(H = H_j|\mathcal{F}^{(1)}_{T^{(1)}}\right),$

and using (7.51). The same holds true for (7.54), as we see from the relationship

$$E\left\{J^\tau_{k+1}(\pi_{k+1})|\mathcal{F}^{(1)}_k\right\} = \int J^\tau_{k+1}(\pi^{(1)}_{k+1})\left[\pi^{(1)}_k q^{(1)}_1(x) + (1-\pi^{(1)}_k)q^{(1)}_0(x)\right]\mathrm{d}x.$$

Regarding J^τ_k and C^τ_k, we have the following result.

PROPOSITION 7.8. *J^τ_k and C^τ_k are concave functions of π on* $[0, 1]$.

Proof: We have that J^τ_τ is the pointwise minimum of two linear functions of π and hence it is concave. Now suppose that J^τ_{k+1} is a concave function of π. Then it can be written as $\inf_i(a_i\pi + b_i)$ for some $a_i, b_i, i = 1, 2, \ldots$. With this representation we have that

$$C^\tau_k(\pi^{(1)}_k) = \int \inf_i(a_i\pi^{(1)}_{k+1} + b_i)\left[\pi^{(1)}_k q^{(1)}_1(x) + (1-\pi^{(1)}_k)q^{(1)}_0(x)\right]\mathrm{d}x$$
$$= \int \inf_i\left[(a_i + b_i)\pi^{(1)}_k q^{(1)}_1(x) + b_i(1-\pi^{(1)}_k)q^{(1)}_0(x)\right]\mathrm{d}x.$$

Consequently, C^τ_k is a concave function of π, as is J^τ_k, and by induction the claim holds true for every $k < \tau$. This establishes the proposition. ■

We further have the following preliminary result.

PROPOSITION 7.9. *Define* $f(\pi_k^{(1)}) = \min\left\{E\left\{C(0,\delta^{(2)};H)|\mathcal{F}_k^{(1)}\right\}, E\left\{C(1,\delta^{(2)};H)|\mathcal{F}_k^{(1)}\right\}\right\}$. *Then the following inequalities hold for all* $k = 0, \ldots, \tau$.

$$f(0) < c + C_k^\tau(0), \tag{7.55}$$

and

$$f(1) < c + C_k^\tau(1). \tag{7.56}$$

Moreover,

$$J_k^\tau(\pi) \leq J_{k+1}^\tau(\pi),\ 0 \leq \pi \leq 1, \tag{7.57}$$

and

$$C_k^\tau(\pi) \leq C_{k+1}^\tau(\pi),\ 0 \leq \pi \leq 1. \tag{7.58}$$

Proof: The proof of (7.55) and (7.56) follows by the definitions of the respective functions and induction. For the monotonicity results of (7.57) and (7.58), we note that each of the left-hand quantities is the infimum over a larger set of stopping times than is the corresponding right-hand quantity. This establishes the proposition. ∎

To solve problem (7.50), we consider the limit $\tau \to \infty$. By following an argument identical to the one used in Section 7.2.1, it follows that the pointwise limit of J_k^τ exists and is independent of k. More specifically we have that

$$J(\pi) = \lim_{\tau\to\infty} J_k^\tau(\pi) \tag{7.59}$$

$$= \inf_{\tau\to\infty} J_k^\tau(\pi),\ 0 \leq \pi \leq 1. \tag{7.60}$$

THEOREM 7.10. *The minimal expected cost to go* $J(\pi)$ *satisfies the Bellman equation*

$$J(\pi) = \min\left\{E\left\{C(0,\delta^{(2)};H)\right\}, E\left\{C(1,\delta^{(2)};H)\right\}, c + C_J(\pi)\right\},\ 0 \leq \pi \leq 1. \tag{7.61}$$

where[6]

$$C_J(\pi) = E\{J(\pi_1)\},\ 0 \leq \pi \leq 1.$$

The optimal stopping time is

$$T_{\text{opt}} = \inf\{k|\pi_k^{(1)} \notin (\pi_{\text{L}}^{(1)}, \pi_{\text{U}}^{(1)})\}, \tag{7.62}$$

where the thresholds of detector 1 are characterized by

$$\pi_{\text{L}}^{(1)} = \sup\left\{0 \leq \pi \leq \frac{1}{2}\middle| c + C_J(\pi) = E\left\{C(0,\delta^{(2)};H)\right\}\right\}, \tag{7.63}$$

and

$$\pi_{\text{U}}^{(1)} = \inf\left\{\frac{1}{2} \leq \pi \leq 1\middle| c + C_J(\pi) = E\left\{C(1,\delta^{(2)};H)\right\}\right\}. \tag{7.64}$$

[6] Note that $C_J(\pi)$ depends on π through the dependence of $E\{\cdot\}$ on π.

Proof: Equation (7.61) follows by taking the limit in (7.54) and using (7.59). The concavity of J follows from the fact that it is the limit of concave functions as Proposition 7.8 asserts. Inequalities similar to (7.55) and (7.56) also hold by the same token. Using these inequalities, the concavity of C_J, and (7.61), it follows that the optimal stopping time is of the threshold type as given in (7.62), where the thresholds are determined by

$$c + C_J(\pi_L^{(1)}) = E\left\{C(0, \delta^{(2)}; H)\right\}\Big|_{\{\pi = \pi_L^{(1)}\}}, \tag{7.65}$$

and

$$c + C_J(\pi_U^{(1)}) = E\left\{C(1, \delta^{(2)}; H)\right\}\Big|_{\{\pi = \pi_U^{(1)}\}}. \tag{7.66}$$

This establishes the theorem. ■

REMARK 7.11. The uniqueness of the limiting value function for problem (7.50) follows by using an argument similar to the one used in the proof of Proposition 7.4. Moreover from (7.65) and (7.66) it is evident that the optimal thresholds $\pi_L^{(1)}$ and $\pi_U^{(1)}$ are coupled, and their determination requires the solution of two simultaneous dynamic programming equations.

In the above, we have derived the optimal local decision rule for S_1 for a fixed value of $(T^{(2)}, \delta^{(2)})$. The globally optimal decision rule is achieved when both sensors simultaneously implement their individually optimal decision rule for each other's optimal decision rule. This globally optimal rule can be found iteratively by successively fixing one sensor's thresholds and optimizing the other's via Theorem 7.10.

Ultimately, the optimal decision rules for the sensors $S_i, i = 1, 2$, are of the following form:

(1) accept H_0 if $\pi_k^{(i)} \leq \pi_L^{(i)}$;
(2) accept H_1 if $\pi_k^{(i)} \geq \pi_U^{(i)}$; or
(3) continue sampling if $\pi_L^{(i)} \leq \pi_k^{(i)} \leq \pi_U^{(i)}$.

The thresholds $(\pi_L^{(i)}, \pi_U^{(i)})$ are related to the thresholds (A_i, B_i) of the per-sensor SPRT's via

$$A_i = \frac{\pi}{1-\pi} \frac{(1 - \pi_U^{(i)})}{\pi_U^{(i)}} \tag{7.67}$$

and

$$B_i = \frac{\pi}{1-\pi} \frac{(1 - \pi_L^{(i)})}{\pi_L^{(i)}}. \tag{7.68}$$

Notice the similarity of (7.67) and (7.68) to (4.46). See [221] for further details of these results.

A similar approach to that described above can be applied to the problem of decentralized quickest detection. In this problem each of the sensors S_j receives sequentially observations $\{Z_k^{(j)}\}$ that have a known marginal density $q_0^{(j)}$ for $k = 1, \ldots, t-1$ and $q_1^{(j)}$ for $k = t, \ldots$. The change point t is assumed to have a geometric distribution with a mass at 0 as in Chapter 5. The sensor observations are assumed to be conditionally independent given the value of the change point. This conditional independence is valid within sensors and across sensors. The problem that each of the sensors face is to optimally select stopping times $T^{(i)}$ (each measurable with respect to their own filtrations $\mathcal{F}^{(i)}$), so as to detect the change point as soon as possible, while at the same time controlling the probability of false alarms. The objective is to minimize $E\left\{C(T^{(1)}, T^{(2)}; t)\right\}$ over all stopping times $T^{(1)}$ and $T^{(2)}$, where

$$C(T^{(1)}, T^{(2)}; t) = 1_{\{T^{(1)}<t\}} 1_{\{T^{(2)}<t\}} + c_1(T^{(1)} - t) 1_{\{T^{(1)}\geq t\}} + c_2(T^{(2)} - t) 1_{\{T^{(2)}\geq t\}}.$$

This problem is treated in [212], where it is shown that the optimal solution to the above problem is

$$T^{(1)} = \inf\{k | P(t \leq k | \mathcal{F}_k^{(1)}) \geq \pi_1^*\},$$

and

$$T^{(2)} = \inf\{k | P(t \leq k | \mathcal{F}_k^{(2)}) \geq \pi_2^*\},$$

where the optimal thresholds π_i^* are coupled through a system of two dynamic programming equations. It is important to note that it is the term $1_{\{T^{(1)}<t\}} 1_{\{T^{(2)}<t\}}$ that appears in the cost function that couples the solution. Just as in (7.50), if we were able to write $C(T^{(1)}, T^{(2)}; t) = C(T^{(1)}; t) + C(T^{(2)}; t)$, the problem faced by each of the sensors would be the typical change-point problem whose solution appears in Chapter 5. Again, a treatment of this problem is found in [212].

7.3 Quickest detection with modeling uncertainty

In Chapters 5 and 6 we considered the quickest detection problem under the assumption that the pre-change and the post-change distributions of the observations are known. In this section we discuss strategies for handling situations in which there is a degree of uncertainty with these distributions. We will begin, in the following subsection, by considering the case in which this uncertainty can be modeled by introducing non-parametric uncertainty classes within which the pre- and post-change distributions must lie. Here, we seek change detection procedures whose performance is robust to uncertainty within these classes. Then in Section 7.3.2, we consider adaptive quickest detection for situations in which the uncertainty can be parametrized.

7.3.1 Robust quickest detection

Consider a sequence of independent real-valued observations $\{Z_k\}$ with the same characteristics as in Section 6.2. In particular, as before, there is an unknown (non-random) change point t, at which there is an abrupt change in the distribution of the given sequence. The measurable space is $(\mathbb{R}^\infty, \mathcal{B}^\infty)$ and, for fixed t, $\{Z_k; k = 1, 2, \ldots\}$ is a sequence of independent random variables with $Z_1, \ldots, Z_{t-1}$ being i.i.d. with marginal distribution Q_0 and $Z_t, \ldots$ being i.i.d. with marginal distribution Q_1. So far, this is exactly the model of Section 6.2. However, now we wish to model the situation in which the distributions Q_0 and Q_1 are not known exactly. Thus, rather than assuming that we know Q_0 and Q_1, we assume instead that they are known only to be within respective classes $\mathcal{Q}_0$ and $\mathcal{Q}_1$ of probability distributions on $(\mathbb{R}, \mathcal{B})$. For simplicity of exposition, we will consider the specific case in which all elements of $\mathcal{Q}_0$ and $\mathcal{Q}_1$ have probability density functions. That is, the Z_k's have a common marginal density q_0 before the change and q_1 after the change. So it is not only the case that we do not know the change point, but that we are also uncertain about which member of $\mathcal{Q}_0$ describes the observations before t and which member of $\mathcal{Q}_1$ does so after t.

Since, as in Chapter 6, we assume that the change point is an unknown constant, the objective is to minimize the worst detection delay (i.e., Lorden's delay penalty), while insisting on a lower bound on the mean time between false alarms, γ. However, we will need to be more specific about what it means to achieve optimal performance when the pair (Q_0, Q_1) is known only to lie in $\mathcal{Q}_0 \times \mathcal{Q}_1$. Naturally, the most desirable design procedure would be the one that minimizes the worst detection delay uniformly over all possible pairs $(Q_0, Q_1) \in \mathcal{Q}_0 \times \mathcal{Q}_1$. Unfortunately, such a procedure does not usually exist.

An alternative way to approach this problem is to seek procedures that minimize the worst-case performance over all pairs of pre- and post-change distribution $(Q_0, Q_1) \in \mathcal{Q}_0 \times \mathcal{Q}_1$; i.e., to adopt a minimax approach, where the maximization used to define $J(T)$ now extends to the unknown distributions Q_0 and Q_1. As in many such formulations (see, e.g. [120]), it turns out that this problem can be solved (at least asymptotically) by choosing stopping times that optimize the performance for a "least favorable" pair of distributions $(Q_{0L}, Q_{1L}) \in \mathcal{Q}_0 \times \mathcal{Q}_1$ for that particular model. Although the performance of such a stopping time may be less than optimal for other pairs of distributions in $\mathcal{Q}_0 \times \mathcal{Q}_1$, the use of procedures of the kind described above guarantees a given level of performance over the entire class under certain conditions (again see, e.g. [120] for a general discussion of such problems).

Let us briefly remind ourselves of the stochastic optimization problem that arises by minimizing Lorden's delay penalty $J(T)$ subject to a given constraint in the mean time between false alarms γ (see Chapter 6), when Q_0 and Q_1 are known. In particular, it was shown in Chapter 6 that Page's CUSUM stopping time is its optimal solution. That is, the stopping time

$$T_h = \inf\{n | S_n \geq h\}, \tag{7.69}$$

with

$$S_n = \max\{0, S_{n-1} + g(Z_n)\}, \; S_0 = 0, \; n = 1, \ldots,$$

where $g(Z_k) = \log L(Z_k) = \log \frac{q_1(Z_k)}{q_0(Z_k)}$ achieves the lowest detection delay $J(T)$ for a given false-alarm constraint γ. Recall also that the performance of this solution can be characterized asymptotically by the quantity, (see Theorem 6.19)

$$\eta = \lim_{\gamma \to \infty} \frac{\log \gamma}{J(T_h)}, \tag{7.70}$$

where η is the Kullback–Leibler divergence of Q_1 with respect to Q_0 is as defined in (6.1). Note that for large γ we have the approximation $J(T) \approx (\log \gamma)/\eta$, and so to minimize $J(T)$ (asymptotically in γ) one needs to maximize η.

The lack of precise knowledge of the pre- and post-change distributions leads us to consider stopping times that will replace the function $L = \log q_1/q_0$ in (7.70) with choices other than L, which we will denote generically by g. An issue that immediately arises is how we can characterize the asymptotic performance of such stopping times. Because of the optimality of the CUSUM, such a stopping time will have a value of η (as in (7.70)) that satisfies $\eta \le D(Q_1 \parallel Q_0)$. It is unclear however, how one could go about calculating η for a general choice of g. To address this problem, Broder [49] has shown that the following quantity is a lower bound on η for a general class of functions g:

$$\widetilde{\eta} = \omega_0 \int_{-\infty}^{+\infty} g(x) q_1(x) \mathrm{d}x, \tag{7.71}$$

where ω_0 is the unique non-zero root of the moment-generating function equality

$$\int_{-\infty}^{\infty} \mathrm{e}^{\omega_0 g(x)} q_0(x) \mathrm{d}x = 1. \tag{7.72}$$

REMARK 7.12. The moment-generating function equality (7.72) has

- one and only one non-zero root $\omega_0 < 0$ if $\int_{-\infty}^{\infty} g(x) q_0(x) \mathrm{d}x > 0$,
- one and only one non-zero root $\omega_0 > 0$ if $\int_{-\infty}^{\infty} g(x) q_0(x) \mathrm{d}x < 0$, or
- no non-zero real root if $\int_{-\infty}^{\infty} g(x) q_0(x) \mathrm{d}x = 0$

provided there exists $1 > \delta > 0$ such that

(1) $Q_0 \left(g(Z_k) > \ln(1+\delta) \right) > 0$; and
(2) $Q_0 \left(g(Z_k) > \ln(1-\delta) \right) > 0$.

For further discussion on this result please refer to the discussion following (4.79), and for a formal proof please refer to [93, p. 103].

With regard to the quantity (7.71), we have the following useful property.

PROPOSITION 7.13. *We have that* $\widetilde{\eta} = \eta = D(Q_1 \parallel Q_0)$ *if and only if* $g = C \cdot L$ *for some* $C > 0$, *where* $L = \log q_1/q_0$.

Proof: Suppose that $g = C \log q_1/q_0$; then we have

$$\begin{aligned} \widetilde{\eta} &= \omega_0 C \int_{-\infty}^{\infty} \log \frac{q_1(x)}{q_0(x)} q_1(x) \mathrm{d}x \\ &= \omega_0 C D(Q_1 \parallel Q_0), \end{aligned}$$

where ω_0 satisfies

$$1 = \int_{-\infty}^{\infty} \exp\left\{\omega_0 C \log \frac{q_1(x)}{q_0(x)}\right\} q_0(x)\mathrm{d}x$$
$$= \int_{-\infty}^{\infty} \left[\frac{q_1(x)}{q_0(x)}\right]^{\omega_0 C} q_0(x)\mathrm{d}x,$$

from which it follows that $\omega_0 = 1/C$. Therefore, $\widetilde{\eta} = \eta = D(Q_1 \parallel Q_0)$. Now suppose that $\widetilde{\eta} = \eta$. We want to show that the function g that solves the optimization problem

$$\max_g \widetilde{\eta} = \omega_0 \int_{-\infty}^{\infty} g(x) q_1(x)\mathrm{d}x \tag{7.73}$$

subject to

$$1 = \int_{-\infty}^{\infty} \exp\{\omega_0 g(x)\} q_0(x)\mathrm{d}x, \tag{7.74}$$

is given by $g = C \log q_1/q_0$ for some $C > 0$.

To solve this problem, it suffices to find g that minimizes

$$\int_{-\infty}^{\infty} \left[\omega_0 g(x) q_1(x) + \lambda \exp\{\omega_0 g(x)\} q_0(x)\right] \mathrm{d}x,$$

where λ is the Lagrange multiplier associated with the constraint (7.74). Suppose that $\widehat{g}$ is the solution to the above problem and consider an arbitrary perturbation of $\widehat{g}$ in the neighborhood of an arbitrary function p. That is, consider $g = \widehat{g} + \epsilon\delta p$. The problem now becomes one of finding a stationary point (in ϵ) of

$$\int_{-\infty}^{\infty} \left(\omega_0 \left[\widehat{g}(x) + \epsilon\delta p(x)\right] q_1(x) + \lambda \exp\left\{\omega_0 \left[\widehat{g}(x) + \epsilon\delta p(x)\right]\right\} q_0(x)\right) \mathrm{d}x.$$

Taking the derivative of the above expression with respect to ϵ, and setting the result equal to 0 we have

$$\omega_0 \int_{-\infty}^{\infty} \delta p(x) \left[q_1(x) + \lambda e^{\omega_0 \widehat{g}(x)} q_0(x)\right] \mathrm{d}x = 0.$$

In order for the above equality to hold for arbitrary δp it is necessary (and sufficient) that

$$\left[q_1(x) + \lambda e^{\omega_0 \widehat{g}(x)} q_0(x)\right] = 0, \text{ for all } x \in \mathbb{R}.$$

Or equivalently

$$\omega_0 \widehat{g}(x) = \log \frac{q_1(x)}{q_0(x)} - \log(-\lambda), \text{ for all } x \in \mathbb{R},$$

where from (7.74) we have $\lambda = -1$. Therefore it follows that

$$\widehat{g}(x) = \frac{1}{\omega_0} \log \frac{q_1(x)}{q_0(x)}.$$

The result now follows by letting $C = 1/\omega_0$, and noticing that the second derivative of

$$\int_{-\infty}^{\infty} \left(\omega_0 \left[\widehat{g}(x) + \epsilon\delta p(x)\right] q_1(x) + \lambda \exp\left\{\omega_0 \left[\widehat{g}(x) + \epsilon\delta p(x)\right]\right\} q_0(x)\right) \mathrm{d}x \bigg|_{\lambda=-1},$$

with respect to ϵ, evaluated at $\widehat{g}$, is negative, which establishes the fact that $\widehat{g}$ is in fact a maximum. This completes the proof of the proposition. ■

Using the above formulation, we can consider an asymptotic max–min stopping time to be one that solves the problem

$$\sup_{g} \inf_{(Q_0,Q_1)\in\mathcal{Q}_0\times\mathcal{Q}_1} \tilde{\eta}(g; Q_0, Q_1), \tag{7.75}$$

where $\tilde{\eta}(g; Q_0, Q_1)$ is defined through (7.71). It is shown in [49], that solving the max–min problem of (7.75) is equivalent to solving the min–max problem

$$\inf_{(Q_0,Q_1)\in\mathcal{Q}_0\times\mathcal{Q}_1} \sup_{g} \tilde{\eta}(g; Q_0, Q_1). \tag{7.76}$$

In view of the above proposition, given any choice of the pair $(Q_0, Q_1) \in \mathcal{Q}_0 \times \mathcal{Q}_1$, the choice of the function g that maximizes $\tilde{\eta}$ is $L = \ln q_1/q_0$ which leads to $\tilde{\eta}(L, Q_0, Q_1) = D(Q_1 \parallel Q_0)$. Hence problem (7.76) can be solved by first solving for a minimizing pair of distributions $(Q_{0L}, Q_{1L}) \in \mathcal{Q}_0 \times \mathcal{Q}_1$, that is, a pair solving

$$\inf_{(Q_0,Q_1)\in\mathcal{Q}_0\times\mathcal{Q}_1} D(Q_1 \parallel Q_0), \tag{7.77}$$

(assuming that such a pair exists), and then choosing $g = \log \frac{\mathrm{d}Q_{1L}}{\mathrm{d}Q_{0L}}$.

By drawing on work of Brandt and Karlin [44], it can be shown that (7.77) has the same solution for a given pair of classes $\mathcal{Q}_0$ and $\mathcal{Q}_1$ as a well-known problem in robust hypothesis testing posed by Huber in [108]. Solutions to the latter problem are known for several useful choices of $\mathcal{Q}_0$ and $\mathcal{Q}_1$, and therefore solutions to (7.77) and (7.76) can be found directly.

As an example, consider the very useful and intuitive ϵ-contaminated model:

Example 7.1: Suppose ϵ_0 and $\epsilon_1 \in (0, 1)$, and define the classes $\mathcal{Q}_0$ and $\mathcal{Q}_1$ in terms of respective classes $\mathcal{C}_0$ and $\mathcal{C}_1$ of density functions:

$$\mathcal{C}_0 = \{f \mid f = (1 - \epsilon_0) f_{n0} + \epsilon_0 h,\ h \in \mathcal{H}\}, \tag{7.78}$$

and

$$\mathcal{C}_1 = \{f \mid f = (1 - \epsilon_1) f_{n1} + \epsilon_1 h,\ h \in \mathcal{H}\}, \tag{7.79}$$

where $\mathcal{H}$ is the class of all probability density functions on $\mathbb{R}$, and where f_{n0} and f_{n1} are nominal densities for the classes $\mathcal{C}_0$ and $\mathcal{C}_1$, respectively.

The least favorable pair of densities in the above ϵ-contaminated classes for robust hypothesis testing are known to be (see [108])

$$f_{0L}(x) = \begin{cases} (1 - \epsilon_0) f_{n0}(x), & f_{n1}(x) < a_0 f_{n0}(x) \\ \frac{1}{a_0}(1 - \epsilon_0) f_{n1}(x), & f_{n1}(x) \ge a_0 f_{n0}(x) \end{cases}$$

and

$$f_{1L}(x) = \begin{cases} (1 - \epsilon_1) f_{n0}(x), & f_{n1}(x) < a_1 f_{n0}(x) \\ a_1(1 - \epsilon_1) f_{n1}(x), & f_{n1}(x) \ge a_1 f_{n0}(x) \end{cases}$$

where a_0, a_1 satisfy $0 < a_1 \le 1 \le a_0$, and are selected so that f_{0L} and f_{1L} are densities. That is, they satisfy:

$$P\left(\frac{f_{n1}(Z_k)}{f_{n0}(Z_k)} < a_0 \middle| f_{n0}\right) + \frac{1}{a_0} P\left(\frac{f_{n1}(Z_k)}{f_{n0}(Z_k)} \ge a_0 \middle| f_{n1}\right) = \frac{1}{1-\epsilon_0}, \tag{7.80}$$

and

$$P\left(\frac{f_{n1}(Z_k)}{f_{n0}(Z_k)} > a_1 \middle| f_{n1}\right) + a_1 P\left(\frac{f_{n1}(Z_k)}{f_{n0}(Z_k)} \le a_1 \middle| f_{n0}\right) = \frac{1}{1-\epsilon_1}, \tag{7.81}$$

where $P(\cdot|f)$ denotes probability assuming that Z_k has probability density f. Note that a_0 and a_1 are unique. (See [108].)

The above f_{0L} and f_{1L} give rise to the following log-likelihood ratio:

$$g_R(x) = \begin{cases} \log a_1 + \log \frac{1-\epsilon_1}{1-\epsilon_0}, & f_{n1}(x) \le a_1 f_{n0}(x) \\ \log \frac{f_{n1}(x)}{f_{n0}(x)} + \log \frac{1-\epsilon_1}{1-\epsilon_0}, & a_1 f_{n0}(x) < f_{n1}(x) < a_0 f_{n0}(x) \\ \log a_0 + \log \frac{1-\epsilon_1}{1-\epsilon_0}, & f_{n1}(x) \ge a_0 f_{n0}(x) \end{cases}$$

which, aside from the (inconsequential) additive constant $\log \frac{1-\epsilon_1}{1-\epsilon_0}$, is simply a censored version of the log-likelihood ratio for the nominal model, $\log \frac{f_{n1}(x)}{f_{n0}(x)}$. For example, in the case in which $\epsilon_0 = \epsilon_1$ and f_{n0} and f_{n1} are $\mathcal{N}(0,1)$ and $\mathcal{N}(\mu,1)$ densities, respectively, g_R is given by

$$g_R(x) = \begin{cases} \log a_1, & \mu x - \frac{1}{2}\mu^2 \le \log a_1 \\ \mu x - \frac{1}{2}\mu^2, & \log a_1 < \mu x - \frac{1}{2}\mu^2 < \log a_0 . \\ \log a_0, & \mu x - \frac{1}{2}\mu^2 \ge \log a_0 \end{cases}$$

When a least-favorable pair of distributions (Q_{0L}, Q_{1L}) (i.e. a minimizing pair in (7.77)) exists for the classes $\mathcal{Q}_0$ and $\mathcal{Q}_1$, then the triple $(g_R; Q_{0L}, Q_{1L})$ with $g_R = \frac{dQ_{1L}}{dQ_{0L}}$ forms a *saddle point* of the max–min game (7.75). That is, not only does g_R maximize $\tilde{\eta}(g; Q_{0L}, Q_{1L})$, but we also have

$$\tilde{\eta}(g_R; Q_{0L}, Q_{1L}) \le \tilde{\eta}(g_R; Q_0, Q_1), \ \forall\, (Q_0, Q_1) \in \mathcal{Q}_0 \times \mathcal{Q}_1. \tag{7.82}$$

Thus, the bound $\tilde{\eta}$ for the stopping time based on g_R is guaranteed to be no smaller for any pair of distributions in the uncertainty classes than it is for the least-favorable pair. This implies further that η itself is uniformly lower-bounded over $\mathcal{Q}_0 \times \mathcal{Q}_1$ for this choice of stopping time, thus leading to the claim of robustness. For further discussion please refer to [49] and [120].

Note that a tacit assumption in the above discussion is that the classes $\mathcal{Q}_0$ and $\mathcal{Q}_1$ represent relatively small degrees of uncertainty in the model; e.g. in the ϵ-contaminated example, ϵ_0 and ϵ_1 should be small. Otherwise the performance guarantee of (7.82) could be very pessimistic. When $\mathcal{Q}_0$ and $\mathcal{Q}_1$ are instead very broad classes, then other approaches, known as non-parametric approaches, are more suitable than the max–min formalism treated here. Some work in the area of non-parametric quickest detection is found in [3,31,35,36,49,63,97,98,158,166].

7.3.2 Adaptive quickest detection

In the previous subsection we have discussed the problem of detecting a change in the case of uncertainty in the pre-change and post-change distributions. In this section we will discuss the problem of detecting a change while at the same time attempting to estimate unknown parameters of the post-change distribution. The difficulties pertaining to achieving this goal are inherent in the uncertainty of the location of the change point.

To consider a specific situation, let us again consider the case of independent discrete-time observations with a non-random change point t, and suppose that the observations are described by a parametric family of distributions whose (real) parameter is $\theta = \theta_0$ before the change, and changes to $\theta = \theta_1$ at the change point, where θ_1 is at least equal to $\theta_0 + \nu_0$ for a fixed positive value of ν_0. In many applications, it is natural to assume that the pre-change distribution is known (or that θ_0 can be estimated) since the pre-change distribution in most practical examples, although not all, corresponds to a normal state of the sequence. Hence, it is typically easy to find historical data and use standard statistical methods to estimate θ_0. The problem of estimating θ_1 however, is much more complex, since we do not know which is the first observation that we should use in estimating it.

A traditional approach to such problems, as described in [139], is to use the generalized CUSUM stopping time defined as

$$T = \inf \left\{ n;\ \max_{0 \le k \le n} \sup_{\theta_1 \ge \theta_0 + \nu_0} \sum_{j=k}^{n} \log L_{\theta_1}(Z_j) \ge a \right\},$$

where L_{θ_1} denotes the likelihood ratio between the post- and pre-change marginal distributions, assuming that $\theta = \theta_1$ after the change. Properties of this stopping time have been studied and its asymptotic optimality as the false-alarm constraint $\gamma \to \infty$ under certain conditions has been established in [77] and in [139]. The main drawback of this stopping time is that it does not admit a recursive implementation. That is, in terms of computation, at each time step all observations must be stored, and the supremum recomputed afresh.

In this section we will present an alternative adaptive algorithm for detecting a change in distribution in the case in which the post-change parameter of the distribution is unknown, following [63]. For simplicity, we treat here the case in which the marginal distribution of the Z_k's is $\mathcal{N}(\theta, \sigma^2)$, where $\sigma^2 > 0$ is known, although the results can be extended to more general cases. (See [63].)

Consider the following stopping time in which L_θ, defined as above, is a monotone function of its argument and $h > 0$ is a fixed threshold.

(1) Initialize $i = 0$, $S_0 = 0$.
(2) $i = i + 1$.
(3) $S_i = \max\{0, S_{i-1} + \log L_{\theta_0+\nu_0}(Z_i)\}$. If $S_i < h$ then repeat 2.
(4) $k = i + 1$.
(5) $i = i + 1$. $S_i = \sum_{j=k}^{i} \log L_{\tilde{\theta}_i}(Z_j)$.

(6) If $a < S_i < b$, repeat 4. If $S_i \leq a$, set $S_i = 0$ go to step 2. Otherwise, proceed to (5).
(7) Declare a disorder at time i.

In step (5), $\widetilde{\theta}_i = \max\{\theta_0 + \nu_0, \hat{\theta}_i\}$ with $\hat{\theta}_i = \frac{1}{i-k+1}\sum_{j=k}^{i} Z_j$. Note that $\hat{\theta}_i$ is an estimate of θ_1 under the assumption that $Z_k, \ldots, Z_i$ come from the post-change distribution

The above algorithm is a combination of a CUSUM tuned to detect the minimum possible change of $\theta_1 = \theta_0 + \nu_0$ and an SPRT with an estimated post-change parameter. The former is used to signal the possibility of a disorder. If an alarm sounds then the second test is initiated. If the SPRT results in a rejection of H_1, then the CUSUM statistic is set to 0 and the process restarts. Otherwise, a change is declared. The selection of h can be done through techniques described in Chapter 6. The determination of a and b can be done through the control of type I and type II errors of the SPRT, which in turn are determined by the oc of the SPRT procedure described above. Both the asn and the oc of the procedure are determined by simulation as no closed form formulas exist. This approach can be extended to other families of distributions for which the CUSUM under step 3 is optimal for the alternative $\theta_1 \geq \theta_0 + \nu_0$ (similarly to the case of Brownian motion with post-change drift parameter $\mu \geq m > 0$ treated in Chapter 6). For further details on this procedure, please refer to [63].

An alternative interesting adaptive approach for detecting a change in the mean in a sequence of observations in the presence of noise is introduced and analyzed in [56]. The objective in this approach focuses more on the sequential estimation of the mean that can change at an unknown point in time rather than on estimating the change point itself. Similarly, an adaptive procedure for detecting an unknown change in the mean for the particular case of Poisson observations in the case of an unknown change in the Poisson arrival rate has been suggested in [65]. Another approach to the detection of signals with unknown shape and unknown magnitude is given in [28].

7.4 Quickest detection with dependent observations

In all previous discussions we have considered only situations in which observations are independent of one another conditioned on the hypothesis or change point. In this section we consider situations in which we have dependent observations. We consider two particular situations, one that extends the optimality of the CUSUM to certain dependent situations, and another that treats more general dependencies using an asymptotic framework.

7.4.1 Quickest detection with independent likelihood ratio sequences

In this section, we generalize the result of Theorem 6.2, which asserts the (Lorden) optimality of the CUSUM for i.i.d. pre- and post-change models. We follow [151]. In particular, consider a general discrete-time observation process $\{Z_k\}$ (not necessarily independent) with the usual filtration $\{\mathcal{F}_k; k = 1, 2, \ldots\}$, and sequences of conditional probability measures $\{Q_0(\cdot|\mathcal{F}_{k-1})\}$ and $\{Q_1(\cdot|\mathcal{F}_{k-1})\}$ that describe the conditional distributions of $\{Z_k\}$, given the past $\mathcal{F}_{k-1}$ before and after the change point t. We assume that, for each k, these probability measures are mutually absolutely continuous. Suppose further that $\{Q_0(\cdot|\mathcal{F}_{k-1})\}$ satisfies the following key condition for all k:

$$Q_0\left(L(Z_k) \leq l|\mathcal{F}_{k-1}\right) = F_0(l), \text{ for all } l \geq 0, \tag{7.83}$$

where F_0 is a cdf, independent of k. Although the observation process can be non-stationary and dependent, the above condition guarantees that the process $\{L(Z_k)\}$, where $L(Z_k) = \frac{\mathrm{d}Q_1(Z_k|\mathcal{F}_{k-1})}{\mathrm{d}Q_0(Z_k|\mathcal{F}_{k-1})}$ is i.i.d. under both Q_0 and Q_1. To see that (7.83) is sufficient to ensure that the process $\{L(Z_k)\}$ is i.i.d., consider the multivariate distribution of the random variables $L(Z_1), \ldots, L(Z_k)$ under the probability distribution induced by the conditional probability measures $\{Q_0(\cdot|\mathcal{F}_{k-1})\}$. Using the fact that $\{L(Z_k) \leq l\} \in \mathcal{F}_k$, for all $l \geq 0$, we have

$$P\left(L(Z_1) \leq l_1, \ldots, L(Z_n) \leq l_n\right) = \prod_{i=1}^{k} Q_0\left(L(Z_i) \leq l_i|\mathcal{F}_{i-1}\right) \tag{7.84}$$

$$= \prod_{i=1}^{k} F_0(l_i), \tag{7.85}$$

which proves that the $\{L(Z_k)\}$ are i.i.d. with corresponding marginal distribution F_0 before the change. Using the absolute continuity of Q_1 with respect to Q_0 it is possible to show that the same holds under the sequence of measures $\{Q_1(\cdot|\mathcal{F}_k)\}$, with

$$F_1(l) = Q_1\left(L(Z_k) \leq l|\mathcal{F}_{k-1}\right), \; l \geq 0, \tag{7.86}$$

where $F_1(l) = \int_0^l z\mathrm{d}F_0(z)$.

On examination of the proof of Theorem 6.2, it can be seen that the optimality of the CUSUM $S_n = \max\{0, S_{n-1}\}L(Z_n)$ does not rely on the independence of the observation sequence, but rather on the independence of the sequence of likelihood ratios. Thus, the CUSUM is (Lorden) optimal for processes satisfying (7.83) and (7.86), a fact noted by Moustakides in [151].

As an important example of a situation in which (7.83) and (7.86) are satisfied, suppose the observation process $\{Z_k\}$ is generated by the equation[7]

$$Z_k = \sum_{i=1}^{p} \beta_i Z_{k-i} + W_k, \tag{7.87}$$

where $\{W_k\}$ is an i.i.d. Gaussian sequence with mean θ and variance σ^2, about which we wish to solve the change-point detection problem in which $\theta = \theta_0$ for all $k = 1, \ldots, t-1$ and $\theta = \theta_1$ for all $k = t, \ldots$

[7] Such a process is known as an autoregressive (AR) process of order p.

It is straightforward to show [19] that the log-likelihood ratio based on n observations from (7.87) takes the form

$$S_n = \sum_{k=1}^{n} \frac{\theta_1 - \theta_0}{\sigma^2} \left(Z_k - \sum_{i=1}^{p} \beta_i Z_{k-i} - \frac{\theta_0 + \theta_1}{2} \right), \tag{7.88}$$

the summands of which are i.i.d. before and after the change regardless of the value of θ.

Therefore, from (7.88), we can see that condition (7.83) is met and thus the CUSUM process is optimal under Lorden's criterion in detecting a change of $\theta = \theta_0$ to $\theta = \theta_1$.

A number of other examples of processes that satisfy (7.83) are given in [151].

It is easy to check however that (7.83) is not satisfied in the case of an autoregressive process in which the hypothesis of change involves changes in the parameters β_1 through β_p. One technique used to detect changes of such type in cases for which the change is small and the distance in the parameters shrinks as the sample size n increases, is the local hypothesis testing approach, to which we turn in the following section.

7.4.2 Locally asymptotically normal distributions

To introduce the notion of a locally asymptotically normal family of distributions let us first review some definitions from statistical theory. Consider a collection $\{Z_k; k = 1, \ldots, n\}$ of (possibly dependent) random variables defined on the measurable space $(\mathbb{R}^n, \mathcal{B}^n)$ and a parametric family of probability measures $\{P_\theta\}_{\theta \in \Theta}$ where Θ is an open subset of $\mathbb{R}^p$ and where P_θ is the joint distribution of the collection $\{Z_k; k = 1, \ldots, n\}$. We assume that the family of measures $\{P_\theta\}_{\theta \in \Theta}$ satisfy certain regularity conditions, given for example on page 45 of [188], among which is that each P_θ has a probability density function p_θ.

For $\theta \in \Theta$, the *efficient score statistic* is defined as

$$E_n = \frac{\partial p_\theta(Z_1, \ldots, Z_n)}{\partial \theta}$$

and the Fisher information matrix is defined to be the $p \times p$ matrix $I_n(\theta)$ with $i - j^{\text{th}}$ element (see, e.g. [178])

$$[I_n(\theta)]_{ij} = \frac{1}{n} E_\theta \left\{ \frac{\partial \ln p_\theta(Z_1, \ldots, Z_n)}{\partial \theta_i} \frac{\partial \ln p_\theta(Z_1, \ldots, Z_n)}{\partial \theta_j} \right\}.$$

Let $\{h_n\}$ be a sequence of points in $\mathbb{R}^p$ such that $h_n \to h \in \mathbb{R}^p$, and set $\theta_n = \theta + h_n \frac{1}{\sqrt{n}}$, so that $\theta_n \in \Theta$ for sufficiently large n since Θ is open and $\theta \in \Theta$. Denote by $L_n(Z_1, \ldots, Z_n)$ the likelihood ratio between the hypotheses

$$H_0 : \text{the distribution of } Z_1, \ldots, Z_n \text{ is } P_\theta$$

versus

$$H_1 : \text{the distribution of } Z_1, \ldots, Z_n \text{ is } P_{\theta_n}$$

Then, the parametric family of distributions $\{P_\theta\}_{\theta\in\Theta}$ is said to be *locally asymptotically normal* (LAN) if we can write

$$\log L_n(Z_1, \ldots, Z_n) = h^T \Delta_n(\theta) - \frac{1}{2} h^T I_n(\theta) h + \kappa(Z_1, \ldots, Z_n, \theta, h), \tag{7.89}$$

where

$$\Delta_n(\theta) = \frac{1}{\sqrt{n}} \frac{\partial \ln q_\theta(Z_1, \ldots, Z_n)}{\partial \theta}$$

and

$$\lim_{n\to\infty} \kappa(Z_1, \ldots, Z_n, \theta, h) = 0 \;\; P_\theta \; a.s.,$$

and where the asymptotic distribution of $\Delta_n(\theta)$ is $\mathcal{N}(0, I(\theta))$ under P_θ.

Members of the LAN family of distributions include i.i.d processes, stationary Markov processes of any given order, stationary Gaussian random processes, a subset of which includes stationary Gaussian autoregressive moving-average (ARMA)[8] processes, etc. (See for example pp. 47–52 [188].)

For the LAN family of distributions, it is also the case that the asymptotic distribution of $\Delta_n(\theta)$ is $\mathcal{N}(I(\theta)h, I(\theta))$ under P_{θ_n}. To see this, consider a continuous random variable X and two hypotheses regarding its probability density, namely assume it has density q_{θ_1} under H_1 and density q_{θ_0} under H_0 for some p vectors $\theta_1, \theta_0 \in \Theta$, where $\Theta \subset \mathbb{R}^p$ and $||\theta_1 - \theta_0||$ is small. The likelihood ratio is $L(X) = \frac{q_{\theta_1}(X)}{q_{\theta_0}(X)}$, and we can write

$$\ln L(X) \approx (\theta_1 - \theta_0)^T \left. \frac{\partial \ln q_\theta(X)}{\partial \theta} \right|_{\theta=\theta_0} + \frac{1}{2}(\theta_1 - \theta_0)^T \left. \frac{\partial^2 \ln q_\theta(X)}{\partial \theta^2} \right|_{\theta=\theta_0} (\theta_1 - \theta_0), \tag{7.90}$$

from which we see that

$$E_{\theta_0}\{\ln L(X)\} \approx -\frac{1}{2}(\theta_1 - \theta_0)^T I(\theta_0)(\theta_1 - \theta_0), \tag{7.91}$$

and

$$E_{\theta_1}\{\ln L(X)\} \approx \frac{1}{2}(\theta_1 - \theta_0)^T I(\theta_1)(\theta_1 - \theta_0) \approx \frac{1}{2}(\theta_1 - \theta_0)^T I(\theta_0)(\theta_1 - \theta_0), \tag{7.92}$$

where the second relationship follows by using (7.90) and interchanging the roles of θ_1 and θ_0. Moreover for small values of $||\theta_1 - \theta_0||$, we have

$$\ln L(X) \approx (\theta_1 - \theta_0)^T \left. \frac{\partial \ln q_\theta(X)}{\partial \theta} \right|_{\theta=\theta_0},$$

[8] ARMA processes generalize the AR model of (7.87) to include a "moving average" piece, as follows:

$$Z_k = \sum_{i=1}^{p} \beta_i Z_{k-i} + \sum_{j=1}^{q} \alpha_j W_{k-j+1},$$

where $\alpha_1, \ldots, \alpha_q$ are constants.

and using (7.92) it follows that

$$E_{\theta_1}\left\{ \frac{\partial \ln q_\theta(X)}{\partial \theta}\bigg|_{\theta=\theta_0} \right\} \approx I(\theta_0)(\theta_1 - \theta_0). \tag{7.93}$$

We note that in the case of a collection of n random variables the expected value of the second-order term appearing in the expansion of $\ln L_n(Z_1, \dots, Z_n)$ is of the order of $nI_n(\theta)$. Hence, it should not come as a surprise that the rate at which $\theta_n = \theta + \frac{h_n}{\sqrt{n}}$ approaches θ is of the order of $1/\sqrt{n}$, as the quantity squared cancels n allowing (together with other regularity conditions and stationarity) the limiting distribution to exist.

Equation (7.93) indicates that in the case of a LAN family, under P_{θ_n}, the distribution of $\Delta_n(\theta)$ is $\mathcal{N}(I(\theta)h, I(\theta))$. Using the normality of $\Delta_n(\theta)$ we can also use (7.91) and (7.92) to deduce that

$$\ln L_n(Z_1, \dots, Z_n) \overset{D}{\to} \mathcal{N}\left(-\frac{1}{2}h^T I(\theta)h, I(\theta)\right) \text{ under } P_\theta, \tag{7.94}$$

and

$$\ln L_n(Z_1, \dots, Z_n) \overset{D}{\to} \mathcal{N}\left(+\frac{1}{2}h^T I(\theta)h, I(\theta)\right) \text{ under } P_{\theta_n}. \tag{7.95}$$

The basic idea behind the LAN family of distribution is that

$$L_n(Z_1, \dots, Z_n) \approx \exp\left[h^T \Delta_n(\theta) - \frac{1}{2}h^T I(\theta)h\right], \tag{7.96}$$

from which it is seen that the likelihood behaves approximately as if it were an exponential family with sufficient statistic $\Delta_n(\theta)$ (see, e.g. [178]).

In the next section we develop a technique used to test sequentially a null hypothesis about the parameter of a parametric family of distributions against local alternatives. We will achieve this through the construction of an appropriate SPRT stopping time. We will examine its optimality properties as well as its asn. Moreover, we will investigate the problem of detecting a change in the case of local alternatives using an appropriately constructed CUSUM stopping time.

7.4.3 Sequential detection (local hypothesis approach)

Continuing with the model above, but with a scalar parameter (i.e. $p = 1$), we are interested in testing the null hypothesis $H_0 : \theta \le \theta_0$ against local alternatives $H_1 : \theta > \theta_0$. We continue to assume that for each $\theta \in \Theta$, P_θ has a probability density function p_θ. In what follows we first assume that $\{Z_k\}$ is an i.i.d. sequence with marginal probability density q_θ, construct an SPRT-type test based on the statistic

$$\sum_{k=1}^{n} \frac{\partial \ln q_\theta(Z_k)}{\partial \theta},$$

and indicate its local optimality properties. We then extend these ideas to situations in which the observations are LAN.

So, we begin with the i.i.d. case. Let (δ, T) be the following SPRT-like stopping time

$$T = \inf\left\{ n \left| \sum_{k=1}^{n} \frac{\partial \ln q_\theta(Z_k)}{\partial \theta} \right|_{\theta=\theta_0} \notin (a, b) \right\}, \tag{7.97}$$

with the associated decision function

$$\delta(Z_1, \ldots, Z_T) = \begin{cases} 1 & ; \text{if } \sum_{k=1}^{T} \left.\frac{\partial \ln q_\theta(Z_k)}{\partial \theta}\right|_{\theta=\theta_0} \geq b \\ \\ 0 & ; \text{if } \sum_{k=1}^{T} \left.\frac{\partial \ln q_\theta(Z_k)}{\partial \theta}\right|_{\theta=\theta_0} \leq a. \end{cases} \tag{7.98}$$

The local performance characteristics of the above sequential test are described by the type I error $\alpha(\delta)$ (compare footnote 2 in Chapter 4), the asn $E\{T\}$ and the local slope of the power function[9] $\beta'(\theta_0)$ at the null hypothesis θ_0.

It can be shown that, within regularity, the test described by (7.97) and (7.98) is locally most powerful. That is, for any other test $(\tilde{\delta}, \tilde{T})$ with characteristics $\left(\alpha(\tilde{\delta}), E\left\{\tilde{T}\right\}, \tilde{\beta}'(\theta_0)\right)$, with

$$E\left\{\tilde{T}\right\} \leq E\{T\},$$

and

$$\alpha(\tilde{\delta}) \leq \alpha(\delta),$$

the following holds

$$\tilde{\beta}'(\theta)|_{\theta=\theta_0} \leq \beta'(\theta)|_{\theta=\theta_0}.$$

For a proof of this result please refer to [33].

To examine the asn of (7.97) for the two simple local hypotheses $H_0 : \theta = \theta_0$ vs. $H_1 : \theta = \theta_0 + h$ as $h \to 0$, let us consider once again a random variable X with density p_θ, with θ as prescribed by H_0 and H_1 and use the Taylor expansion of the log-likelihood ratio. We have

$$L(X) = h \left.\frac{\partial \ln p_\theta(X)}{\partial \theta}\right|_{\theta=\theta_0} + \frac{1}{2}h^2 \left.\frac{\partial^2 \ln p_\theta(X)}{\partial \theta^2}\right|_{\theta=\theta_0} + o(h^2).$$

Taking expectations of both sides under H_0 and using the fact that $E_{\theta_0}\left\{ \left.\frac{\partial \ln p_\theta(X)}{\partial \theta}\right|_{\theta=\theta_0} \right\}$ $= 0$, we obtain

$$E_{\theta_0}\{L(X)\} = -\frac{1}{2}I(\theta_0)h^2 + o(h^2)$$

and

$$E_{\theta_1}\{L(X)\} = \frac{1}{2}I(\theta_1)h^2 + o(h^2),$$

[9] The power function is defined as $\beta(\theta) = P_\theta(\delta_T = 1)$.

where $I(\theta)$ denotes the Fisher information when the parameter is θ. Using the definition of the Kullback–Leibler divergence of Section 6.1, we obtain

$$\begin{aligned} D(P_{\theta_0}||P_{\theta_1}) &= \frac{1}{2}I(\theta_0)h^2 + o(h^2), \\ D(P_{\theta_1}||P_{\theta_0}) &= \frac{1}{2}I(\theta_1)h^2 + o(h^2), \\ E_\theta\{L(X)\} &= h\left(\tilde{h} - \frac{1}{2}h\right)I(\theta) + o(h^2), \end{aligned}$$

for $\theta \in [\theta_0, \theta_1]$ and $\tilde{h} = \theta - \theta_0$. Using the analysis and the results of Section 4.2, we thus obtain

$$E_{\theta_0}\{T\} \approx \frac{(1-\alpha)\ln\frac{1-\alpha}{\gamma} + \alpha\ln\frac{1-\gamma}{\alpha}}{\frac{1}{2}h^2 I(\theta_0) + o(h^2)}, \tag{7.99}$$

$$E_{\theta_1}\{T\} \approx \frac{(1-\gamma)\ln\frac{1-\gamma}{\alpha} + \gamma\ln\frac{1-\alpha}{\gamma}}{\frac{1}{2}h^2 I(\theta_1) + o(h^2)}, \tag{7.100}$$

and

$$E_\theta\{T\} \approx \frac{-a\, oc(\theta) + b(1 - oc(\theta))}{h\left(\tilde{h} - \frac{1}{2}h\right)I(\theta) + o(h^2)}, \tag{7.101}$$

where oc(θ) is the operating characteristic.

For further details on the local approach for calculating oc(θ) please refer to [19].

Before considering the local approach for hypothesis testing in the case of processes with more involved dependence structures than the i.i.d. case (such as the case of LAN distributions), it is important to mention two results concerning the most powerful test of fixed size in the case of LAN distributions.

Let $\{P_\theta\}_{\theta\in\Theta}$, $\Theta \subset \mathbb{R}$ be a LAN family of distributions with scalar parameter θ. Consider the hypotheses $H_0 : \theta \le \theta_0 = \theta^* + \frac{h_0}{\sqrt{n}}$, vs. $H_1 : \theta > \theta_1 = \theta^* + \frac{h_1}{\sqrt{n}}$, and let

$$C_\alpha = \left\{\delta \,\middle|\, \limsup_{n\to\infty} \sup_{\theta\le\theta_0} E_\theta\{\delta(Z_1, \ldots, Z_n)\} \le \alpha\right\}.$$

The test with critical region

$$\Delta_n \ge K, \tag{7.102}$$

where $\Delta_n = \Delta_n(\theta^*)$ as defined in (7.89), is asymptotically uniformly most powerful (AUMP) in the class C_α. This is a result of [42].

That (7.102) is AUMP means that as $n \to \infty$ the test based on Δ_n is the best among all tests with Type I error less than or equal to α (i.e. all α-level tests). (For a more detailed discussion on the AUMP test please refer to [188].) Essentially, this result follows because the likelihood ratio depends asymptotically on the quantity

$$\Delta_n = \frac{1}{\sqrt{n}} \frac{\partial \ln p_\theta(Z_1, \ldots, Z_n)}{\partial \theta}\bigg|_{\theta=\theta^*}$$
$$= \frac{1}{\sqrt{n}} \sum_{k=1}^{n} \frac{\partial \ln p_\theta(Z_k | Z_1, \ldots, Z_{k-1})}{\partial \theta}\bigg|_{\theta=\theta^*} \tag{7.103}$$

via (7.89).

Given the above information, we can discuss the construction of tests for the following hypotheses:

$$H_0 : \theta = \theta_0$$
$$\text{vs.} \tag{7.104}$$
$$H_1 : \theta = \theta_0 + h/\sqrt{n},$$

and

$$H_0 : \theta_0 = \theta^* - h/2\sqrt{n}$$
$$\text{vs.} \tag{7.105}$$
$$H_1 : \theta_1 = \theta^* + h/2\sqrt{n},$$

where in both of the above cases we assume $\Theta \subset \mathbb{R}$.

The obvious candidate for the construction of an SPRT-type test for both types of hypotheses is the stopping time

$$T = \inf\left\{ n \bigg| \frac{1}{\sqrt{n}} \sum_{k=1}^{n} \frac{\partial \ln p_\theta(Z_k | Z_1, \ldots, Z_{k-1})}{\partial \theta}\bigg|_{\theta=\theta^*} \notin (a, b) \right\}, \tag{7.106}$$

with the associated decision function

$$\delta_T(Z_1, \ldots, Z_T) = \begin{cases} 1 & ; \text{if } \Delta_T \geq b \\ & \\ 0 & ; \text{if } \Delta_T \leq a. \end{cases} \tag{7.107}$$

Let us begin with the first set of hypotheses. To calculate the asn of this stopping time we can follow the procedure described above or use the fact that asymptotically it behaves as in the Gaussian i.i.d case to obtain

$$E_{\theta_0}\{T\} \approx \frac{(1-\alpha)\ln(1-\alpha)/\gamma + \alpha \ln(1-\gamma)/\alpha}{\frac{1}{2}h^2 I(\theta_0) + o(h^2)}, \tag{7.108}$$

$$E_{\theta_1}\{T\} \approx \frac{(1-\alpha)\ln(1-\alpha)/\gamma + \alpha \ln(1-\gamma)/\alpha}{\frac{1}{2}h^2 I(\theta_1) + o(h^2)}, \tag{7.109}$$

and

$$E_\theta\{T\} \approx \frac{-a\, oc(\theta) + b(1 - \mathrm{oc}(\theta))}{h(\tilde{h} - \frac{1}{2}h)I(\theta) + o(h^2)},$$

for $\theta \in [\theta_0, \theta_1]$. We can then use the following central limit theorem which appears in [103] and concerns the limiting distribution of

$$\Delta_{n,s}(\theta^*) = \frac{1}{\sqrt{n}} \sum_{k=1}^{[ns]} \left. \frac{\partial \ln p_\theta(Z_k|Z_1, \ldots, Z_{k-1})}{\partial \theta} \right|_{\theta=\theta^*}, \tag{7.110}$$

with $s \in [0, 1]$ and $\theta \in \Theta \subset \mathbb{R}$.

We have

(1) under P_{θ^*} : $I_n(\theta^*)^{-\frac{1}{2}} \Delta_{n,s}(\theta^*) \xrightarrow{D} W_s$; and
(2) under $P_{\theta^* + \frac{h}{\sqrt{n}}}$: $(I_n(\theta^*)^{-\frac{1}{2}} \left(\Delta_{n,s}(\theta^*) - I_n(\theta^*)hs\right) \xrightarrow{D} W_s$,

where $\{W_s; 0 \le s \le 1\}$ is a standard Brownian motion.

REMARK 7.14. In the above two results, convergence in distribution actually means weak convergence of the stochastic process $\{\Delta_{n,s}; s \in [0, 1]\}$ to the limiting Brownian motion.

We can then obtain approximations to both oc(θ), and asn(θ) by using results on the hitting times of Brownian motion.

In the case of the second pair of hypotheses (7.105), notice that

$$\ln \frac{p_{\theta_1}(Z_1, \ldots, Z_n)}{p_{\theta^*}(Z_1, \ldots, Z_n)} \approx \frac{h}{2\sqrt{n}} \left. \frac{\partial p_\theta(Z_1, \ldots, Z_n)}{\partial \theta} \right|_{\theta=\theta^*} - \frac{h^2}{8} I(\theta^*), \tag{7.111}$$

while,

$$\ln \frac{p_{\theta^*}(Z_1, \ldots, Z_n)}{p_{\theta_0}(Z_1, \ldots, Z_n)} \approx \frac{h}{2\sqrt{n}} \left. \frac{\partial p_\theta(Z_1, \ldots, Z_n)}{\partial \theta} \right|_{\theta=\theta^*} + \frac{h^2}{8} I(\theta^*). \tag{7.112}$$

Therefore, by combining (7.111) and (7.112), we obtain

$$\ln \frac{p_{\theta_1}(Z_1, \ldots, Z_n)}{p_{\theta_0}(Z_1, \ldots, Z_n)} \approx \frac{h}{\sqrt{n}} \left. \frac{\partial p_\theta(Z_1, \ldots, Z_n)}{\partial \theta} \right|_{\theta=\theta^*} = h\Delta_n. \tag{7.113}$$

Using the same central limit theorem as above, with convergence in the sense of Remark 7.14, it follows that:

(1) under H_0: $I_n(\theta^*)^{-\frac{1}{2}} \left(\Delta_{n,s}(\theta^*) + \frac{1}{2} I_n(\theta^*)hs\right) \xrightarrow{D} W_s$; and
(2) under H_1: $I_n(\theta^*)^{-\frac{1}{2}} \left(\Delta_{n,s}(\theta^*) - \frac{1}{2} I_n(\theta^*)hs\right) \xrightarrow{D} W_s$.

Therefore, again the approximation of oc(θ) and asn(θ) reduces to answering the same questions in the case of a Brownian motion with negative drift and positive drift, respectively. The questions are easily answered in this case of a scalar parameter. (See for example [118].)

Given the result of Berk [33] discussed at the beginning of the section, which applies to local scalar alternatives and the asymptotic behavior of an LAN as an exponential family with sufficient statistic Δ_n, one could potentially characterize (although no such result exists in the literature) the SPRT-type test described by (7.106) and (7.107) as a locally asymptotically most powerful sequential test for both the first and the second pair of hypotheses.

The extension of the above results to a p-dimensional parameter set is straightforward at least in the second pair of hypotheses (7.105) as long as the hyperplane separating the parameter sets Θ_0 and Θ_1 under the two hypotheses can be approximated by the hyperplane $h^T I(\theta^*)(\theta - \theta^*) = 0$. This is true in the case in which the following two inequalities hold for all $k \geq 1$:

$$E_{\theta_0}\left\{h^T \left.\frac{\partial \ln q_\theta(Z_k|Z_1, \ldots, Z_{k-1})}{\partial \theta}\right|_{\theta=\theta^*}\right\} < 0 \;\; \forall\, \theta \in \Theta_0,$$

and

$$E_{\theta_1}\left\{h^T \left.\frac{\partial \ln q_\theta(Z_k|Z_1, \ldots, Z_{k-1})}{\partial \theta}\right|_{\theta=\theta^*}\right\} > 0 \;\; \forall\, \theta \in \Theta_1.$$

It is important to mention that one is guaranteed that the SPRT described by (7.106) and (7.107) is closed in the second pair of hypotheses (that is, it will terminate) even in the case in which we do not have a LAN family of distributions or the limit theorem of convergence to Brownian motion does not hold. All that is necessary is that the asymptotic expansions (7.111) and (7.112) be valid.

In what follows we will discuss how to construct a CUSUM type of algorithm in the case of a LAN family of distributions to detect local changes of the type appearing in the second hypotheses pair (7.104).

7.4.4 Quickest detection (local hypothesis approach)

We would like to detect a change in the parameter determining the statistical behavior of a process $\{Z_k\}$ whose distribution before and after the change point t belongs to a LAN family of distributions $\{P_\theta\}_{\theta\in\Theta}$, $(\Theta \subset \mathbb{R}^p)$ with parameter θ. It is assumed that before the change point t, hypothesis H_0 is true regarding the value of the parameter θ and after the change point, hypothesis H_1 is true regarding the value of the parameter θ. In this section we will consider the local hypotheses $H_0 : \theta_0 = \theta^* - (h/\sqrt{n})$ vs. $H_1 : \theta_1 = \theta^* + (h/\sqrt{n})$, with $\Theta \subset \mathbb{R}$.

REMARK 7.15. Notice that the dependence on n of the hypotheses treated in this subsection becomes relevant in the convergence of the distribution of the statistic (7.103) through the mechanism explained in the paragraph following (7.93). Intuitively, one can consider n to be much larger than the change point t when a change occurs, so that the asymptotics can be applied both pre- and post-change.

As is evident from (7.108) and (7.109), we have that

$$E_{\theta_0}\left\{\left.\frac{\ln \partial p_\theta(Z_k|Z_1, \ldots, Z_{k-1})}{\partial \theta}\right|_{\theta=\theta^*}\right\} < 0,$$

and

$$E_{\theta_1}\left\{\frac{\ln \partial p_\theta(Z_k|Z_1,\ldots,Z_{k-1})}{\partial\theta}\bigg|_{\theta=\theta^*}\right\} > 0.$$

This suggests that a CUSUM type of stopping time based on the first passage time of the process (7.71) with g replaced by

$$g_k(Z_k) = \frac{\ln \partial p_\theta(Z_k|Z_1,\ldots,Z_{k-1})}{\partial\theta}\bigg|_{\theta=\theta^*}, \tag{7.114}$$

will terminate. That is, $g_k(Z_k)$ is directly related to the statistic (7.103). Therefore, asymptotic results regarding the LAN family of distributions and in particular the statistic (7.110) become relevant here in the computation of the first moment of the first passage time of the process (7.71) with g_k of (7.114). Such a result is given in [103], where it is shown that $I(\theta^*)^{-\frac{1}{2}}\Delta_{n,s}(\theta^*)$ of (7.110) converges in the sense of Remark 7.14 to

$$\mathrm{d}B_s = -1_{\{s\le t\}}\frac{1}{2}I(\theta^*)h\mathrm{d}s + 1_{\{s\ge t\}}\frac{1}{2}I(\theta^*)h\mathrm{d}s + I(\theta^*)^{\frac{1}{2}}\mathrm{d}W_s, \tag{7.115}$$

where $\{W_s; s\in[0,1]\}$ is a standard Brownian motion, (in the case of a vector parameter it is a multi-dimensional Brownian motion) and the change point t is any fixed point in the interval [0, 1] (see pp. 358–360 of [19]).

The extension of the above to the p-dimensional case, that is to the case in which $H_0 : \theta = \theta^* - (h/\sqrt{n})$ is true for $Z_1, \ldots, Z_{t-1}$ and $H_1 : \theta = \theta^* + (h/\sqrt{n})$ is true thereafter, (h_n is given at the beginning of Section 7.4.2). Here, we use the CUSUM-like stopping time of (7.69) with the g of (7.70) replaced by

$$g_k(Z_k) = h^T\,\frac{\ln \partial p_\theta(Z_k|Z_1,\ldots,Z_{k-1})}{\partial\theta}\bigg|_{\theta=\theta^*}, \tag{7.116}$$

since

$$E_{\theta_0}\left\{h^T\,\frac{\ln \partial p_\theta(Z_k|Z_1,\ldots,Z_{k-1})}{\partial\theta}\bigg|_{\theta=\theta^*}\right\} < 0,$$

and

$$E_{\theta_1}\left\{h^T\,\frac{\ln \partial p_\theta(Z_k|Z_1,\ldots,Z_{k-1})}{\partial\theta}\bigg|_{\theta=\theta^*}\right\} > 0.$$

In the case in which $H_0 : \theta = \theta_0$ is true before the change and $H_1 : \theta = \theta_0 + (h/\sqrt{n})$ is true after the change, a practical algorithm arises from the approximation

$$\max_{1\le r\le k}\sup_{\theta\in\Theta_1}\sum_{i=r}^{k}\ln\frac{p_\theta(Z_i|Z_1,\ldots,Z_{i-1})}{p_{\theta_0}(Z_i|Z_1,\ldots,Z_{i-1})} \approx \max_{1\le r\le k}(\Delta_r)^T I(\theta_0)^{-1}(\Delta_r). \tag{7.117}$$

where

$$\Delta_r = \frac{1}{\sqrt{n-r+1}}\sum_{i=r}^{n}\ln\frac{\partial p_\theta(Z_i|Z_1,\ldots,Z_{i-1})}{\partial\theta}\bigg|_{\theta=\theta_0} \tag{7.118}$$

whose asymptotic distribution, in the LAN family, is χ^2 with $n - r + 1$ degrees of freedom.[10] The crux of this result lies in the fact that $\ln \left.\frac{\partial p_\theta(Z_i|Z_1,\ldots,Z_{i-1})}{\partial \theta}\right|_{\theta=\theta_0}$ can be treated as being i.i.d. in the LAN family. Hence the quadratic form that appears on the right-hand side of (7.117) has a χ^2 distribution.

Other approaches to problems with dependent observations, include the work of Bansal and Papantoni-Kazakos [13], who proved an asymptotic optimality result equivalent to the one of Lorden [139] discussed in Chapter 6 for the CUSUM stopping time of (7.70) with g in (7.71) replaced by

$$g(Z_k) = \frac{p_{\theta_1}(Z_k|Z_1,\ldots,Z_{k-1})}{p_{\theta_0}(Z_k|Z_1,\ldots,Z_{k-1})}. \tag{7.119}$$

Other related works that investigate the asymptotic optimality properties of the CUSUM stopping time and the Shiryaev–Roberts stopping time (compare Remark 6.18) in the case of dependence in the observations include [16,34,88,89,105,126–128,145,222].

[10] A continuous random variable X has a χ^2 distribution with p degrees of freedom ($p \in \{1, 2, \ldots\}$) if its probability density function is given by

$$f_X(x) = \frac{1}{\Gamma(\frac{p}{2})2^{\left(\frac{p}{2}\right)}} x^{\left(\frac{p}{2}\right)} e^{-\frac{x}{2}},\ x \geq 0,$$

where Γ is the gamma function.

Bibliography

1. E. E. Alvarez and D. K. Dey, Bayesian isotonic changepoint analysis, *Annals of the Institute of Statistical Mathematics,* to appear.
2. T. W. Anderson, *The Statistical Analysis of Time Series*. Series in Probability and Mathematical Statistics, (Wiley, New York), 1971.
3. E. Andersson, On-line detection of turning points using non-parametric surveillance: The effect of the growth after the turn, *Statistics & Probability Letters*, Vol. 73, No. 4, pp. 433–439, 2005.
4. E. Andersson, D. Bock and M. Frisén, Detection of turning points in business cycles, *Journal of Business Cycle Measurement and Analysis*, Vol. 1, No. 1, pp. 93–108, 2004.
5. E. Andersson, D. Bock and M. Frisén, Some statistical aspects of methods for detection of turning points in business cycles, *Journal of Applied Statistics*, Vol. 33, No. 3, pp. 257–278, 2006.
6. R. Andre-Obrecht, A new statistical approach for the automatic segmentation of continuous speech signals, *IEEE Transactions on Acoustics, Speech and Signal Processing*, Vol. 36, No. 1, pp. 29–40, 1988.
7. E. Andreou and E. Ghysels, The impact of sampling frequency and volatility estimators on change-point tests, *Journal of Financial Econometrics*, Vol. 2, No. 2, pp. 290–318, 2004.
8. D. W. K. Andrews, I. Lee and W. Ploberger, Optimal changepoint tests for normal linear regression, *Journal of Econometrics*, Vol. 70, No. 1, pp. 9–38, 1996.
9. D. Assaf, A dynamic sampling approach for detecting a change in distribution, *Annals of Statistics*, Vol. 16, No. 1, pp. 236–253, 1988.
10. D. Assaf, M. Pollak, Y. Ritov and B. Yakir, Detecting a change of a normal mean by dynamic sampling with a probability bound on a false alarm, *Annals of Statistics*, Vol. 21, No. 3, pp. 1155–1165, 1993.
11. D. Assaf and Y. Ritov, Dynamic sampling procedures for detecting a change in the drift of Brownian motion: A non-Bayesian model, *Annals of Statistics*, Vol. 17, No. 2, pp. 793–800, 1989.
12. B. Azimi-Sadjadi and P. S. Krishnaprasad, A particle filtering approach to change detection for non-linear systems, *Journal on Applied Signal Processing*, Vol. 15, No. 15, pp. 2295–2305, 2004.
13. R. K. Bansal and P. Papantoni, An algorithm for detecting a change in a stochastic process, *IEEE Transactions on Information Theory*, Vol. 32, No. 2, pp. 227–235, 1986.
14. G. Barnard, Control charts and stochastic processes, *Journal of the Royal Statistical Society B*, Vol. 39, No. 2, pp. 239–271, 1959.
15. T. P. Barnett, D. W. Pierce and R. Schnur, Detection of anthropogenic climate change in the world's oceans, *Science*, Vol. 292, No. 5515, pp. 270–274, 2001.

16. M. Baron and A. G. Tartakovsky, Asymptotic Bayesian change-point detection theory for general continuous-time models, *Sequential Analysis*, Vol. 25, No. 3, pp. 257–296, 2006.
17. M. Basseville, Edge detection using sequential methods for change in level – Part II: Sequential detection of a change in mean, *IEEE Transactions on Acoustics, Speech and Signal Processing*, Vol. 29, No. 1, pp. 29–40, 1981.
18. M. Basseville, *et al.*, *In situ* damage monitoring in vibration mechanics: Diagnostics and predictive maintenance, *Mechanical Systems and Signal Processing*, Vol. 7, No. 5, pp. 401–423, 1993.
19. M. Basseville and I. Nikiforov, *Detection of Abrupt Changes: Theory and Applications*, (Prentice-Hall, Englewood Cliffs, NJ), 1993.
20. J. A. Bather, On a quickest detection problem, *Annals of Mathematical Statistics*, Vol. 38, No. 3, pp. 711–724, 1967.
21. E. Bayraktar and S. Dayanik, Poisson disorder problem with exponential penalty for delay, *Mathematics of Operations Research*, Vol. 31, No. 2, pp. 217–233, 2006.
22. E. Bayraktar, S. Dayanik and I. Karatzas, The standard Poisson disorder problem revisited, *Stochastic Processes and their Applications*, Vol. 115, No. 9, pp. 1437–1450, 2004.
23. E. Bayraktar, S. Dayanik and I. Karatzas, The adaptive Poisson disorder problem, *Annals of Applied Probability*, Vol. 16, No. 3, pp. 1190–1261, 2006.
24. E. Bayraktar and H. V. Poor, Quickest detection of a minimum of two Poisson disorder times, *SIAM Journal on Control and Optimization*, Vol. 46, No. 1, pp. 308–331, 2007.
25. M. Beibel, A note on Ritov's Bayes approach to the minimax property of the CUSUM procedure, *Annals of Statistics*, Vol. 24, No. 2, pp. 1804–1812, 1996.
26. M. Beibel, Bayes problems in change-point models for the Wiener process, in *Change Point Problems*, E. Carlstein, H.-G. Müller and D. Siegmund, eds, (Institute of Mathematical Statistics, Hayward, CA), pp. 1–6, 1994.
27. M. Beibel, Sequential change-point detection in continuous time when the post-change drift is unknown, *Bernoulli*, Vol. 3, No. 4, pp. 457–478, 1997.
28. M. Beibel, Sequential detection of signals with unknown shape and unknown magnitude, *Statistica Sinica*, Vol. 10, pp. 715–729, 2000.
29. M. Beibel, A note on sequential detection with exponential penalty for the delay, *Annals of Statistics*, Vol. 28, No. 6, pp. 1696–1701, 2000.
30. M. Beibel and H. R. Lerche, A new look at optimal stopping problems related to mathematical finance, *Statistica Sinica*, Vol. 7, pp. 93–108, 1997.
31. C. Bell, L. Gordon and M. Pollak, An efficient nonparametric detection scheme and its application to surveillance of a Bernoulli process with unknown baseline, in *Change Point Problems*, E. Carlstein, H.-G. Müller and D. Siegmund, eds, (Institute of Mathematical Statistics, Hayward, CA), pp. 7–23, 1994.
32. A. Benveniste, M. Basseville and G. Moustakides, The asymptotic local approach to change detection and model validation, *IEEE Transactions on Automatic Control*, Vol. 32, No. 2, pp. 583–592, 1987.
33. R. H. Berk, Locally most powerful sequential tests, *Annals of Statistics*, Vol. 3, No. 2, pp. 373–381, 1975.
34. I. Berkes, E. Gombay, L. Horváth and P. Kokoszka, Sequential change-point detection in GARCH(p,q) models, *Econometric Theory*, Vol. 20, No. 6, pp. 1140–1167, 2004.
35. P. K. Bhattacharya, Some aspects of change-point analysis, *Change Point Problems*, E. Carlstein, H.-G. Müller and D. Siegmund, eds, (Institute of Mathematical Statistics, Hayward, CA), pp. 28–56, 1994.

36. P. K. Bhattacharya and H. Zhou, A rank-CUSUM procedure for detecting small changes in a symmetric distribution, *Change Point Problems*, E. Carlstein, H.-G. Müller and D. Siegmund, eds, (Institute of Mathematical Statistics, Hayward, CA), pp. 57–65, 1994.
37. P. Billingsley, *Probability and Measure*, (Wiley, New York), 1979.
38. D. A. Blackwell and M. A. Girshik, *Theory of Games and Statistical Decisions.* (Wiley, New York), 1954.
39. R. Blum, S. A. Kassam and H.V. Poor, Distributed signal detection Part II: Advanced topics, *Proceedings of the IEEE*, Vol. 85, No.1, pp. 64–79, 1997.
40. T. Bojdecki, Probability maximizing approach to optimal stopping and its application to a disorder problem, *Stochastics*, Vol. 3, No. 1–4, pp. 61–71, 1979.
41. T. Bojdecki and J. Hosza, On a generalized disorder problem, *Stochastic Processes and Their Applications*, Vol. 18, No. 2, pp. 349–359, 1984.
42. A. A. Borovkov, *Theory of Mathematical Statistics – Estimation and Hypotheses Testing.* (Russian) (Naouka, Moscow), 1984. Translated in French under the title *Statistique Mathématique – Estimation et Tests d' Hypothèses.* (Mir, Paris), 1987.
43. A. A. Borovkov, Asymptotically optimal solutions in the change-point problem, *Theory of Probability and its Applications*, Vol. 43, No. 4, pp. 539–561, 1998.
44. R. N. Brandt and S. Karlin, On the design and comparison of certain dichotomous experiments, *Annals of Mathematical Statistics*, Vol. 26, No. 2, pp. 390–409, 1956.
45. S. Braun, *Mechanical Signature Analysis – Theory and Applications.* (Academic Press, New York), 1986.
46. L. Breiman, *Probability.* (Addison-Wesley, Reading, MA), 1968.
47. P. Brémaud, *Point Processes and Queues: Martingale Dynamics.* (Springer-Verlag, New York), 1981.
48. P. J. Brockwell and R. A. Davis, *Time Series: Theory and Methods.* (Springer Verlag, New York), 1991.
49. B. Broder, *Quickest Detection Procedures and Transient Signal Detection*, Ph.D. Thesis, Department of Electrical Engineering, Princeton University, Princeton, NJ, 1990.
50. B. E. Brodsky and B. S. Darkhovsky, *Nonparametric Methods in Change-Point Problems.* (Kluwer, Dordrecht), 1993.
51. L. D. Broemling and H. Tsurumi, *Econometrics and Structural Change.* (Marcel Dekker, New York), 1987.
52. T. N. Bui, V. Krishnamurthy and H. V. Poor, On-line Bayesian activity detection in DS/CDMA networks, *IEEE Transactions on Signal Processing*, Vol. 53, No. 1, pp. 371–375, 2005.
53. A. A. Cardenas, J. S. Baras and V. Ramezani, Distributed change detection for worms, DDoS and other network attacks, *Proceedings of the 2004 American Control Conference*, Vol. 2, Boston, MA, June 30–July 2, 2004, pp. 1008–1013.
54. E. Carlstein, H.-G. Müller and D. Siegmund, eds., *Change Point Problems.* (Institute of Mathematical Statistics, Hayward, CA), 1994.
55. R. K. C. Chang, Defending against flooding-based distributed denial-of-service attacks: A tutorial, *IEEE Communications Magazine*, Vol. 40, No. 10, pp. 42–51, 2002.
56. H. Chernoff and S. Zacks, Estimating the current mean of a normal distribution which is subjected to changes in time, *Annals of Mathematical Statistics*, Vol. 35, No. 3, pp. 999–1018, 1964.
57. Y. S. Chow, H. Robbins and D. Siegmund, *Great Expectations: The Theory of Optimal Stopping.* (Houghton-Mifflin, Boston), 1971.

58. K. L. Chung, *A Course in Probability Theory.* (Academic Press, New York), 1968.
59. A. Cohen, *Biomedical Signal Processing.* (CRC Press, Boca Raton, FL), 1987.
60. D. Commenges, J. Seal and F. Pinatel, Inference about a change point in experimental neurophysiology, *Mathematical Biosciences*, Vol. 80, No. 1, pp. 81–108, 1986.
61. N. Cressie and P. B. Morgan, The VSPRT: A sequential testing procedure dominating the SPRT, *Econometric Theory*, Vol. 9, No. 3, pp. 431–450, 1993.
62. N. Cressie, J. Biele and P. B. Morgan, Sample-size optimal sequential testing, *Journal of Statistical Planning and Inference*, Vol. 39, No. 2, pp. 305–327, 1994.
63. R. W. Crow, *Robust, Distributed and Adaptive Quickest Detection Procedures*, Ph.D. Thesis, Department of Electrical Engineering, Princeton University, Princeton, NJ, 1995.
64. R. W. Crow and S. C. Schwartz, Quickest detection for sequential decentralized decision systems, *IEEE Transactions on Aerospace and Electronic Systems*, Vol. 32, No. 1, pp. 267–283, 1996.
65. R. W. Crow and S. C. Schwartz, An adaptive procedure for detecting jump changes with Poisson observables, *Technical Report*, Department of Electrical Engineering, Princeton University, Princeton, NJ, 1996.
66. N. J. Davey, P. H. Ellaway and R. B. Stein, Statistical limits for detecting change in the cumulative sum derivative of the peristimulus time histogram, *Journal of Neuroscience Methods*, Vol. 17, Nos. 2 & 3, pp. 153–166, 1986.
67. M. H. A. Davis, A note on the Poisson disorder problem, in *Mathematical Control Theory*, Vol. 1, (Banach Centre Publications, Warsaw), pp. 65–72, 1978.
68. S. Dayanik, C. Goulding and H. V. Poor, Bayesian sequential change diagnosis, *Mathematics of Operations Research*, Vol. 33, No. 2, pp. 475–496, 2008.
69. S. Dayanik, H. V. Poor and S. O. Sezer, Multisource Bayesian sequential change detection, *Annals of Applied Probability*, Vol. 18, No. 2, pp. 552–559, 2008.
70. S. Dayanik, H. V. Poor and S. O. Sezer, Sequential multi-hypothesis testing for compound Poisson processes, *Stochastics*, Vol. 80, No. 1, pp. 19–50, 2008.
71. S. Dayanik and S. O. Sezer, Compound Poisson disorder problem, *Mathematics of Operations Research*, Vol. 31, No. 4, pp. 649–672, 2006.
72. S. Dayanik and S. O. Sezer, Sequential testing of simple hypotheses about compound Poisson processes, *Stochastic Processes and Their Applications*, Vol. 116, No. 12, pp. 1892–1919, 2006.
73. J. DeLucia and H. V. Poor, Performance analysis of sequential tests between Poisson processes, *IEEE Transactions on Information Theory*, Vol. 41, No. 1, pp. 221–238, 1997.
74. J. DeLucia and H. V. Poor, The moment generating function of the stopping time for a linearly stopped Poisson process, *Communications in Statistics – Stochastic Models*, Vol. 13, No. 2, pp. 275–292, 1997.
75. D. K. Dey and S. Purkayastha, Bayesian approach to change point problems, *Communications in Statistics – Theory and Methods*, Vol. 26, No. 8, pp. 2035–2047, 1997.
76. J. L. Doob, *Stochastic Processes.* (Wiley, New York), 1953.
77. V. P. Dragalin, Optimality of the generalized CUSUM procedure in quickest detection problem, *Proceedings of the Steklov Institute of Mathematics*, Vol. 202, No. 4, pp. 107–119, 1994.
78. V. P. Dragalin, The design and analysis of 2-CUSUM procedure, *Communications in Statistics – Simulations*, Vol. 26, No. 1, pp. 67–81, 1997.
79. L. E. Dubins and H. Teicher, Optimal stopping when the future is discounted, *Annals of Mathematical Statistics*, Vol. 38, No. 2, pp. 601–605, 1967.

80. D. Duffie, *Dynamic Asset Pricing Theory.* (Princeton University Press, Princeton, NJ), 1992.
81. A. Dvorestsky, J. Kiefer and J. Wolfowitz, Sequential decision problems for processes with continuous time parameter. Testing hypotheses, *Annals of Mathematical Statistics*, Vol. 24, No. 2, pp. 254–264, 1953.
82. E. A. Feinberg and A. N. Shiryaev, Quickest detection of drift change for Brownian motion in generalized Bayesian and minimax settings, *Statistics & Decisions*, Vol. 24, No. 4, pp. 445–470, 2006.
83. S. Fotopoulos and V. Jandhyalab, Maximum likelihood estimation of a change-point for exponentially distributed random variables, *Statistics & Probability Letters*, Vol. 51, No. 4, pp. 423–429, 2001.
84. T. Friede, F. Miller, W. Bischoff and M. Kieserc, A note on change point estimation in dose–response trials, *Computational Statistics & Data Analysis*, Vol. 37, No. 2, pp. 219–232, 2001.
85. M. Frisén, Evaluations of methods for statistical surveillance, *Statistics in Medicine*, Vol. 11, No. 11, pp. 1489–1502, 1992.
86. M. Frisén, Statistical surveillance: Optimality and methods, *International Statistical Review*, Vol. 71, No. 2, pp. 403–434, 2003.
87. M. Frisén, Properties and use of the Shewhart method and its followers, *Sequential Analysis*, Vol. 26, No. 2, pp. 171–193, 2007.
88. M. Frisén and C. Sonesson, Optimal surveillance based on exponentially weighted moving averages, *Sequential Analysis*, Vol. 25, No. 4, pp. 379–403, 2006.
89. C. D. Fuh, SPRT and CUSUM in hidden Markov models, *Annals of Statistics*, Vol. 31, No. 3, pp. 942–977, 2003.
90. C. D. Fuh, Asymptotic operating characteristics of an optimal change point detection in hidden Markov models, *Annals of Statistics*, Vol. 32, No.5, pp. 2305–2339, 2004.
91. W. A. Fuller, *Introduction to Statistical Time Series.* (John Wiley & Sons, New York), 1976.
92. L. I. Galchuk and B. L. Rozovskii, The disorder problem for a Poisson process, *Theory of Probability and its Applications*, Vol. 15, No. 4, pp. 729–734, 1971.
93. P. V. Gapeev, The disorder problem for compound Poisson processes with exponential jumps, *Annals of Applied Probability*, Vol. 15, No. 1A, pp. 487–499, 2005.
94. B. K. Ghosh, *Sequential Tests of Statistical Hypotheses*, (Addison-Wesley, Reading, MA), 1970.
95. J. Giron, J. Ginebra and A. Riba, Bayesian analysis of a multinomial sequence and homogeneity of literary style, *The American Statistician*, Vol. 59, No.1, pp. 19–30, 2005.
96. E. Gombay, Sequential change-point detection with likelihood ratios, *Statistics & Probability Letters*, Vol. 49, No. 2, pp. 195–204, 2000.
97. L. Gordon and M. Pollak, An efficient sequential nonparametric scheme for detecting a change of distribution, *Annals of Statistics*, Vol. 22, No. 2, pp. 763–804, 1994.
98. L. Gordon and M. Pollak, A robust surveillance scheme for stochastically ordered alternatives, *Annals of Statistics*, Vol. 23, No. 4, pp. 1350–1375, 1995.
99. L. Gordon and M. Pollak, Average run length to false alarm for surveillance schemes designed with partially specified pre-change distribution, *Annals of Statistics*, Vol. 25, No. 3, pp. 1284–1310, 1997.
100. O. Hadjiliadis, Optimality of the 2-CUSUM drift equalizer rules for detecting two-sided alternatives in the Brownian motion model, *Journal of Applied Probability*, Vol. 42, No. 4, pp. 1183–1193, 2005.

101. O. Hadjiliadis and G. V. Moustakides, Optimal and asymptotically optimal CUSUM rules for change point detection in the Brownian motion model with multiple alternatives, *Theory of Probability and its Applications*, Vol. 50, No. 1, pp. 131–144, 2006.
102. O. Hadjiliadis and H. V. Poor, On the best 2-CUSUM stopping rule for quickest detection of two-sided alternatives in a Brownian motion model, *Theory of Probability and Its Applications*, Vol. 53, No. 3, pp. 610–622, 2008.
103. P. Hall and C. C. Heyde, *Martingale Limit Theory and its Application.* Probability and Mathematical Statistics, A Series of Monographs and Textbooks. (Academic Press, New York), 1980.
104. C. Han, P. K. Willett and D. A. Abraham, Some methods to evaluate the performance of Page's test as used to detect transient signals, *IEEE Transactions on Signal Processing*, Vol. 47, No. 8, pp. 2112–2127, 1999.
105. T. Herberts and Y. Jensen, Optimal detection of a change point in a Poisson process for different observation schemes, *Scandinavian Journal of Statistics*, Vol. 31, No. 3, pp. 347–366, 2004.
106. A. J. Heunis, *Notes on Stochastic Calculus*, Lecture Notes on Stochastic Calculus, Course E&CE 784/STAT 902 (http://www.control.uwaterloo.ca/heunis/notes.902.pdf, University of Waterloo, Waterloo, Canada), 2008.
107. L. Horváth, Change-point detection in long-memory processes, *Journal of Multivariate Analysis*, Vol. 78, No. 2, pp. 218–234, 2001.
108. P. J. Huber, A robust version of the probability ratio test, *Annals of Mathematical Statistics*, Vol. 36, No. 6, pp. 1753–1758, 1965.
109. A. Irle, Extended optimality of sequential probability ratio tests, *Annals of Statistics*, Vol. 12, No. 1, pp. 380–386, 1984.
110. E. Jaärpe, Surveillance of the interaction parameter of the Ising model, *Communications in Statistics – Theory and Methods*, Vol. 28, No. 12, pp. 3009–3027, 1999.
111. E. Jaärpe and P. Wessman, Some power aspects of methods for detecting different shifts in the mean, *Communications in Statistics – Simulation and Computation*, Vol. 29, No. 2, pp. 633–646, 2000.
112. J. Jacod, *Calcul Stochastique et Problemes de Martingale*, Lecture Notes in Mathematics 714, (Springer-Verlag, Berlin), 1979.
113. B. James, K. L. James and D. Siegmund, Conditional boundary crossing probabilities with applications to change-point problems, *Annals of Probability*, Vol. 16, No. 2, pp. 825–839, 1988.
114. D. Jarušková, Change-point detection in meteorological measurement, *Monthly Weather Review*, Vol. 124, No. 7, pp. 1535–1543, 1996.
115. D. Jarušková, Some problems with application of change-point detection methods to environmental data, *EnvironMetrics*, Vol. 8, No. 5, pp. 469–483, 1997.
116. T. Kailath and H. V. Poor, Detection of stochastic processes, *IEEE Transactions on Information Theory*, Vol. 44, No. 6, pp. 2230–2231, 1998.
117. I. Karatzas, A note on Bayesian detection of change-points with an expected miss criterion, *Statistics and Decisions*, Vol. 21, No. 1, pp. 3–14, 2003.
118. I. Karatzas and S. Shreve, *Brownian Motion and Stochastic Calculus – Second Edition*, (Springer-Verlag, New York), 1991.
119. R. J. Karunamuni and S. Zhang, Empirical Bayes detection of a change in distribution, *Annals of the Institute of Statistical Mathematics*, Vol. 48, No. 2, pp. 229–246, 1996.

120. S. A. Kassam and H. V. Poor, Robust techniques for signal processing: A survey, *Proceedings of the IEEE*, Vol. 23, No. 3, pp. 433–481, 1985.
121. D. P. Kennedy, Martingales related to cumulative sum tests and single-server queues, *Stochastic Processes and Their Applications*, Vol. 4, No. 3, pp. 261–269, 1976.
122. F. Kerestecioğlu, *Change Detection and Input Design in Dynamical Systems.* (Research Studies Press, Taunton, UK), 1993.
123. R. A. Khan, Wald's approximations to the average run length in CUSUM procedures, *Journal of Statistical Planning and Inference*, Vol. 2, No. 1, pp. 63–77, 1978.
124. H. J. Kim, Likelihood ratio and cumulative sum tests for a change-point in linear regression, *Journal of Multivariate Analysis*, Vol. 51, No. 1, pp. 54–70, 1994.
125. H. Kim, B. L. Rozovskii and A. G. Tartakovsky, A nonparametric multichart CUSUM test for rapid detection of DOS attacks in computer networks, *International Journal of Computing & Information Sciences*, Vol. 2, No. 3, pp. 149–158, 2004.
126. E. Koda, Scene-change-point detecting method and moving-picture editing/displaying method, US Patent No. 6,408,030, Issued June 18, 2002.
127. B. Krishnan, *Nonlinear Filtering and Smoothing: An Introduction to Martingales, Stochastic Integrals and Estimation.* (Wiley, New York), 1984.
128. T. L. Lai, Sequential change-point detection in quality control and dynamical systems (with discussion), *Journal of the Royal Statistical Society B*, Vol. 57, No. 4, pp. 613–658, 1995.
129. T. L. Lai, Information bounds and quick detection of parameter changes in stochastic systems, *IEEE Transactions on Information Theory*, Vol. 44, No. 7, pp. 2917–2929, 1998.
130. T. L. Lai, Sequential analysis: Some classical problems and new challenges, *Statistica Sinica*, Vol. 11, No. 2, pp. 303–408, 2001.
131. L. Lai, Y. Fan and H. V. Poor, Quickest detection in cognitive radio: A sequential change detection framework, *Proceedings of the 2008 IEEE Global Communications Conference*, New Orleans, LA, November 30 – December 4, 2008.
132. E. L. Lehmann, *Testing Statistical Hypotheses.* (Wiley, New York), 1968.
133. J. P. Lehoczky, Formulas for stopped diffusion processes with stopping times based on the maximum, *Annals of Probability*, Vol. 5, No. 4, pp. 601–607, 1977.
134. D. Lelescu and D. Schonfeld, Statistical sequential analysis for real-time video scene change detection on compressed multimedia bitstream, *IEEE Transactions on Multimedia*, Vol. 5, No. 1, pp. 106–117, 2003.
135. R. S. Liptser and A. N. Shiryaev, *Statistics of Random Processes II.* (Springer-Verlag, Berlin), 1979.
136. Y. Liu and S. D. Blostein, Quickest detection of an abrupt change in a random sequence with finite change-time, *IEEE Transactions on Information Theory*, Vol. 40, No. 6, pp. 1985–1993, 1994.
137. C. R. Loader, A log-linear model for a Poisson process change point, *Annals of Statistics*, Vol. 20, No. 3, pp. 1391–1411, 1992.
138. A. Løkka, Detection of disorder before an observable event, *Stochastics*, Vol. 79, Nos. 3 & 4, pp. 219–231, 2007.
139. G. Lorden, Procedures for reacting to a change in distribution, *Annals of Mathematical Statistics*, Vol. 42, No. 6, pp. 1897–1908, 1971.
140. G. Lorden and M. Pollak, Nonanticipating estimation applied to sequential analysis and changepoint detection, *Annals of Statistics*, Vol. 33, No. 3, pp. 1422–1454, 2005.
141. C. A. Macken and H. M. Taylor, On deviations from the maximum in a stochastic process, *Journal on Applied Mathematics*, Vol. 32, No. 1, pp. 96–104, 1977.

142. M. Marcus and J. Rosen, *Markov Processes, Gaussian Processes, and Local Times.* (Cambridge University Press, New York), 2006.
143. M. Marcus and P. Swerling, Sequential detection in radar with multiple resolution elements, *IEEE Transactions on Information Theory*, Vol. 8, No. 3, pp. 237–245, 1962.
144. Y. Mei, Comments on "A note on optimal detection of a change in distribution," by Benjamin Yakir, *Annals of Statistics*, Vol. 34, No. 3, pp. 1570–1576, 2006.
145. Y. Mei, Information bounds and quickest change detection in decentralized decision systems, *IEEE Transactions on Information Theory*, Vol. 51, No. 7, pp. 2669–2681, 2005.
146. Y. Mei, Sequential change-point detection when unknown parameters are present in the pre-change distribution, *Annals of Statistics*, Vol. 34, No. 1, pp. 92–122, 2006.
147. Y. Mei, Suboptimal properties of Page's CUSUM and Shiryaev–Roberts procedures in change-point problems with dependent observations, *Statistica Sinica*, Vol. 16, No. 4, 2006.
148. P. Misra and P. Enge, *Global Positioning System: Signals, Measurements and Performance.* (Ganga-Jumuna Press, Lincoln, MA), 2001.
149. U. Mitra and H. V. Poor, Activity detection in a multi-user environment, *Wireless Personal Communications – Special Issue on Signal Separation and Cancellation for Personal, Indoor and Mobile Radio Communications*, Vol. 3, Nos. 1–2, pp. 149–174, 1996.
150. G. V. Moustakides, Optimal stopping times for detecting changes in distributions, *Annals of Statistics*, Vol. 14, No. 4, pp. 1379–1387, 1986.
151. G. V. Moustakides, Quickest detection of abrupt changes for a class of random processes, *IEEE Transactions on Information Theory*, Vol. 44, No. 5, pp. 1965–1968, 1998.
152. G. V. Moustakides, Optimality of the CUSUM procedure in continuous time, *Annals of Statistics*, Vol. 32, No. 1, pp. 302–315, 2004.
153. G. V. Moustakides, Decentralized CUSUM change detection, *Proceedings of the 9th International Conference on Information Fusion*, Florence, Italy, July 10–13, 2006, pp. 1–6.
154. G. V. Moustakides, *CUSUM Techniques for Sequential Change Detection*, Lecture Notes in Probability, Course 8201, (Columbia University, New York), June 2007.
155. G. V. Moustakides, Sequential change detection revisited, *Annals of Statistics*, Vol. 36, No. 2, pp. 787–807, 2008.
156. N. Mukhopadhyay, *Probability and Statistical Inference.* (Marcel-Dekker, Basel, Switzerland), 2000.
157. J. Neveu, *Discrete Parameter Martingales.* (North-Holland, Amsterdam), 1975.
158. X. L. Nguyen, M. J. Wainwright and M. I. Jordan, Nonparametric decentralized detection using kernel methods, *IEEE Transactions on Signal Processing*, Vol. 53, No. 11, pp. 4053–4066, 2005.
159. I. V. Nikiforov, I. V. Varavva and V. Kireichikov, Application of statistical fault detection algorithms for navigation systems monitoring, *Automatica*, Vol. 29, No. 5, pp. 1275–1290, 1993.
160. B. Oksendal, *Stochastic Differential Equations.* (Springer-Verlag, Berlin), 1998.
161. T. Oskiper and H. V. Poor, On-line activity detection in a multiuser environment using the matrix CUSUM algorithm, *IEEE Transactions on Information Theory*, Vol. 46, No. 2, pp. 477–493, 2002.
162. T. Oskiper and H. V. Poor, Quickest detection of a random signal in background noise using a sensor array, *EURASIP Journal on Applied Signal Processing – Special Issue on Advances in Sensor Array Processing Technology*, Vol. 2005, No. 1, pp. 13–24, 2005.
163. E. S. Page, Continuous inspection schemes, *Biometrika*, Vol. 41, Nos. 1–2, pp. 100–115, 1954.

164. P. Papantoni-Kazakos, Algorithms for monitoring changes in the quality of communication links, *IEEE Transactions on Communications*, Vol. 27, No. 4, pp. 682–693, 1976.
165. B. E. Parker, Jr. and H. V. Poor, Fault diagnostics using statistical change detection in the bispectral domain, *Mechanical Systems and Signal Processing*, Vol. 14, No. 4, pp. 561–570, 2000.
166. M. Pawlak, E. Rafajowicz and A. Steland, On detecting jumps in time series: Nonparametric setting, *Journal of Nonparametric Statistics*, Vol. 16, Nos. 3 & 4, pp. 329–347, 2004.
167. L. Pelkowitz, The general discrete time disorder problem, *Stochastics*, Vol. 20, No. 2, pp. 89–110, 1987.
168. G. Peskir and A. N. Shiryaev, Sequential testing problems for Poisson processes, *Annals of Statistics*, Vol. 28, No. 3, pp. 837–859, 2000.
169. G. Peskir and A. N. Shiryaev, Solving the Poisson disorder problem, in *Advances in Finance and Stochastics*. K. Sandmann and P. Schönbucher, eds, (Springer-Verlag, Berlin), pp. 295–312, 2002.
170. M. Pettersson, Monitoring a freshwater fish population: Statistical surveillance of biodiversity, *Environmetrics*, Vol. 9, No. 2, pp. 139–150, 1998.
171. M. Petzold, C. Sonesson, E. Bergman and H. Kieler, Surveillance in longitudinal models: Detection of intrauterine growth restriction, *Biometrics*, Vol. 60, No. 4, pp. 1025–1033, 2004.
172. V. F. Pisarenko, A. F. Kushnir and I. V. Savin, Statistical adaptive algorithms for estimation of onset moments of seismic plates, *Physics of the Earth and Planetary Interiors*, Vol. 47, pp. 4–10, 1987.
173. M. Pollak, Optimal detection of a change in distribution, *Annals of Statistics*, Vol. 13, No. 1, pp. 206–227, 1985.
174. M. Pollak, Non-parametric detection of a change, *Proceedings of the 56th Session of the International Statistical Institute (ISI)*, Lisbon, Portugal, August 22–29, 2007.
175. M. Pollak and D. Siegmund, Approximations to the expected sample sizes of certain sequential tests, *Annals of Statistics*, Vol. 3, No. 6, pp. 1267–1282, 1975.
176. M. Pollak and D. Siegmund, A diffusion process and its application to detecting a change in the drift of Brownian motion, *Biometrika*, Vol. 72, No. 2, pp. 267–280, 1985.
177. M. Pollak and A. G. Tartakovsky, On optimality properties of the Shiryaev-Roberts procedure, *Annals of Statistics*, submitted.
178. H. V. Poor, *An Introduction to Signal Detection and Estimation.* (Springer-Verlag, New York), 1994.
179. H. V. Poor, Quickest detection with exponential penalty for delay, *Annals of Statistics*, Vol. 26, No. 6, pp. 2179–2205, 1998.
180. P. Protter, *Stochastic Integration and Differential Equations.* (Springer-Verlag, New York), 1985.
181. M. R. Reynolds, Jr., Approximations to the average run length in cumulative sum control charts, *Technometrics*, Vol. 17, No. 1, pp. 65–71, 1975.
182. A. Riba and J. Ginebra, Diversity of vocabulary and homogeneity of literary style, *Journal of Applied Statistics*, Vol. 33, No. 7, pp. 729–741, 2006.
183. Y. Ritov, Decision theoretic optimality of the CUSUM procedure, *Annals of Statistics*, Vol. 18, No. 3, pp. 1464–1469, 1990.
184. H. E. Robbins and E. Samuel, An extension of a lemma of Wald, *Journal of Applied Probability*, Vol. 3, No. 1, pp. 272–273, 1966.

185. S. W. Roberts, Control chart tests based on geometric moving average, *Technometrics*, Vol. 1, No. 1, pp. 239–250, 1959.
186. S. W. Roberts, A comparison of some control chart procedures, *Technometrics*, Vol. 8, No. 2, pp. 411–430, 1966.
187. S. M. Ross, *Stochastic Processes.* (Wiley, New York), 1996.
188. G. G. Roussas, *Contiguity of Probability Measures: Some Applications in Statistics.* (Cambridge University Press, Cambridge, UK), 1972.
189. I. R. Savage, Contributions to the theory of rank order statistics – the two-sample case, *Annals of Mathematical Statistics*, Vol. 27, No. 2, pp. 590–615, 1956.
190. A. N. Shiryaev, *Statistical Sequential Analysis.* (American Mathematical Society, Providence, RI), 1973.
191. A. N. Shiryaev, *Optimal Stopping Rules.* (Springer-Verlag, New York), 1978.
192. A. N. Shiryaev, Minimax optimality of the method of cumulative sums (CUSUM) in the continuous case, *Russian Mathematical Surveys*, Vol. 51, No. 4, pp. 750–751, 1996.
193. A. N. Shiryaev, Quickest detection problems in the technical analysis of financial data, *Mathematical Finance – Bachelier Congress, 2000 (Paris).* (Springer, Berlin) pp. 487–521, 2002.
194. A. N. Shiryaev, On optimum methods in quickest detection problems, *Theory of Probability and Its Applications*, Vol. 8, No. 1, pp. 22–46, 1963.
195. A. N. Shiryaev. Some exact formulas in a 'disorder' problem, *Theory of Probability and its Applications*, Vol. 10, pp. 348–354, 1965.
196. A. Shwartz and A. Weiss, *Large Deviations for Performance Analysis.* (Chapman and Hall, London), 1995.
197. D. Siegmund, Corrected diffusion approximations in certain random walk problems, *Advances in Applied Probability*, Vol. 11, No. 4, pp. 701–719, 1979.
198. D. Siegmund, *Sequential Analysis.* (Springer-Verlag, New York), 1985.
199. D. Siegmund, The 1984 Wald Memorial Lectures. Boundary crossing probabilities and statistical applications, *Annals of Statistics*, Vol. 14, No. 2, pp. 361–404, 1986.
200. D. L. Snyder and M. I. Miller, *Random Point Processes in Time and Space.* (Springer-Verlag, New York), 1991.
201. C. Sonesson and D. Bock, A review and discussion of prospective statistical surveillance in public health, *Journal of the Royal Statistical Society A*, Vol. 166, No. 1, pp. 5–21, 2003.
202. M. Stoto, *et al.*, Evaluating statistical methods for syndromic surveillance. In *Statistical Methods in Counterterrorism: Game Theory, Modeling, Syndromic Surveillance, and Biometric Authentication*, A. G. Wilson, G. D. Wilson and D. H. Olwell, eds. (Springer, New York), 2006.
203. S. Tantaratana and H. V. Poor, Asymptotic efficiencies of truncated sequential tests, *IEEE Transactions on Information Theory*, Vol. 28, No. 6, pp. 911–923, 1982.
204. A. G. Tartakovsky, Asymptotically minimax multi-alternative sequential rule for disorder detection, *Statistics and Control of Random Processes: Proceedings of the Steklov Institute of Mathematics*, (American Mathematical Society, Providence, RI), Vol. 202, No. 4, pp. 229–236, 1994.
205. A. G. Tartakovsky, Asymptotic properties of CUSUM and Shiryaev's procedures for detection a change in a nonhomogeneous Gaussian process, *Mathematical Methods of Statistics*, Vol. 4, No. 4, pp. 389–404, 1995.

206. A. G. Tartakovsky, Asymptotic performance of a multichart CUSUM test under false alarm probability constraint, *Proceedings of the 44th IEEE Conference on Decision and Control*, Seville, Spain, December 12–15, 2005, pp. 320–325.
207. A. G. Tartakovsky, Asymptotic optimality in Bayesian change-point detection problems under global false alarm probability contraint, *Theory of Probability and its Applications*, submitted.
208. A. G. Tartakovsky, B. L. Rozovskii, R. B. Blazek and H. Kim, A novel approach to detection of intrusions in computer networks via adaptive sequential and batch-sequential change-point detection methods, *IEEE Transaction on Signal Processing*, Vol. 54, No. 9, pp. 3372–3382, 2006.
209. A. G. Tartakovsky and V. V. Veeravalli, General asymptotic Bayesian theory of quickest change detection, *Theory of Probability and its Applications*, Vol. 49, No. 3, pp. 458–497, 2005.
210. H. M. Taylor, A stopped Brownian motion formula, *Annals of Probability,* Vol. 3, No. 2, pp. 234–246, 1975.
211. D. Teneketzis and Y.-C. Ho, The decentralized Wald problem, *Information and Computation*, Vol. 73, No. 1, pp. 23–44, 1987.
212. D. Teneketzis and P. Varaiya, The decentralized quickest detection problem, *IEEE Transactions on Automatic Control*, Vol. 29, No. 7, pp. 641–644, 1984.
213. M. Thottan and C. Ji, Anomaly detection in IP networks, *IEEE Transactions on Signal Processing*, Vol. 51, No. 8, pp. 2191–2204, 2003.
214. R. Trivedi and R. Chandramouli, Secret key estimation in sequential steganography, *IEEE Transactions on Signal Processing*, Vol. 53, No. 2, Part 2, pp. 746–757, 2005.
215. J. N. Tsitsiklis, Decentralized detection, in *Advances in Statistical Signal Processing – Vol. 2*, H. V. Poor and J. B. Thomas, eds. (JAI Press, Greenwich CT), pp. 297–344, 1993.
216. H. Tucker, *A Graduate Course in Probability.* (Academic Press, New York), 1967.
217. J. Ushiba, Y. Tomita, Y. Masakado and Y. Komune, A cumulative sum test for a peri-stimulus time histogram using the Monte Carlo method, *Journal of Neuroscience Methods*, Vol. 118, No. 2, pp. 207–214, 2002.
218. V. V. Veeravalli, Sequential decision fusion: Theory and applications, *Journal of the Franklin Institute*, Vol. 336, No. 2, pp. 301–322, 1999.
219. V. V. Veeravalli, Decentralized quickest change detection, *IEEE Transactions on Information Theory*, Vol. 47, No. 4, pp. 1657–1665, 2001.
220. V. V. Veeravalli, T. Basar and H. V. Poor, Decentralized sequential detection with a fusion center performing the sequential test, *IEEE Transactions on Information Theory*, Vol. 37, No. 2, pp. 433–442, 1993.
221. V. V. Veeravalli, T. Basar and H. V. Poor, Decentralized sequential detection with sensors performing sequential tests, *Mathematics of Control, Signals and Systems*, Vol. 7, No. 4, pp. 292–305, 1994.
222. M. H. Vellekoop and J. M. C. Clark, Optimal speed of detection in generalized Wiener disorder problems, *Stochastic Processes and their Applications*, Vol. 95, No. 1, pp. 25–54, 2001.
223. A. Wald, *Sequential Analysis.* (Wiley, New York), 1947.
224. G. B. Wetherhill and D. W. Brown, *Statistical Process Control.* (Chapman and Hall, London), 1991.
225. E. Wong and B. Hajek, *Stochastic Processes in Engineering Systems.* (Springer-Verlag, New York), 1985.

226. B. Yakir, Optimal detection of a change in distribution when the observations form a Markov chain with a finite state space, in *Change Point Problems*, E. Carlstein, H.-G. Müller and D. Siegmund, eds. (Institute of Mathematical Statistics, Hayward, CA), pp. 346–359, 1994.
227. B. Yakir, Dynamic sampling policy for detecting a change in distribution, with a probability bound on false alarm, *Annals of Statistics*, Vol. 24, No. 5, pp. 2199–2214, 1996.
228. B. Yakir, A note on optimal detection of a change in distribution, *Annals of Statistics*, Vol. 25, No. 5, pp. 2117–2126, 1997.
229. M. C. Yang, Y. Y. Namgung, R. G. Marks, I. Magnusson and W. B. Clark, Change detection on longitudinal data in periodontal research, *Journal of Periodontal Research*, Vol. 28, No. 2, pp. 152–160, 1993.
230. E. Yashchin, Change-point models in industrial applications, *Nonlinear Analysis*, Vol. 30, No. 7, pp. 3997–4006, 1997.
231. M. Yoshida, Probability maximizing approach to a detection problem with continuous Markov processes, *Stochastics*, Vol. 11, Nos. 1–2, pp. 173–189, 1984.
232. A. J. Yu, Optimal change-detection and spiking neurons, in *Advances in Neural Information Processing Systems 19*. (MIT Press, Cambridge, MA), 2006.

Index

Note: footnotes are indicated by suffix 'n'